# Interdisciplinary Biotechnological Advances

**Series Editors**

Jayanta Kumar Patra, Research Institute of Integrative Life Sciences,
Dongguk University, Ilsandong, Goyang, Kyonggi-do, Korea (Republic of)
Gitishree Das, Research Institute of Integrative Life Sciences,
Dongguk University, Goyang, Korea (Republic of)

This series is an authoritative resource that explains recent advances and emerging directions in biotechnology, reflecting the forefront of research clearly and reliably, without excessive hype. Each volume is written by authors with excellent reputations and acknowledged expertise in the topic under discussion. The volumes span the entire field from an interdisciplinary perspective, covering everything from biotechnology principles and methods to applications in areas including genetic engineering, transgenic plants and animals, environmental problems, genomics, proteomics, diagnosis of disease, gene therapy, and biomedicine. The significance of these applications for the achievement of UN Sustainable Development Goals is highlighted. The series will be highly relevant for Master's and PhD students in Biotechnology, Nanochemistry, Biochemical Engineering, and Microbiology, medical students, academic and industrial researchers, agricultural scientists, farmers, clinicians, industry personnel, and entrepreneurs.

Alok Das Mohapatra • Priyadarshi S. Sahu
Editors

# Advances in Immunology and Immuno-techniques

## Updated Concepts and Recent Applications

Springer

*Editors*
Alok Das Mohapatra
Pre-Clinical In Vivo Pharmacology Division
TScan Therapeutics
Waltham, MA, USA

Priyadarshi S. Sahu
Department of Microbiology &
Immunology, School of Medicine
Medical University of the Americas
Nevis, St. Kitts and Nevis

ISSN 2730-7069 ISSN 2730-7077 (electronic)
Interdisciplinary Biotechnological Advances
ISBN 978-981-97-5510-3 ISBN 978-981-97-5508-0 (eBook)
https://doi.org/10.1007/978-981-97-5508-0



This Springer imprint is published by the registered company Springer Nature Singapore Pte Ltd.
The registered company address is: 152 Beach Road, #21-01/04 Gateway East, Singapore 189721, Singapore

*This book is dedicated to all our teachers and mentors without whom we would not have been at this stage today.*
*—Editors*

# Preface

The major focus of this book is to facilitate a clear understanding of the emerging concepts and applications in different fields of immunology. The chapters included in this book are not part of any conventional textbook on Immunology. Each chapter in this book has a chapter outline in the beginning and a chapter glossary at the end, for better comprehension of the topic. Hence, this book is expected to be an interesting read for students of immunology as well as immunologists from academics, research, and industry.

Waltham, MA, USA — Alok Das Mohapatra
Nevis, St. Kitts and Nevis — Priyadarshi S. Sahu

# About This Book

The book introduces comprehensive chapters on emerging concepts and applications in the field of immunology, which are yet to find their place in conventional immunology textbooks. Each chapter in this book focuses on: (1) clarifying the basic concepts and principles underlying the subject and (2) listing out the unanswered research questions for potential future research directions. Thus, this book will serve as a supplementary reference manual for students (from various disciplines including life science, biochemistry, biomedical science, bio-technology, medicine, and veterinary sciences) as well as researchers of immunology (both academia and industry). Moreover, this book will provide conceptual clarity on the promising recent developments in immunology and arouse curiosity among the readers to pursue further research in related areas.

# Contents

# Editors and Contributors

## About the Editors

**Alok Das Mohapatra** is an Immunologist, with more than 20 years of research experience in the broad field of immunology, in premier academic research institutions of the USA and India. While in India, Dr. Das Mohapatra has almost 7 years of teaching experience in the field of Microbiology and Immunology, at both undergraduate and postgraduate level. After completing his Master's degree in "Microbiology" and PhD in the field of "Cellular immunology" from India, Dr. Das Mohapatra came to the USA as a postdoctoral scientist and worked in the Center for Immunotherapy of Cancer and Infectious Diseases, Department of Immunology at UConn Health Center, Farmington, CT, USA. After working there for nearly 6 years, he moved to the Center for International Health Research, Department of Pathology and Laboratory Medicine, Brown University, Providence, RI, USA, as a Research Scientist and then joined the division of Infectious Diseases and Immunology at Umass Medical School, Worcester, MA, USA. He has many research publications to his credit and has published in top tier international journals including Nature, Cancer Immunology Research, *Journal of Infectious Diseases*, and many others. After working in academia for more than a decade, Dr. Das Mohapatra moved into industry. Currently, Dr. Das Mohapatra is working as a principal scientist on cutting-edge research in the field of cell therapy in cancer. Overall, Dr. Das Mohapatra has gathered knowledge and expertise in diverse areas of immunology, including innate and adaptive immunity, basic and translational immunology, clinical immunology and experimental immunology, in the context of both cancer and infectious diseases caused by unicellular as well as multicellular parasites. As evidence of his expertise in the field, Dr. Das Mohapatra has reviewed close to 100 research articles for publication in more than 12 different peer-reviewed international journals. Dr. Das Mohapatra has received a certificate of recognition for "outstanding contribution to reviewing" from the renowned "Elsevier" publication. Dr. Das Mohapatra was elected as a full member of Sigma Xi, the Scientific

Research Honor Society. Dr. Das Mohapatra has also received a travel award from the American Association of Immunology. Dr. Das Mohapatra realized that there is a pressing need for a comprehensive book on immunology in the market that will disseminate the complex concepts of immunology in ways that are easy to understand by the general readers and at the same time it should also inspire the intellectual curiosity of researchers in the field.

**Priyadarshi S. Sahu** is a Medical Microbiologist and Immunologist having long years of experience in academics, research, laboratory services, and consultancy. This Indian born professor had a strong motivation towards science from his early education; understanding body's defense mechanisms to infections has always been fascinating him. He completed his master's in Microbiology followed by PhD under Faculty of Medicine from a nationally reputed postgraduate medical education center in India. Then he relocated to the USA and started working as a post-doctoral fellow in a neuroimmunology laboratory at the University of Texas at San Antonio. Subsequently, he accepted a career to teach and has taken various academic positions so far having worked in different countries in Asia and the Americas. He has been teaching microbiology, immunology, parasitology, and infectious disease epidemiology to students at both undergraduate and postgraduate levels under Medicine as well as other health science course programs *viz.*, Dentistry, Nursing, Biomedical Sciences, Medical Laboratory Technology. Currently, he is working as a Full-time Professor in Microbiology and Immunology teaching Pre-clinical Medicine at Medical University of the Americas, Nevis. He advocates the benefits of self-guided learning methods as well as appropriate learning materials for students in professional and higher education courses. An effective source of information is needed to empower to understand the concepts and to excel. He is keen to develop effective learning materials and also to explore newer methods of teaching in both small and large groups in Medicine in order to integrate the basic sciences knowledge into clinical applications and contribute to the learning process of the student through his expertise and experiences. He has guided multiple PhD as well as postgraduate research projects and has been actively involved in collaborative research in the field of infectious diseases, immunology, and tropical medicine.

## Contributors

**Nandini Acharya** Pelotonia Institute of Immuno-Oncology, The Ohio State University Comprehensive Cancer Center, Columbus, OH, USA

**Rishabh Agrahari** Amity Institute of Microbial Technology (AIMT), Amity University, Noida, Uttar Pradesh, India

**Janak Raj Awasthi** National Institute of Drug Abuse, National Institute of Health, Baltimore, MD, USA

Department of Neuroscience, University of Medicine and Health Sciences, Basseterre, St. Kitts & Nevis

**Subrat Kumar Bhattamisra** Department of Pharmacy, School of Health Sciences, Central University of South Bihar, Gaya, India

**Rafael Oliveira Brandão** Biotech Research and Innovation Centre, University of Copenhagen, Copenhagen, Denmark

**Cory Alexander Brennick** Yale School of Medicine, Yale University, New Haven, CT, USA

**Mariam Mathew George** Weill Cornell Medicine, New York, NY, USA

**Marc Andrew Gillig** Enable Life Sciences, Worcester, MA, USA

**Bapi Gorain** Department of Pharmaceutical Sciences, Birla Institute of Technology, Ranchi, Jharkhand, India

**Xin Jiang** Biotech Research & Innovation Centre (BRIC), Biomedical Institute, University of Copenhagen, Copenhagen, Denmark

**Varnita Karmakar** Department of Pharmaceutical Sciences, Birla Institute of Technology, Ranchi, Jharkhand, India

**Sunil Kumar Jada** Sri Venkata Sai Medical College and Hospital, Affiliated to KNR University of Health Sciences, Warangal, Telangana, India

**K. S. Kumar** School of Science, GSFC University, Vadodara, India

**Rahul Kumar** Department of Environmental Sciences, H.N.B. Garhwal University, Srinagar, Uttarakhand, India

**Subrat Kumar** Infection Biology Laboratory, School of Biotechnology, KIIT University, Bhubaneswar, Odisha, India

**Ramchander Merugu** Department of Biochemistry, Mahatma Gandhi University, Nalgonda, Telangana, India

**Snigdha Misra** Jeffrey Cheah School of Medicine & Health Sciences, Monash University Malaysia, Bandar Sunway, Selangor DE, Malaysia

**Debapriya Mohanty** National Institute of Foot and Mouth Disease, Indian Council of Agricultural Research, Jatni, Odisha, India

**Soumitra Mohanty** Department of Microbiology, Tumor and Cell Biology, Karolinska Institutet, Stockholm, Sweden

Division of Clinical Microbiology, Karolinska University Hospital, Stockholm, Sweden

**Alok Das Mohapatra** Pre-Clinical In Vivo Pharmacology Division, TScan Therapeutics, Waltham, MA, USA

**Purusottam Mohapatra** Department of Biotechnology, National Institute of Pharmaceutical Education and Research (NIPER) Guwahati, Kamrup, Assam, India

**Swati Mohapatra** Department of Infection Biology, School of Medicine, Wonkwang University, Iksan, Republic of Korea

**Manisha Pandey** Department of Pharmaceutical Sciences, Central University of Haryana, Mahendragarh, Haryana, India

**Pooja** Department of Pharmaceutical Sciences, Central University of Haryana, Mahendragarh, Haryana, India

**Swaminathan Ravichandran** Department of Pathology, School of Medicine, Medical University of the Americas, Nevis, St. Kitts & Nevis

**Biswa Mohan Sahoo** School of Pharmacy and Life Sciences, Centurion University of Technology & Management, Bhubaneswar, Odisha, India

**Priyadarshi S. Sahu** Department of Microbiology & Immunology, School of Medicine, Medical University of the Americas, Nevis, St. Kitts and Nevis

**Sudhagar Selvaraju** Department of Biotechnology, National Institute of Pharmaceutical Education and Research (NIPER) Guwahati, Kamrup, Assam, India

**Dhirodatta Senapati** KIIT School of Pharmacy, KIIT (Deemed to be University), Bhubaneswar, Odisha, India

**Arpit Kumar Srivastava** Atal Bihari Vajpayee Government Medical College, Vidisha, India

Department of Biotechnology, Devi Ahilya Vishwavidyalaya, Indore, India

**Subhra Subhadra** Regional Medical Research Center, Bhubaneswar, Odisha, India

**Toru Takumi** Department of Physiology and Cell Biology, Kobe University School of Medicine, Kobe, Japan

**Saranya Vilvanathan** Department of Pharmacology, School of Medicine, Medical University of the Americas, Nevis, St. Kitts & Nevis

# Immunologic Basis of Immuno-therapeutics and Vaccines: Present Approaches and Future Perspectives

**Alok Das Mohapatra and Priyadarshi S. Sahu**

**Abstract** Immunotherapy is a rapidly evolving branch of biomedical science that deals with the modulation or manipulation of host immune response to treat various immune-related diseases including cancer, infectious diseases, allergies, autoimmune and inflammatory diseases, etc. The process of immunotherapy often exploits the natural function of native or engineered versions of effector molecules (i.e., antibodies) or immune cells (T cells, NK cells, dendritic cells, etc.) to specifically target disease-causing molecules, abnormal cells, or pathogens conferring therapeutic benefits. The present chapter which is primarily aimed at a broad scientific audience emphasizes the general immunological principles and different approaches used in immunotherapy of common infectious and non-infectious diseases with a futuristic approach; newer vaccines, vaccine development, vaccine delivery strategies, and overall applications against infectious agents, as well as non-infectious conditions (e.g., cancer, allergy, autoimmune and inflammatory disorders, etc.) are discussed.

**Keywords** Immunotherapy · Vaccine · CAR-T cell · TCR-T cell · Antibody · Cancer therapy

---

A. Das Mohapatra
Pre-Clinical In Vivo Pharmacology Division, T-Scan Therapeutics, Waltham, MA, USA

P. S. Sahu (✉)
Department of Microbiology & Immunology, School of Medicine, Medical University of the Americas, Nevis, St. Kitts and Nevis
e-mail: p.sahu@mua.edu



A. Das Mohapatra, P. S. Sahu (eds.), *Advances in Immunology and Immuno-techniques*, Interdisciplinary Biotechnological Advances,
https://doi.org/10.1007/978-981-97-5508-0_1

## 1 Introduction

The human immune system comprises a complex network of innate and adaptive immune cells (i.e., dendritic cells, macrophages, NK cells, T cells, B cells, etc.), effector molecules (i.e., cytokines, chemokines, antibodies, complements, etc.), and lymphoid organs (i.e., thymus, bone marrow, spleen, lymph nodes, and other lymphoid tissues). All the above components of immune system work in coordination to protect us from infectious diseases caused by various pathogens like bacteria, viruses, parasites, fungi, and harmful foreign substances, as well as abnormal cells including infected or cancerous cells (Marshall et al. 2018; Paludan et al. 2021). In addition to recognizing and eliminating external threats, the immune system also maintains tolerance to our own normal cells and tissues. However, sometimes due to lapses in the functioning of one or more component of the immune system, individuals suffer from infectious diseases, cancer, autoimmune diseases, inflammatory diseases, allergies, etc. These diseases are primarily attributed to failure of the immune system in eliminating pathogens, killing abnormal cells, and maintaining tolerance against normal cells or tissues, and hyper-reaction of immune system to specific agents, etc. respectively (Marshall et al. 2018; Paludan et al. 2021). However, our in-depth understanding of the functioning of different components of our immune system has allowed us to therapeutically manipulate our immune system to prevent or treat such diseases effectively, by following different approaches. All these approaches are called immunotherapy, and the therapeutic agents used here are referred to as immunotherapeutic.

## 2 Immunotherapy of Different Diseases

Immunotherapy is a form of biotherapy that involves the modulation or manipulation of the immune system for the treatment of various immunological diseases including, immunodeficiencies, hypersensitivity/allergic reactions, autoimmune diseases, tissue, and organ transplantations, inflammatory disorders, infectious diseases, as well as cancer (Justiz Vaillant et al. 2024) to achieve an improvement in the quality and/or life expectancy of individuals (Cantarovich et al. 2008; Xu et al. 2013; Miller et al. 2018; Thangamathesvaran et al. 2018; Justiz Vaillant and Qurie 2023, Klein et al. 2024). Following are certain diseases and conditions where immunotherapy finds applications.

## 2.1 *Immunotherapy of Cancer*

The immune system is capable of specifically recognizing abnormal cancer cells and killing them. However, sometimes our immune system fails to recognize and kill the cancer cells as the cancer cells downregulate their HLA (human leukocyte antigens, which display the cancer antigen recognized by cytotoxic T cells) to prevent their detection and recruit various immune suppressive cells like regulatory T cells and immune suppressive cytokines like TGFβ, IL-10, etc. There are many different types of immunotherapies for cancer (Rosenberg et al. 1986; Waldman et al. 2020; Oiseth and Aziz 2017; Baulu et al. 2023; Shalhout et al. 2023; Klein et al. 2024) including monoclonal antibodies, bi-specific antibodies, cancer vaccines, CAR-T cell therapy, TCR-T cell therapies, oncolytic virus therapy, etc. that are commonly used to fight cancer.

### 2.1.1 Cancer Vaccines

Vaccines to treat cancer are still under development; however, most commonly they are used to challenge the immune system of an individual with cancer-specific antigens in various forms. This results in an increased number of cancer antigen-specific cytotoxic T cells in the immunized individual that can destroy the cancer cells. An emerging concept in the field of cancer vaccines is neo-antigen-based personalized vaccines (Xie et al. 2023). The neo-antigens refer to cancer antigens which arise due to tumor-specific mutations.

### 2.1.2 Monoclonal Antibodies

These are designed to target specific antigens present on cancer cells to kill them directly by inducing programmed cell death or by ADCC (antibody-dependent cell-mediated cytotoxicity). Recently, monoclonal antibodies have been successfully used as check point inhibitors. These check point inhibitors target the co-inhibitory receptors (i.e., CTLA-4, PD-1, LAG-3, TIM-3, TIGIT, etc.) expressed on T cells and restore their tumor cell killing activity (Schnell et al. 2020) (Fig. 1).

### 2.1.3 Bi-specific Antibodies

Unlike monoclonal antibodies, the bi-specific antibodies have binding specificities to two different antigens (Klein et al. 2024) which are present on two different target cells. They are usually designed to have specificities for both the T cells and specific cancer cells. By binding to the T cells, they activate the T cells, and by simultaneously interacting with the cancer cells, they can bring the two cells together, to mediate effective killing of the cancer cells by activated T cells.

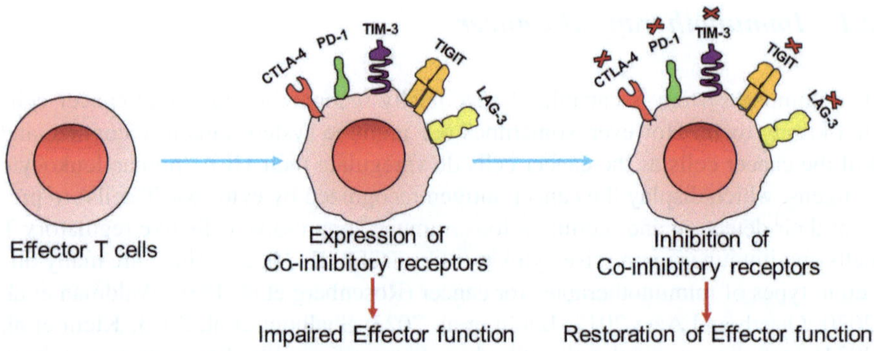


**Fig. 1** Inhibition or blocking of co-inhibitory receptors by corresponding monoclonal antibodies restores the normal effector function of T cells

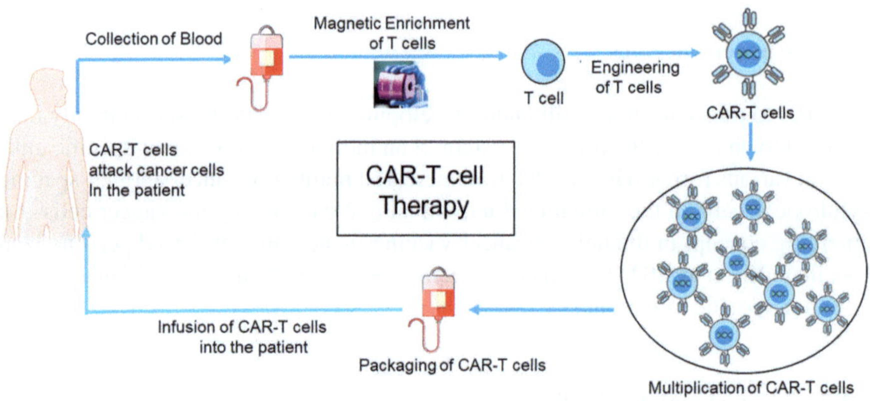


**Fig. 2** Schematic representation of chimeric antigen receptor (CAR)-T cell therapy

### 2.1.4 T Cell Therapy

There are two types of T cell therapy, CAR-T cell therapy (Sterner and Sterner 2021) and TCR-T cell therapy (Baulu et al. 2023). In the case of both the CAR-T cell therapy and TCR-T cell therapy, the T cells are harvested from the blood of cancer patients, engineered, and expanded in the laboratory before transferring them back to the cancer patients to fight cancer. In the case of CAR-T cells, the T cells are engineered to express a chimeric antigen receptor that specifically recognizes the specific antigens on cancer cells, not necessarily presented in the context of host HLA molecules. In the case of TCR-T cell therapy, the T cells are engineered to express modified T cell receptor which specifically recognizes specific cancer antigen expressed in the context of host HLA molecules (Fig. 2).

#### 2.1.5 Oncolytic virus therapy

Oncolytic virus therapy uses viruses that have been engineered in a laboratory to specifically infect and destroy cancer cells (Shalhout et al. 2023).

### 2.2 *Immunotherapy of Autoimmune and Chronic Inflammatory Disorders*

In normal conditions, the effector cells of the immune system protect host individuals from infectious diseases and cancer. In this process, the B cells target antigens derived from invading pathogens and the T cells attack the infected or abnormal host cells in an individual, respectively. However, in the case of autoimmune disorders, certain abnormal B cells produce antibodies against self-antigens, and some abnormal T cells attack normal host cells expressing self-antigens, thereby causing harm to the host. Immunotherapies against autoimmune (Caspi 2008) and inflammatory disorders (McInnes and Gravallese 2021) have made appreciable progress. All the immunotherapies that target cancer cells can be potentially manipulated to specifically target and eliminate the abnormal B cells and T cells or specifically inhibit the activity of these abnormal B cells or T cells in an individual to treat the autoimmune disorders. Besides, conventional immunotherapies aim to suppress or modulate the immune response by using corticosteroids, disease modifying anti-rheumatic drugs (DMRDS), and biologic agents targeting specific immune pathways mediating inflammation in inflammatory disorders. Most common treatment modalities involve inhibition of TNF, IL-6, IL-17, and IL-23 for autoimmune diseases, such as rheumatoid arthritis, ankylosing spondylitis, and psoriasis. Anti-CD20 mAbs aimed at depleting B cells has shown great promise in patients having neuro-inflammatory diseases. Further, inhibition of B cell survival factors has been approved for treatment of SLE (systemic lupus erythematosus).

Chronic inflammatory diseases like inflammatory bowel disease (IBD), multiple sclerosis (MS), and psoriasis can be treated by targeting specific components of the immune system mediating inflammation and tissue damage.

### 2.3 *Immunotherapy of Infectious Diseases*

In the case of infectious diseases, immunotherapy is primarily aimed at protecting the host against infection and/or disease and alleviating disease symptoms (Wallis et al. 2023; Sangeetha et al. 2023). Here, based on their mode of action, immunotherapies can be classified into passive immunotherapies or active immunotherapies. In passive immunotherapies, immune cells or effector molecules including antibodies, and cytokines produced, ex vivo are used. On the other hand, active immunotherapy triggers the host immune system to produce or recruit its own effector cells and molecules to fight infection and/or disease. The different immunotherapeutic approaches used in the case of infectious diseases can be broadly categorized into the following.

### 2.3.1 T Cell Engineering

The T cells used in this approach are commonly isolated from the blood of the patient and called autologous T cells. After a phase of in vitro expansion and engineering, the engineered T cells are infused back into the patients to fight the infection and/or disease. Based on the engineering approach, the cell therapy can be of the following two types.

### 2.3.2 TCR T Cell Therapy

Here the T cells are engineered to express a specific T cell receptor that recognizes specific antigens of the pathogen expressed on the surface of an infected host cell in the context of specific HLA (human leukocyte antigen). Because of their high specificity, the engineered T cells specifically target and eliminate the infected cells and spare the normal host cells.

### 2.3.3 CAR-T Cell Therapy

Here the T cells are engineered to express a chimeric antigen receptor (CAR) that is composed of four distinct regions including an antigen recognition domain, an extracellular hinge region, a transmembrane domain, and an intracellular T cell signaling domain. The extracellular antigen recognition domain of CAR-T cells is specific to specific pathogen-derived antigens expressed on the surface of infected host cells. Unlike the TCR T cells, the antigens recognized by CAR-T cells are not HLA-restricted.

### 2.3.4 Modulation of T Cell and B Cell Activity—Vaccination

Here the host is immunized with killed or live-attenuated pathogens, or pathogen-derived, or artificially synthesized antigens specific to a pathogen. Following immunization, the antigens are taken up, processed, and presented by antigen-presenting cells like dendritic cells/macrophages to naïve T cells and B cells to elicit pathogen-specific cellular or humoral immune response.

### 2.3.5 Checkpoint Inhibition

During the immune response against pathogens, the T cells are activated, but over time the activated T cells express different co-inhibitory receptors (i.e., CTLA-4, PD-1, LAG-3, TIM-3, TIGIT, etc.) which act as checkpoints to dampen the effector function of T cells. *Mycobacterium tuberculosis* is known to induce the expression of ligands for PD-1, CTLA-4, TIM3, and LAG3 on the surfaces of infected macrophages, thereby inhibiting T cell activation. Using monoclonal antibodies targeting either the co-inhibitory receptors on the T cell or corresponding ligands on antigen-presenting cells (APCs) can disrupt the interaction between the co-inhibitory receptors and their respective ligands, resulting in the restoration of the effector function of the T cells.

### 2.3.6 Therapeutic Monoclonal Antibodies

Monoclonal antibodies are being considered for the treatment of viral, bacterial, and protozoan infections. These antibodies can block the invasion of host cells by binding to the receptors on host cells or corresponding ligands on the pathogens. Further, by binding to the antigens expressed on infected host cells or on the pathogen itself, the monoclonal antibodies can eliminate the infected host cells and neutralize pathogens, by different direct (induction of apoptotic cell death) or indirect (ADCC—antibody-dependent cellular cytotoxicity) mechanisms. In ADCC, the antibody-bound target cells are subsequently recognized by host NK cells and macrophages for elimination. Additionally, monoclonal antibodies targeting inflammatory cytokines IL-6 have been used to alleviate the disease symptoms during the cytokine storm in case of COVID-19 viral infections.

### 2.3.7 Antibody-Drug Conjugates

Here the monoclonal antibody-conjugated drugs can undergo receptor-mediated internalization to deliver specific drugs in infected target cells, resulting killing of the pathogens or infected host cells. This approach has been successfully used against *Trypanosoma brucei,* in vitro where monoclonal antibody targeting human haptoglobin hemoglobin receptor (HpHbR) conjugated to pyrrolobenzodiazepine toxin was used.

### 2.3.8 Bi-specific Antibodies

These are synthetic antibodies having specificity to two different target antigens expressed on two different target cells. The bi-specific antibodies commonly used in case of infectious diseases, commonly target the CD3 on T cells and specific antigen on pathogen-infected host cells. Administration of these antibodies into an individual achieves two purposes simultaneously: By binding to CD3, it activates the T cells and by interacting with the antigen of pathogen-infected cells, it brings the activated T cells and pathogen-infected T cells in close proximity to mediate effective killing of infected host cells.

#### 2.3.9 Plasma Therapy

Immune individuals who recover from an infection harbor protective antibodies against the corresponding infection in their blood. Plasma of such individuals containing protective antibodies against specific pathogens is referred to as convalescent plasma. Such convalescent plasma can be used to passively transfer protective antibodies to infected individuals to give a therapeutic benefit. This convalescent plasma therapy has been successfully used in pandemic influenza, Lassa virus, Ebola virus, and more recently in the case of the COVID-19 pandemic.

#### 2.3.10 Cytokine Therapy

A better understanding of the biological roles of individual cytokines during the manifestation of different diseases has resulted in their therapeutic use to ameliorate the disease conditions. The role of GM-CSF (granulocyte-macrophage colony-stimulating factor) in increasing the dendritic cell population responsible for inducing a protective immune response against Mycobacterium tuberculosis has been demonstrated. Additionally, recombinant human interleukin-2 (rhIL2) is being evaluated in a clinical trial to treat multi-drug resistant TB patients (NCT03069634).

### 2.4 *Immunotherapy of Allergic Diseases*

Immunotherapy of allergic reactions to allergens (Durham and Shamji 2023) like pollens, bee venom, house dust mites, etc. involves the administration of gradually increasing doses of the specific allergen to an individual. The incremental exposure of an individual to a specific allergen results in the production of allergen-specific blocking antibodies which reduces the symptoms of allergy upon encounter of the same allergen in the future. Inflammation characterized by rhinitis and asthmatic conditions can also be improved by immunotherapy.

## 2.5 *Immunotherapy of Organ Transplantation*

In organ transplantation, the recipient is usually transplanted with a tissue or an organ from the donor. Although HLA typing is done between the donor and the recipient to find the best match before transplantation, it is difficult to get a 100% match between the donor and the recipient. This leaves the scope of rejection of the transplant by the recipient's immune system. The primary objective of immunotherapy in organ transplantation is to prevent rejection of the transplant by modulating the host's immune response while minimizing the risk of infection and other adverse effects on the recipient.

### 2.5.1 Immunosuppressive Drugs

These drugs are most used in the field of immunotherapy for transplantation. Their mode of action involves suppression of the recipient's immune response to reduce the risk of rejection of the transplant. However, this makes the recipient susceptible to various infections. The most commonly used immunosuppressive drugs include corticosteroids, calcineurin inhibitors (i.e., tacrolimus, cyclosporine), mTOR inhibitors (i.e., sirolimus, everolimus), and anti-metabolites (i.e., azathioprine and mycophenolate mofetil).

### 2.5.2 Induction Therapy

Here potent immunosuppressive drugs are used during transplantation to prevent rejection of transplant at an early stage. Common examples include monoclonal antibodies such as basiliximab or anti-thymocyte globulin (ATG).

### 2.5.3 Maintenance Therapy

After the organ transplantation, the rejection of transplant at the late stage is prevented by maintenance therapy where immune suppressive drugs are administered considering the patient's immune status, transplant type, and historical medical conditions.

### 2.5.4 Immunomodulatory Therapy

In traditional immune suppressive therapies, the drugs induce general immune suppression. However, recently biologic agents targeting specific immune cells or molecules involved in the rejection are being explored.

### 2.5.5 Tolerance Induction

This is another emerging research area of immunotherapy where the objective is to induce immune tolerance in the recipient to the transplanted organ so that the recipient's immune system considers the transplanted organ as "self." Multiple approaches are being explored in this field including mixed chimerism, Treg (regulatory T cell) therapy, hematopoietic stem cell (HSC) transplantation, etc.

### 2.5.6 Risk Management

During immunotherapy of organ transplant patients, it is critical to monitor any side effects of therapy in them including their increased susceptibility to infections, metabolic disorders, adverse off-target effects, etc.

# 3 Conventional Vaccines

Normally, vaccines help to protect us from infectious diseases and have a major contribution to global health. Vaccination helps economic growth through its tremendous impact in lowering morbidity and mortality due to infections, thereby saving millions of lives globally (Andre et al. 2008). Further, recent efforts are ongoing to develop effective vaccines against cancer, autoimmune diseases, and allergies. Future potential targets of vaccine-based treatment include lifestyle diseases like diabetes, hypertension, addiction, and Alzheimer's disease (Greenwood 2014).

## *3.1 What Is a Vaccine?*

A vaccine is a biological product or formulation of an antigen that is often administered along with an adjuvant to confer protection from a specific infection and/or disease by eliciting an effective immune response in the immunized subject (Pollard and Bijker 2021). The antigen component of the vaccine usually includes any of the following, i.e., a live and attenuated pathogen, killed pathogen, purified proteins, polysaccharides, synthetic recombinant proteins, peptides, nucleic acids, or inactive toxins (toxoids) derived from the pathogen (Fig. 3).

The adjuvants are specific compounds that typically activate the antigen-presenting cells like dendritic cells and amplify the immune response. Conventional vaccines have successfully prevented infections by pathogens expressing relatively conserved antigens through the generation of specific cellular (T cell-mediated) and humoral (antibody-mediated) immune responses (De Gregorio and Rappuoli 2012). In clinical trials, the effectiveness of a vaccine is usually measured in terms of

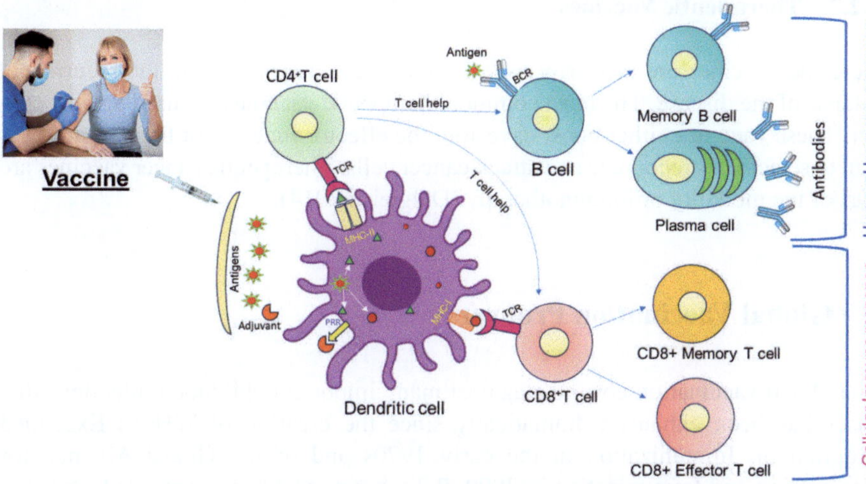


**Fig. 3** Schematic representation of the mechanism of action of vaccines

prevention from infection and or disease, reduced disease severity, or a decreased rate of hospitalization.

## 3.2 *Types of Vaccine*

Depending upon their timing of administration and ultimate objective, the vaccines can be broadly categorized into two types, i.e., prophylactic vaccines and therapeutic vaccines. Further, depending upon the nature of the antigen used for immunization, the vaccines can be of different types, i.e., protein vaccine, peptide vaccine, nucleic acid vaccine (i.e., recombinant DNA-based vaccines, mRNA based vaccines), anti-idiotype vaccines (use of anti-idiotype antibodies as vaccines), etc.

### 3.2.1 Prophylactic Vaccine

Here the vaccines aim to prevent a disease and or infection, and hence usually administered before the onset of infection and or disease. These vaccines mostly target infectious diseases caused by unicellular pathogens or multicellular parasites.

### 3.2.2 Therapeutic Vaccines

Here the vaccines are primarily used to treat a disease or prevent the recurrence/relapse of the disease. The most common therapeutic vaccines are used against cancer. These vaccines either boost or restore the effector function of the immune system to selectively recognize and attack cancer cells. Therapeutic cancer vaccines are part of the modality of immunotherapy (Dalgleish 2014).

## 4 Global Vaccination Program

The global vaccination coverage against many important childhood infectious diseases has been enhanced dramatically since the creation of WHO's Expanded Program on Immunization in the early 1970s and of the Global Alliance for Vaccination and Immunization in 2000. Polio has almost been eradicated, and success in controlling measles makes this infection another potential target for eradication (Greenwood 2014). Immunization currently prevents 3.5–5 million deaths every year from diseases like diphtheria, tetanus, pertussis, influenza, and measles (WHOWeblink1 n.d.). This has contributed to the appreciable decline in the mortality of children below 5 years of age from 93 deaths per 1000 live births in 1990 to 38 deaths per 1000 live births in 2018 (WHOWeblink2 n.d.). Vaccines not only prevent infections, but also prevent severe disease manifestations, and transmission of infections. The recent success of the COVID-19 vaccine in restoring normal life in the context of one of the dreaded global pandemics in the twenty-first century has reiterated the importance of vaccination programs in global public health. However, there are still many infections that are not preventable with vaccination, which represent a major cause of mortality worldwide. Developing vaccines against global pandemics such as HIV, malaria, and tuberculosis have been obstructed by several challenges (Nakaya and Pulendran 2015). Also, we do not have safe and effective vaccines against emerging infectious diseases that continually pose new threats (Pezeshki et al. 2019) [PMID: 30700167].

## 5 Factors Determining Vaccine-Induced Protection

The level of protection afforded by vaccination is affected by many genetic and environmental factors, including age, maternal antibody levels, prior antigen exposure, vaccine type/schedule, and vaccine dose.

## *5.1 Age of Vaccination*

The highest burden of and mortality from infectious disease occur in the first 5 years of life, with the youngest infants being most affected. For this reason, immunization programs have largely focused on this age group where there is the greatest benefit from vaccine-induced protection. It is somewhat inconvenient from an immunological perspective as the induction of strong immune responses in the first year of life is challenging. Indeed, vaccination of older children and adults would induce stronger immune responses.

## *5.2 Host Genotype*

The genetic makeup of host individuals plays a crucial role in modulating the immune response to vaccine antigens. Genetic polymorphism of different immune mediators can result in different outcomes of vaccination in different individuals.

## *5.3 Schedule of Vaccination*

For most vaccines that are used in the first year of life, 3–4 doses are administered by 12 months of age. Conventionally, in human vaccinology, "priming" doses are all administered at less than 6 months of age and the "booster" dose is given at 9–12 months of age.

A lifelong protection seems to be the rule following a single dose with some of the live-attenuated viral vaccines, such as the yellow fever vaccine.

An important parental concern is that vaccines might overwhelm their children's immune systems (Gellin et al. 2000). Despite the public impression that vaccines are associated with specific safety concerns, the existing data indicate that vaccines are remarkably safe as interventions to defend human health.

## *5.4 Incubation Period*

The memory response may be sufficient to protect against disease if there is a long incubation period between pathogen exposure and the onset of symptoms to allow for the 3–4 days required for memory B cells to generate antibody titers above the protective threshold. The memory response may not be sufficient to protect against disease if the pathogen has a short incubation period and there is a rapid onset of symptoms before antibody levels have reached the protective threshold (Pollard and Bijker 2021).

## 6 Immune Memory is an Important Feature of Vaccine-Induced Protection

Vaccines have been developed over the past centuries to provide direct protection for the immunized individual through the B- and T-cell-dependent mechanisms. As our immunological understanding of vaccines has developed, it has become apparent that this protection is largely manifested through the production of antibodies. Another important feature of vaccine-induced protection is the induction of immune memory (Pollard and Bijker 2021). Thus, the major objective of vaccination/immunization is to generate high levels of memory cells. The memory response may be insufficient to control a few infections, and sustained immunity for individual protection through vaccination can be difficult to achieve.

In some cases, antibody levels after primary vaccination remain above the protective threshold and can provide lifelong immunity (Pollard and Bijker 2021). If the antibody levels in the circulation wane after primary vaccination, often to a level below that required for protection, there is a solution to this; the provision of booster doses of vaccine through childhood (as is the case, for example, for diphtheria, tetanus, pertussis, and polio vaccines), in an attempt to sustain antibody levels above the protective threshold.

Whether immune memory can protect against a future pathogen encounter depends on the incubation time of the infection, the quality of the memory response, and the level of antibodies induced by memory B cells (Pollard and Bijker 2021).

## 7 "Herd Immunity" or "Herd Protection"

Herd immunity is an important feature of vaccine-induced protection. Vaccines cannot protect every individual in a population directly, as some individuals are not vaccinated for various reasons and others do not mount an immune response despite vaccination. Fortunately, however, if enough individuals in a population are vaccinated, and if vaccination prevents not only the development of disease but also infection itself (discussed in more detail below), the transmission of the pathogen can be interrupted and the incidence of disease can fall further than would be expected, because of the indirect protection of individuals who would otherwise be susceptible.

## 8 Novel Approaches for Vaccine Design

The identification of correlates of protection is helpful in vaccine development as they can be used to compare products and to predict whether the use of an efficacious vaccine in a new population (e.g., a different age group, medical background, or geographical location) is likely to provide the same protection as that observed in the original setting (Pollard and Bijker 2021).

The availability of novel vaccine technologies, together with the knowledge of the distinct human immune responses required to prevent different types of infection, should help design effective vaccines where conventional approaches have failed rationally (De Gregorio and Rappuoli 2012).

In silico approaches to vaccine design include designing and optimizing interfering peptides against protein–protein interactions (Hashemi et al. 2021). In silico tools are used for:

- Structure-based and off-structure approaches of peptide design.
- 3D peptide modeling
- Peptide optimization strategies like cyclization.
- Peptide binding energy evaluation.

New approaches to vaccine designs and their principles of action are stated in the following section.

### *8.1 Virus-Like Particles (VLPs)*

Self-assembling high-density viral capsid proteins formulate this vaccine type without infectious nucleic acid of the virus. It provides virus-like particles (VLPs), which is a safer alternative to traditional attenuated vaccines because of their inability to replicate. VLPs are relatively simpler in composition, allowing rapid production of vaccines; VLPs take a very short time, around 3–12 weeks, for production after a new strain is sequenced, compared to traditional vaccines, which take around 10 years (Wallis et al. 2019). Furthermore, VLPs are very useful for highly mutating pathogens like influenza.

### *8.2 Conjugate Vaccines*

A non-protein antigen or weak immunogen is conjugated with a protein carrier or a strong immunogen that enhances the overall immunogenicity of the vaccine preparation. The conjugation necessitates a careful pairing and orientation between the helper and target portion. Conjugate vaccines are low price and simple to manufacture, resulting in less chances of contamination than traditional vaccines. The most commonly used approved conjugated vaccine is the Hib vaccine (Droppa-Almeida et al. 2018).

### *8.3 Nucleic Acid Vaccines*

Nucleic acid-based vaccines including DNA vaccines and mRNA vaccines are revolutionary techniques in the field of immunization that generate immune responses like live, attenuated vaccines. Endogenous production of viral proteins with native conformation, glycosylation patterns, and other posttranslational changes occurs after nucleic acid vaccines are administered, simulating antigens produced during natural viral infection. Nucleic acid vaccines are also being explored in the case of cancer to elicit both cytotoxic T-lymphocyte and antibody response against a variety of protein antigens. Ease of the vector, its distribution, and expression are the major advantages of these vaccines (Lin et al. 2022).

### *8.4 Cellular Vaccines*

These vaccines are given to elicit effective cellular- and antibody-mediated responses, particularly for antigens generated only by cancer cells, leading to immunologic targeting of cancer cells. Due to its fine specificity and lack of incidental tissue harm observed with other traditional medications such as chemotherapy, the possibility to engage the adaptive immune system as a focused tool against cancer appeals. However, finding immunogenic tumor-specific antigen targets, choosing a platform to deliver the antigens, and enhancing the immune-stimulatory environment in which the vaccines are administered are a few challenges faced during the development of cellular vaccines (Yan et al. 2020).

## 9 Systems Vaccinology Approach

The role of systems biology approaches in determining molecular signatures for developing more effective vaccines (Pezeshki et al. 2019) [PMID: 30700167].

Recent advances in the application of the systems biology approach to analyze immune responses to vaccination in humans are beginning to yield new insights into mechanisms of vaccine immunity and to define molecular signatures induced rapidly after vaccination that correlate with and predict vaccine-induced immunity (Nakaya and Pulendran 2015) [PMID: 25964458].

It is important to discuss the potential of the systems vaccinology approach in defining novel correlates of protection in clinical trials and in infection-induced “experimental challenge models” in humans (Nakaya and Pulendran 2015) [PMID: 25964458].

## 10 Novel Approaches of Vaccine Delivery

Liposomes, polymeric particles, inorganic particles, plant-like material, infectious material, outer membrane vesicles, immune-stimulating complexes, and emulsions are the novel approaches used for vaccine delivery.

Most often, liposomes are utilized in vaccines as an adjuvant or a delivery system. When used in conjunction with liposomes, immunostimulatory chemicals such as cytokines or toll-like receptor agonists can be delivered to target immune cells, reducing systemic exposure to these adjuvant substances. In a diphtheria toxin vaccination, liposomes were used for the first time in 1974. Since then, hepatitis A (Epaxal) and influenza vaccinations using liposomes have been (Inflexal V) developed.

Due to their slow rate of biodegradation, polymeric particles, which come in a variety of natural and synthetic varieties, may either entrap or adsorb antigen for delivery to cells or they can allow for antigen release over time. In clinical trials, Advax, an insulin-derived polymeric microparticle, was employed as an adjuvant for hepatitis B, influenza, and insect-sting allergy vaccinations.

To boost the immune response, inorganic particles are used as adjuvants and antigen delivery systems. The most often utilized inorganic particles are calcium phosphate, aluminum, silica, and gold.

Because these vaccines are inexpensive and simple to administer orally, plant cells are also a desirable vaccine delivery system. Bacteria, viruses, and other infectious agents can be genetically altered by another pathogen. These bacterial and viral strains are used to make vaccines, which are thought to be safe. Outer membrane vesicles provide a different kind of platform for vaccine administration (Kim et al. 2021).

## 11 Newest Vaccines

Major progress has been made with the COVID-19 vaccination response, and it is critical to continue the progress, particularly for those most at risk of disease. During the COVID-19 pandemic, COVAX aimed to accelerate the development and manufacture of COVID-19 vaccines. COVAX was a historic multilateral effort co-led by Gavi, the Vaccine Alliance, the Coalition for Epidemic Preparedness Innovations (CEPI), the World Health Organization (WHO), and UNICEF from 2020 through 2023. Other vaccines against COVID-19 include Moderna COVID-19 (mRNA-1273) vaccine, Nuvaxovid (NVX-CoV2373), and Covovax (NVX-CoV2373) vaccine.

Malaria is an illness caused by *Plasmodium* parasites transmitted to humans by infected mosquitoes. Of the five species that infect humans, *Plasmodium falciparum* exacts the highest toll in terms of humans. RTS, S/AS01 (Mosquirix) malaria vaccine targets the circumsporozoite protein on the sporozoite surface and targets *P. falciparum* parasites before they infect liver cells, it is considered as a pre-erythrocytic vaccine. Mosquirix induces antibodies against circumsporozoite protein; thus, limiting the ability of the parasites to mature in the liver and cause clinical disease. The new vaccine, R21/Matrix-M vaccine is the second malaria vaccine recommended by WHO, following the RTS, S/AS01 vaccine, which received a WHO recommendation in 2021. Both vaccines are shown to be safe and effective in preventing malaria in children and, when implemented broadly, are expected to have a high public health impact (WHO- Web link-3).

Future potential targets of vaccine-based treatment include lifestyle diseases like diabetes, hypertension, addiction, and Alzheimer's disease (Greenwood 2014).

## 12 Challenges in Developing New Generation Vaccines

Developing a successful vaccine to control infectious diseases and/or cancer is typically driven by decades of fundamental research into pathogen biology, the host response to infection (Bansal et al. 2022), and cancer biology (Dalgleish 2014). The higher cost associated with the development of vaccines is a major challenge. Besides, scientific challenges in developing successful vaccines often include difficulties in discovering protective antigens, especially in the context of genetic polymorphisms and the natural evolution of the pathogens over time. Further, the genetic background/genotype of the host, age, and several other factors strongly modulate the immune response of the host and hence play a determining role in predicting the effectiveness of vaccines in individuals. Another major challenge is to determine the quality of the protective immune response required to confer protection against an infection as sometimes either the cellular or humoral immune response in isolation is not enough, we might be required to elicit both the cellular and humoral immune response to realize true lasting benefits. Further, pathogens and cancer cells often adopt different escape mechanisms to evade host immune response. Hence, a comprehensive understanding of the mechanism of immune response (i.e., which immunological parameters are stimulated by vaccination—correlates of immunogenicity, or are associated with protection against subsequent infection, as determined in clinical trials—correlates of protection, escape mechanisms adopted by pathogens or cancer cells, etc.) is crucial in designing effective vaccines.

In the case of cancer, there are often many antigens shared by both the normal and cancer cells, and sometimes the same type of cancer in different individuals might be expressing different antigens making it difficult to select potential vaccine antigens for universal application. The heterogeneity of solid tumors (i.e., simultaneous expression of more than one antigen) adds further complexity to the field of cancer vaccines and necessitates the inclusion of multiple antigens in a single vaccine formulation to effectively target solid tumors. All these difficulties have led to the emergence of the concept of personalized vaccines for cancer.

The final challenge is to devise strategies to induce protective immunity that is long-lasting and with minimal side effects. Adjuvants are often used to amplify the magnitude of immune response during vaccination, by activating the antigen-presenting cells. However, sometimes due to overactivation, the use of adjuvants might have serious side effects and prove detrimental; in other cases, some adjuvants might have suboptimal activity. Hence, the selection of suitable adjuvants is another inherent challenge in the field of vaccination.

## 13 Conclusion

Immunotherapy is finding successful application in a wide variety of diseases including cancer, infectious diseases, allergies, organ transplantation, inflammatory and autoimmune diseases, etc. because of its target specificity, and minimal/manageable side effects. It is exciting to note here that the scope of immunotherapy is now being extended to include metabolic and lifestyle disorders including cardiovascular diseases (Lutgens et al. 2019; Rurik et al. 2021; Khambhati et al. 2018) and diabetes (Rapini et al. 2020), etc. Future research in the field is going to focus on improved versions of existing immunotherapeutic approaches and exploring approaches to have long-lasting benefits with minimal side effects.

## Glossary

**Adjuvant** Specific compounds which are typically used along with antigens during vaccination to activate the antigen-presenting cells (i.e., dendritic cells) and amplify the immune response.

**Bi-specific antibodies** Antibodies having binding specificities to specific epitopes of two different antigens which are usually present on two different target cells.

**CAR-T cell therapy** In this process, T cells are isolated and engineered to express a chimeric antigen receptor (CAR) that specifically recognizes an antigen on cancer cells and selectively eliminate those cancer cells.

**Cellular immune response** Immune response where the T cells are activated and the effector function is mediated by these activated T cells.

**Check point inhibition** Restoration of the effector function of exhausted T cells by blocking different co-inhibitory receptors expressed on them by using specific monoclonal antibodies.

**Humoral immune response** Immune response where the B cells are activated and the effector function is mediated by antibodies produced from activated B cell-derived plasma cells.

**Immunotherapy** Treatment of different immune-related diseases by manipulating the immune system or modulating the immune response.

**Monoclonal antibodies** Antibodies having binding specificity to single epitope of an antigen.

**TCR T cell therapy** In this process, T cells are isolated and engineered to express a T cell receptor (TCR) that specifically recognizes specific epitopes presented by specific HLA (human leukocyte antigen) on cancer cells to selectively kill those cancer cells.

**Vaccine** A formulation of antigen that is used to immunize an individual to elicit specific cellular and/or humoral immune response conferring protection against specific infection and/or disease.

## References

Andre FE, Booy R, Bock HL, Clemens J, Datta SK, John TJ, Lee BW, Lolekha S, Peltola H, Ruff TA, Santosham M, Schmitt HJ (2008) Vaccination greatly reduces disease, disability, death and inequity worldwide. Bull World Health Organ 86(2):140–146. https://doi.org/10.2471/blt.07.040089

Bansal S, Prakash A, Medhi B (2022) New generation vaccine: novel approaches of vaccine design and delivery and current challenges of vaccine development. Indian J Pharmacol. https://doi.org/10.4103/ijp.ijp_555_22

Baulu E, Gardet C, Chuvin N, Depil S (2023) TCR-engineered T cell therapy in solid tumors: State of the art and perspectives. Sci Adv. 9(7):eadf3700. https://doi.org/10.1126/sciadv.adf3700. Epub 2023 Feb 15

Cantarovich D, Vistoli F, Soulillou JP (2008) Immunosuppression minimization in kidney transplantation. Front Biosci. 13:1413–1432

Caspi RR (2008) Immunotherapy of autoimmunity and cancer: the penalty for success. Nat Rev Immunol 8(12):970–976

Dalgleish AG (2014) Vaccines versus immunotherapy: overview of approaches in deciding between options. Hum Vaccin Immunother 10(11):3369–3374. https://doi.org/10.4161/21645515.2014.980707

De Gregorio E, Rappuoli R (2012) Vaccines for the future: learning from human immunology. Microb Biotechnol 5(2):149–155. https://doi.org/10.1111/j.1751-7915.2011.00276.x. Epub 2011 Aug 31

Droppa-Almeida D, Franceschi E, Padilha FF (2018) Immune-informatic analysis and design of peptide vaccine from multi-epitopes against corynebacterium pseudotuberculosis. Bioinform Biol Insights 12:1177932218755337. https://doi.org/10.1177/1177932218755337

Durham SR, Shamji MH (2023) Allergen immunotherapy: past, present and future. Nat Rev Immunol 23:317–328

Gellin BG, Maibach EW, Marcuse EK (2000) Do parents understand immunizations? A national telephone survey. Pediatrics 106:1097–1102

Greenwood B (2014) The contribution of vaccination to global health: past, present and future. Philos Trans R Soc Lond Ser B Biol Sci 369(1645):20130433. https://doi.org/10.1098/rstb.2013.0433

Hashemi ZS, Zarei M, Fath MK, Ganji M, Farahani MS, Afsharnouri F, Pourzardosht N, Khalesi B, Jahangiri A, Rahbar MR, Khalili S (2021) In silico approaches for the design and optimization of interfering peptides against protein-protein interactions. Front Mol Biosci 8:669431. https://doi.org/10.3389/fmolb.2021.669431

Justiz Vaillant AA, Qurie A (Jun 26, 2023) Immunodeficiency. In: StatPearls [Internet]. StatPearls Publishing, Treasure Island (FL)

Justiz Vaillant AA, Nessel TA, Zito PM (2024 Jan) Immunotherapy. [Updated 2023 Jun 5]. In: StatPearls [Internet]. StatPearls Publishing, Treasure Island (FL). Available from: https://www.ncbi.nlm.nih.gov/books/NBK519046/

Khambhati J, Engels M, Allard-Ratick M, Sandesara PB, Quyyumi AA, Sperling L (2018) Immunotherapy for the prevention of atherosclerotic cardiovascular disease: promise and possibilities. Atherosclerosis 276:1–9

Kim D, Wu Y, Kim YB, Oh YK (2021) Advances in vaccine delivery systems against viral infectious diseases. Drug Deliv Transl Res 11(4):1401–1419. https://doi.org/10.1007/s13346-021-00945-2. Epub 2021 Mar 10

Klein C, Brinkmann U, Reichert JM et al (2024) The present and future of bispecific antibodies for cancer therapy. Nat Rev Drug Discov 23:301–319

Lin MJ, Svensson-Arvelund J, Lubitz GS et al (2022) Cancer vaccines: the next immunotherapy frontier. Nat Cancer 3:911–926

Lutgens E, Atzler D, Döring Y, Duchene J, Steffens S, Weber C (2019) Immunotherapy for cardiovascular disease. Eur Heart J 40:3937–3946. https://doi.org/10.1093/eurheartj/ehz283

Marshall JS, Warrington R, Watson W et al (2018) An introduction to immunology and immunopathology. Allergy Asthma Clin Immunol 14(Suppl 2):49

McInnes IB, Gravallese EM (2021) Immune-mediated inflammatory disease therapeutics: past, present and future. Nat Rev Immunol 21:680–686

Miller HK, Hanley PJ, Lang H, Lazarski CA, Chorvinsky EA, McCormack S, Roesch L, Albihani S, Dean M, Hoq F, Adams RH, Bollard CM, Keller MD (2018) Antiviral T cells for adenovirus in the pretransplant period: a bridge therapy for severe combined immunodeficiency. Biol Blood Marrow Transplant 24(9):1944–1946

Nakaya HI, Pulendran B (2015) Vaccinology in the era of high-throughput biology. Philos Trans R Soc Lond Ser B Biol Sci 370(1671):20140146. https://doi.org/10.1098/rstb.2014.0146

Oiseth SJ, Aziz MSA-E (2017) Cancer immunotherapy: a brief review of the history, possibilities, and challenges ahead. J Cancer Metastasis Treat 3:250–261

Paludan SR, Pradeu T, Masters SL et al (2021) Constitutive immune mechanisms: mediators of host defence and immune regulation. Nat Rev Immunol 21:137–150

Pezeshki A, Ovsyannikova IG, McKinney BA, Poland GA, Kennedy RB (2019) The role of systems biology approaches in determining molecular signatures for the development of more effective vaccines. Expert Rev Vaccines 18(3):253–267. https://doi.org/10.1080/14760584.2019.1575208. Epub 2019 Feb 11

Pollard AJ, Bijker EM (2021) A guide to vaccinology: from basic principles to new developments. Nat Rev Immunol 21:83–100

Rapini N, Schiaffini R, Fierabracci A (2020) Immunotherapy strategies for the prevention and treatment of distinct stages of type 1 diabetes: an overview. Int J Mol Sci 21:2103. https://doi.org/10.3390/ijms21062103

Rosenberg SA, Spiess P, Lafreniere R (1986) A new approach to the adoptive immunotherapy of cancer with tumor-infiltrating lymphocytes. Science 233(4770):1318–1321. https://doi.org/10.1126/science.3489291

Rurik JG, Aghajanian H, Epstein JA (2021) Immune cells and immunotherapy for cardiac injury and repair. Circ Res 128:1766–1779

Sangeetha VP, Xavier J, Valappil MP (2023) A review of immune modulators and immunotherapy in infectious diseases. Mol Cell Biochem. https://doi.org/10.1007/s11010-023-04825-w

Schnell A, Bod L, Madi A et al (2020) The yin and yang of co-inhibitory receptors: toward antitumor immunity without autoimmunity. Cell Res 30:285–299

Shalhout SZ, Miller DM, Emerick KS et al (2023) Therapy with oncolytic viruses: progress and challenges. Nat Rev Clin Oncol 20:160–177

Sterner RC, Sterner RM (2021) CAR-T cell therapy: current limitations and potential strategies. Blood Cancer J 11:69

Thangamathesvaran L, Shah R, Verma R, Mahmoud O (2018 Apr) Immune checkpoint inhibitors and radiotherapy-concept and review of current literature. Ann Transl Med 6(8):155

Waldman AD, Fritz JM, Lenardo MJ (2020) A guide to cancer immunotherapy: from T cell basic science to clinical practice. Nat Rev Immunol 20:651–668

Wallis J, Shenton DP, Carlisle RC (2019) Novel approaches for the design, delivery and administration of vaccine technologies. Clin Exp Immunol. 196(2):189–204. https://doi.org/10.1111/cei.13287. Epub 2019 Apr 8

Wallis RS, O'Garra A, Sher A et al (2023) Host-directed immunotherapy of viral and bacterial infections: past, present and future. Nat Rev Immunol 23:121–133

WHOWeblink1 (n.d.). https://www.who.int/health-topics/vaccines-and-immunization#tab=tab_1

WHOWeblink2 (n.d.). https://www.who.int/data/gho/data/themes/topics/topic-details/GHO/child-mortality-and-causes-of-death

Xie N, Shen G, Gao W et al (2023) Neoantigens: promising targets for cancer therapy. Signal Transduct Target Ther 8:9

Xu Y, Zhang Q, Zhan X, Xie D, Dai G, Yang H (2013) [Preparation and immunological evaluation of oral solution of egg yolk-derived hepatitis B virus-specific transfer factor]. Nan Fang Yi Ke Da Xue Xue Bao 33(12):1827–1830

Yan Y, Zeng S, Gong Z, Xu Z (2020) Clinical implication of cellular vaccine in glioma: current advances and future prospects. J Exp Clin Cancer Res 39(1):257

# Blocking Co-inhibitory Receptors: A Promising Cancer Immunotherapy strategy

**Nandini Acharya**

**Abstract** The quality of T cell response, which is initiated through antigen recognition by the T cell receptor (TCR), is regulated by a balance between co-stimulatory and co-inhibitory signals. Under normal physiological conditions, co-inhibitory molecules are instrumental in preventing activation-induced cell death in T cells, maintaining self-tolerance, and reducing collateral damage to healthy tissue from the cytopathic effects of T cells and aberrant inflammation. Cancer cells can usurp these inhibitory pathways that the immune system uses to maintain self-tolerance and evade the anti-tumor immune responses. Given that CD8 + T cells have the capacity to recognize and kill the cognate antigen-expressing cell types, T cells have been the cornerstone of anti-tumor therapeutic interventions. Several co-inhibitory receptors have been identified and studied in cancer in past decades, including but not limited to PD-1, CTLA-4, LAG3, TIM3, TIGIT, and BTLA. They are named "immune checkpoints" referring to molecules that act as gatekeepers of immune responses. Co-inhibitory receptors are heavily expressed on a subset of T cells described as exhausted T cells (Tex) or dysfunctional T cells (Tdys). Dysfunctional T cells have reduced effector function and are generated in place of memory T cells when the T cells are chronically activated as in case of viral infection or in the tumor microenvironment (TME). Dysfunctional T cells in the TME prevent the establishment of durable immunological memory and permit tumor progression or relapse. Thus, targeting these dysfunctional T cells with antibody against co-inhibitory receptors is a promising strategy of cancer immunotherapy.

**Keywords** Co-inhibitory receptors · Checkpoint blockade · PD-1 · CTLA4 · LAG3 · TIM3 · TIGIT

N. Acharya (✉)
Pelotonia Institute of Immuno-Oncology, The Ohio State University Comprehensive Cancer Center, Columbus, OH, USA
e-mail: acharya.108@osu.edu



A. Das Mohapatra, P. S. Sahu (eds.), *Advances in Immunology and Immuno-techniques*, Interdisciplinary Biotechnological Advances,
https://doi.org/10.1007/978-981-97-5508-0_2

# 1 Introduction

The importance of immune cells in malignant progression has been appreciated since the early nineteenth century wherein a New York-based surgeon, William Coley, correlated the occurrence of post-operative infection with improved clinical outcomes in cancer patients. However, the scientific interest in immuno-oncology was reinvigorated over the past decade with the advent of checkpoint blockade therapy. Based on therapeutic gains being made in two fields: chimeric antigen receptor (CAR)-modified T cells and immune modulation using antibodies that block immune regulatory checkpoints, cancer immunotherapy was hailed as the breakthrough of the year in 2013.

Agonists of co-stimulatory receptors or antagonists of co-inhibitory molecules both of which result in the augmentation of antigen-specific T cell responses are the primary agents in current clinical testing. Indeed, the therapeutic impact of Immune Checkpoint Blockade (ICB) is tremendous and is now a widely used therapeutic modality. ICB can induce durable responses across multiple types of cancer, which has enabled the oncology community to begin to envision potentially curative therapeutic approaches. There is a growing number of cancers for which single-agent checkpoint blockade or a combination of checkpoint inhibitors is a US Food and Drug Administration (FDA)-approved treatment. Considering the mechanisms of action of inhibitors of various immune checkpoints, it is crucial to appreciate the diversity of immune functions that they regulate. For example, the two immune checkpoint receptors that have been most actively studied in the context of clinical cancer immunotherapy, cytotoxic T-lymphocyte-associated antigen 4 (CTLA4; also known as CD152) and programmed cell death protein 1 (PD1; also known as CD279)—which are both inhibitory receptors—regulate immune responses at different levels and by different mechanisms. The clinical activity of antibodies that block either of these receptors implies that anti-tumor immunity can be enhanced at multiple levels and the combinatorial strategies can be intelligently designed, guided by mechanistic considerations and preclinical models. However, the major bottleneck of immune checkpoint blockade therapy is its low response rate in most cancers, with a range of 10–30% (Ribas and Wolchok 2018). Mechanisms of non-responsiveness have been extensively studied, and many factors have been found to be relevant, such as tumor mutational burden, PD-L1 expression level, IFN signaling, and MHC-I loss (Le et al. 2015; Samstein et al. 2019; Topalian et al. 2012; Gao et al. 2016). However, biomarkers that faithfully predict efficacy are still lacking. A better understanding of checkpoint biology is therefore urgently needed to design next-generation therapies and to improve the clinical protocols of current therapies.

## 2 Checkpoint Blockade in the Clinic

Ipilimumab, an anti-CTLA4 monoclonal antibody has been approved by the FDA for the treatment of several indications (Table 1). Interestingly, however, a phase III randomized clinical trial comparing tremelimumab, another anti-CTLA4 mAb, with standard-of-care chemotherapy in 655 patients with advanced metastatic melanoma could not demonstrate a significant survival advantage; this therapy failed in multiple late-phase clinical trials (Sade-Feldman et al. 2017). Tremelimumab is not currently approved for cancer therapy. Ipilimumab is an IgG1 mAb, whereas tremelimumab is an IgG2 mAb: these two isotypes have different ADCC and CDC activities (Ribas et al. 2013). The clinical failure of tremelimumab highlights the complexities associated with creating an effective therapeutic antibody.

Additionally, six monoclonal antibodies targeting the PD1–PDL1 or PD1–PDL2 pathway have been approved by the FDA for the treatment of numerous cancer types (Table 1). Recently, a combination of nivolumab (anti-PD1) and relatlimab (anti-lag3) has been FDA-approved for metastatic melanoma (Table 1).

Of note, despite dramatic positive responses observed in a subset of patients, the overall response rate of patients to these monotherapies remained modest. For example, the objective response rate (ORR) for ipilimumab (anti-CTLA-4) was about 10–16%, and for nivolumab (anti-PD-1) was 30–40% in patients with metastatic melanoma. Indeed, combining ipilimumab and nivolumab did increase the ORR and survival rates of patients to about 60% in various cancers (Furness et al. 2014; Rotte 2019; Callahan et al. 2017). However, a significant fraction of patients developed severe irAEs even with monotherapies (Table 1) and was even worse with combined CTLA-4 and PD-1 blockade (Wolchok et al. 2017). Importantly, some tumors are refractory or acquire tumor resistance to these therapies preventing clinical success (Myers 2018). Further investigations are required to achieve the full potential of ICB in different types of cancers.

### *2.1 CTLA4*

CTLA4, the first immune checkpoint receptor to be clinically targeted, is expressed primarily in T cells where it regulates the amplitude of the early stages of T cell activation. In resting T cells, CTLA-4 is an intracellular protein; however, after T cell receptor engagement and a co-stimulatory signal through CD28, CTLA-4 translocates to the cell surface where it outcompetes CD28 for binding to critical co-stimulatory molecules (CD80, CD86) and mediates inhibitory signaling into the T cell, resulting in the arrest of both proliferation and activation (Sharma et al. 2017). One of the proposed mechanisms via which CTLA4 dampens immune responses is due to the active removal of co-stimulatory molecules such as CD80 and CD86 from the surface of professional antigen-presenting cells such as dendritic cells and thus sequestrating CD80 and CD86 from CD28 engagement (Chambers et al. 2001)

**Table 1** Co-inhibitory receptors as therapeutic targets in cancer and autoimmunity

| | Features | CTLA-4 | LAG-3 | PD-1 | TIM3 | TIGIT |
|---|---|---|---|---|---|---|
| 1. | Discovery | 1987 | 1990 | 1992 | 2002 | 2009 |
| 2. | Clinical trial in Cancer | Anti-CTLA 4 antibody is approved as a monotherapy or in combination with PD-1. Since its discovery it has been used in around 40 clinical trials in different cancers including melanoma, prostate cancer, pancreatic cancer, breast cancer, colorectal cancer, Non small cell lungs cancer etc. | Anti-LAG-3 antibody is also approved as a monotherapy or in combination with PD-1. Since its discovery it has been used in around 20 clinical trials in different cancers including renal cancer, metastatic breast cancer, pancreatic cancer or hematologic malignancies etc. | Anti-PD-1 antibody is approved as a monotherapy or in combination with antibodies targeting other co-inhibitors i.e. CTLA-4, LAG-3, T1GIT, and anti-PD-L1 etc. Since its discovery anti-PD1 antibody has been used in about 1000 clinical trials in many different types of cancer. | Anti-TIM-3 antibody has been approved for use as a monotherapy or in combination with anti-PD-1 antibodies. At least, 10 clinical trials are evaluating anti-TIM-3 alone or in combination with anti-PD1 antibodies. | Anti-TIGIT antibody has been approved for use as a monotherapy or in combination with anti-PD-1 antibodies. At least, 10 clinical trials are evaluating anti-TIGIT alone or in combination with anti-PD1 antibodies. |
| 3. | Clinical trial in Autoimmunity | CTLA-4 Ig, Abatacept has been approved by FDA for auto immune diseases. It has been used in about 15 clinical trials for different auto immune diseases like Rheumatoid arthritis, Type 1 Diabetes, and multiple sclerosis etc. | A depleting anti-LAG-3 antibody (targeting activated T cells) is under clinical trial in autoimmune diseases (# GSK2831781) Another preclinical agonist anti-LAG-3 antibody (#IMP761) has shown promise in skin inflammation. | Anti-PD1 antibody is not commonly used in clinical trials against autoimmunity. However, apparently, a PD-1 agonist antibody (#CC90006) has been tried in autoimmune disease. | Application of anti-TIM-3 antibodies in autoimmune diseases are mostly in preclinical stage and showing great promise in auto immune disorders like EAEin animal models | Preclinical studies have demonstrated encouraging results in mouse models of EAE,SLE etc. in animal models |

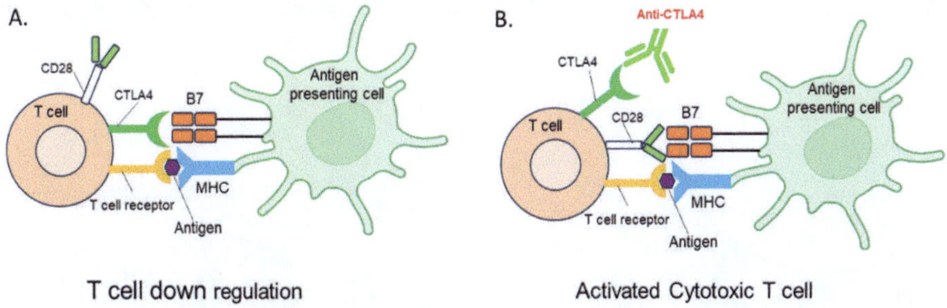


**Fig. 1** (**a**) Interaction of CTLA4 and B7 inhibits T cell activation. (**b**) Blocking of CTLA-4 by anti-CTLA4 antibody results in activation of T cells

given the fact that CTLA4 has a much higher overall affinity for both ligands, compared to CD28 (Fig. 1).

Additionally, the other proposed mechanism includes CTLA4-mediated blocking of T cell activation via activation of the protein phosphatases such as SHP2 (also known as PTPN11) and PP2A that are important in counteracting kinase signals that are induced by TCR and CD28 (Qureshi et al. 2011). The central and non-redundant role of CTLA4 as a co-inhibitory pathway is dramatically demonstrated by the lethal systemic immune hyperactivation phenotype of *Ctla4*-knockout mice (Rudd et al. 2009; Tivol et al. 1995). CTLA4 plays a fundamental role in the functionality of Tregs where it is expressed constitutively. CTLA4 engagement on Tregs enhances their suppressive function. *CTLA4* is a target gene of the forkhead transcription factor FOXP3, the signature transcription factor of Tregs. Importantly, Tregs specific blockade of CTLA4 significantly inhibits their ability to regulate both autoimmunity and anti-tumor immunity (Waterhouse et al. 1995).

## 2.2 *PD1*

PD1 is expressed on activated T cells, and it is known to limit the activity of T cells in the peripheral tissues at the time of an inflammatory response to infection and to limit autoimmunity (Peggs et al. 2009; Ishida et al. 1992; Freeman et al. 2000). However, this major co-inhibitory pathway is coopted by the tumor microenvironment and constitutes a major immune resistance mechanism (Nishimura et al. 1999). The two ligands of PD1 are PDL1 (PDL1; also known as B7-H1 and CD274) (Ishida et al. 1992) and PDL2 (also known as B7-DC and CD273) (Dong et al. 2002). PD-L1 is broadly expressed on hematopoietic and non-hematopoietic parenchymal tissue cells, and PD-L2 is only expressed on DCs and some subsets of myeloid cells. These ligands belong to the B7 family members, and they share 37% sequence homology (Fig. 2).

When engaged by one of its ligands, PD1 inhibits downstream kinases that are involved in T cell activation through the phosphatase SHP2. Additionally, PD1

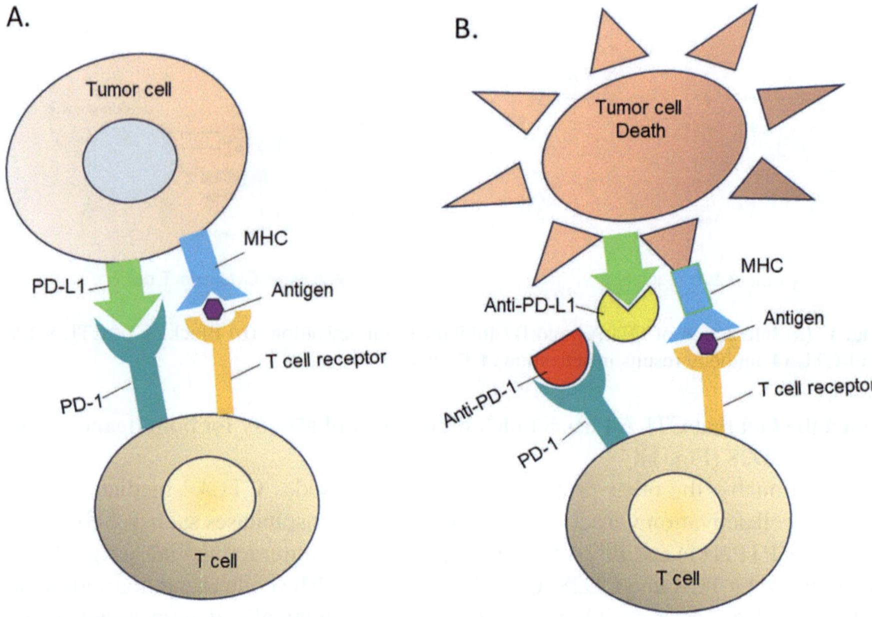


**Fig. 2** (**a**) Interaction of PDL-1 with PD-1 inhibits T cell mediated killing of tumor cell. (**b**) Blocking of PDL-1 or PD-1 results in killing of tumor cell by T cells

engagement may also regulate the activity of T cells by modifying the duration of T cell–APC or T cell–target cell contact. Chronic activation of T cells results in the production of a pro- inflammatory cytokine such as IFN-γ, which is the strongest stimulator of PD-L1 expression (Velu et al. 2009; Baumeister et al. 2016) and continuous PD-1 signaling in T cells induces an epigenetic program of T cell exhaustion (Latchman et al. 2001).

PD1 is also highly expressed on Treg cells, and PD-1 blockade augments the proliferation and suppressive activity of Tregs (Sen et al. 2016; Fife et al. 2009) . Because many tumors are highly infiltrated with $T_{reg}$ cells, the efficacy of anti-PD1 could possibly be enhanced by combining it with treatment strategies that deplete or reduce the function of Tregs.

Additionally, PD1 is also expressed on other activated non-T cell subsets, including B cells and natural killer (NK) cells. Thus, the tumor-controlling effect of anti-PD1 could also partially be mediated through altered antibody production on $PD1^+$ B cells (Tan et al. 2021). (Baumeister et al. 2016; Garcia-Diaz et al. 2017) Furthermore, the relatively mild phenotypes of *PD1* (also known as *Pdcd1*), *PDL1* (also known as *Cd274* and *Pdcd1lg1*), and *PDL2* (also known as *Cd273* and *Pdcd1lg2*) knockout mice suggest that blockade of this pathway would result in less collateral immune toxicity than for CTLA4 blockade, which seems to be the case in clinical trials. A PD1 antibody would block PD1 from interacting with both PDL1 and PDL2 but not the interaction between PDL1 and CD80. By contrast, most PDL1 antibodies block the interaction between PDL1 and CD80 and between PDL1 and PD1 but would not block PD1 from interacting with PDL2. Thus, it is possible that, depending on which interactions dominate in particular cancer, PD1 and PDL1 antibodies might not have redundant activity.

### *2.3 LAG-3*

LAG-3 (CD223) is a cell surface molecule expressed on activated T cells (Kamada et al. 2019) (Francisco et al. 2009), NK cells (Francisco et al. 2009), B cells (Velu et al. 2009), and plasmacytoid dendritic cells (Baumeister et al. 2016). Lag3 interacts with major histocompatibility complex (MHC) class II. Although the mechanism of action of LAG-3 is incompletely understood, its interaction with MHC class II causes downregulation of T cell cytokine production, CD4 and CD8 T cell expansion, and favors Treg phenotype (Garcia-Diaz et al. 2017). Lag3 is highly expressed on exhausted CD8+ T cells along with other co-inhibitory receptors. In contrast to mice deficient in other co-inhibitory receptors, such as CTLA-4 and PD-1, mice deficient in LAG-3 are not susceptible to autoimmunity unless bred to a permissive genetic background. LAG-3 deficiency in B6.SJL mice result in higher susceptibility to Hg-induced autoimmunity (Huard et al. 1994). In addition, NOD mice deficient for LAG-3 display accelerated T1D with 100% incidence (Triebel et al. 1990). Anti-Lag3 antibody was FDA-approved for metastatic melanoma in 2022.

## 3 Immune-Related Adverse Events (irAE)

Single-agent PD-1 pathway blockade has a relatively favorable toxicity profile, with toxicities requiring medical intervention (grade 3–4) in the range of 10–15% in the majority of series (Kisielow et al. 2005; Workman et al. 2009). About 5% of the patients have to discontinue the therapy due to toxicities. The most common treatment-related adverse events of any grade are fatigue, diarrhea, rash, and pruritus in 20–15% of patients. In a smaller percentage of patients, toxicities are more serious and include several endocrinopathies, where the immune system infiltrates a hormone-producing gland and leads to permanent dysfunction requiring life-long substitutive hormonal therapy, such as thyroid disorders, hypophysitis, adrenal gland disorders, and type I diabetes. Serious visceral organ inflammatory toxicities are uncommon (~1%) but can affect any organ including the brain (encephalopathy), meninges (meningitis), lung (pneumonitis), heart (myocarditis), gastrointestinal tract (esophagitis, colitis), liver (hepatitis), kidney (nephritis), muscles (myositis), and joints (arthritis). Frequently, the clinically relevant toxicities with both PD-1 and CTLA-4 blockade therapy are managed with immune suppressive therapies such as with high doses of corticosteroids (Goldberg and Drake 2010). Although recent studies suggest that steroids should be administered to ICB-treated patients with caution as it may impair the effector CD8+ T cell responses (Jha et al. 2014).

## 4 Mechanism of Action of Checkpoint Therapy

Although the mechanisms behind the ability of ICB to mitigate tumor growth are still unclear comparable trends in the functionality of T cells of those patients who show objective responses to checkpoint blockade treatment have been observed. A positive clinical outcome correlates with increased T cell infiltration into the TME and increased activation in terms of the production of pro-inflammatory cytokines and cytotoxic molecules such as GranzymeB and CD107a (Bettini et al. 2011). In addition to the changes in the CD8+ T cell compartment, profound changes in the CD4+ T cells have also been observed; these changes include an increase in the number of T helper 1 cells (Th1 cells) and a concomitant reduction in the number of immunosuppressive regulatory T cells (Tregs). An emerging body of literature indicates that exhausted T cells or dysfunctional T cells are terminally differentiated with a distinct chromatin landscape and hence have limited capacity for functional plasticity.

Owing to the terminal differentiation state and unique epigenetic signature of dysfunctional T cells (Blackburn et al. 2009; Robert et al. 2015; Garon et al. 2015), it is plausible that ICB may not be acting via rejuvenation of the exhausted T cells even though it may potentially rescue some of the effector functions of the T cells but rather by affecting the TME in such a way that it subsequently leads to recruitment of new T cells into the TME. Interestingly, recent studies of T cell dynamics indicate that the T cell response to ICB may originate and rely on peripheral T cell recruitment (Daud et al. 2016; Pauken et al. 2016; Philip et al. 2017). Combination therapies that recruit new effector cells, in addition to preventing T cell dysfunction, may be a critical approach in generating long-lasting anti-tumor immune responses.

Of note, TCF1+ T cells display a stem-like or progenitor phenotype, undergo a proliferative burst upon PD1 blockade, and provide tumor control (Mognol et al. 2017; Acharya and Anderson 2019; Yost et al. 2019). Importantly, it has been shown that patients that respond to ICB have higher frequency of TCF1+ T cells in the TME compared to the non-responsive patients (Im et al. 2016).

In addition, the nuclear factor TOX may be critical for T cell dysfunction caused by persistent antigen exposure (Miller et al. 2019). Not only did the deletion of TOX reduce the expression of co-inhibitory receptors on T cells but it also resulted in the loss of antigen-specific T cells within the TME. TOX expression is driven by TCR-dependent NFAT activation, so, given that antigen-specific T cell persistence is lacking in a tumor, TOX may help dampen the robust tonic TCR signal that drives exhaustion but provides protection against activation-induced cell death (Miller et al. 2019).

# 5 Second Generation of Co-inhibitory Receptors

Inhibitors for a number of second-generation immune checkpoint targets are either entering the clinic or are under active development. Several clinical trials are currently investigating the use of TIM-3 and TIGIT and as targets for monotherapy or in combination with anti-PD-1 and have shown promise for cancer therapy.

## 5.1 *TIM-3*

T cell immunoglobulin and mucin domain-containing protein 3 (TIM-3) is a member of the TIM family of immunoregulatory proteins. Although it was originally identified as a molecule expressed by interferon-γ (IFNγ)-producing $CD4^+$ and $CD8^+$ T cells, many other cell types, myeloid cells, natural killer (NK) cells, have been shown to express TIM3 (Siddiqui et al. 2019). Thus, therapeutic targeting of TIM-3 likely modulates immune responses by acting on multiple cell types.

In contrast to the more "classic" checkpoint receptors such as programmed cell death 1 (PD1) and TIGIT, TIM3 is unique as it lacks known inhibitory signaling motifs in its cytoplasmic tail. The main ligands of Tim3 include galectin-9 (Gal9), phosphatidylserine, HMGB1, and carcinoembryonic antigen-related cell adhesion molecule (CEACAM)-1 (Siddiqui et al. 2019). Tim3 is highly expressed on dysfunctional CD8+ T cells. High TIM-3 levels have correlated with poor prognosis in various malignancies (e.g., prostate, renal cell, colon, cervical) (Sade-Feldman et al. 2018). Furthermore, it has been shown that Tim3 plays a fundamental role in regulating the function of intra-tumoral DCs and regulating anti-tumor immune responses (Scott et al. 2019; Acharya et al. 2020). Given its expression in various T cell subsets and in the myeloid cells, the combination of anti-PD-1 and anti-TIM-3 has become a particularly attractive target (Siddiqui et al. 2019) . Given the wide expression pattern of TIM3, therapeutic targeting of TIM-3 likely modulates immune responses by acting on multiple immune cell types. Three monoclonal antibodies under investigation include MBG453, Sym023, and TSR-022.

## 5.2 *TIGIT*

T cell immunoglobulin and ITIM domain (TIGIT) (also called WUCAM, Vstm3, VSIG9) is a receptor of the Ig superfamily, which plays a critical role in limiting adaptive and innate immunity (Johnston et al. 2014). TIGIT is expressed by activated $CD8^+$ T and $CD4^+$ T cells, natural killer (NK) cells, regulatory T cells (Tregs), and follicular T helper cells in humans (de Mingo Pulido et al. 2021). The ligands of TIGIT include CD155 (PVR/NECL-5) and CD112 (Nectin-2/PVRL2). Interestingly, much like the CD28/CTLA-4/CD80/CD86 pathway, for which inhibitory and co-stimulatory receptors compete for binding to the same ligands, TIGIT also has one competing co-stimulatory receptor, DNAM-1/CD226. CD226 is widely expressed by immune cells, including T cells, NK cells, monocytes, and platelets. Of note, in sharp contrast with *CTLA-4* $^{-/-}$ mice, *TIGIT*$^{-/-}$ mice do not develop autoimmunity however compared with wild-type mice, *Tigit*$^{-/-}$ mice develop more severe experimental autoimmune encephalitis when immunized with myelin oligodendrocyte glycoprotein (Dixon et al. 2021).

In a cancer setting, TIGIT is coexpressed with other co-inhibitory molecules such as PD-1, TIM3 on the CD8$^{+}$ tumor-infiltrating lymphocytes (TILs) in mice and humans. In addition to the CD8+ T cells, the expression of TIGIT on Tregs is further upregulated in the TME.

TIGIT potently inhibits innate and adaptive immunity through multiple mechanisms. TIGIT affects the function of DCs via CD155, wherein it induces CD155 phosphorylation and triggers a signaling cascade leading to the induction of tolerogenic DCs with decreased production of interleukin (IL)-12 and increased production of IL-10 (Anderson et al. 2016). TIGIT can also exhibit directly dampen the function of T cells as shown by studies where the agonistic anti-TIGIT antibodies inhibit T cell proliferation and function by attenuating T cell receptor (TCR)-driven activation signals (Dixon et al. 2021; Levin et al. 2011). As previously mentioned, TIGIT has a competitive co-stimulatory molecule CD226, thus it has been shown that TIGIT impedes CD155-mediated CD226 activation as TIGIT binds CD155 with higher affinity than CD226 (Dixon et al. 2021). TIGIT is highly expressed by a subset of Tregs in both mice and humans.

Additionally, TIGIT acts on Tregs to augment immunosuppressive function and stability (Yu et al. 2009), and TIGIT$^{+}$ Tregs are more suppressive than TIGIT$^{-}$ Tregs (Joller et al. 2011).

Preclinical studies involving the use of dual therapy such as anti-PDL1/PD1+ anti-TIGIT have shown promising results (Hung et al. 2018; Kurtulus et al. 2015 and Fuhrman et al. 2015). The anti-TIGIT mAbs that are currently being used in various clinical trials include: MTIG7192, MK-7684/, OP-313 M32 BMS-98620, AB-154 and ASP8374.

## 6 Conclusion

Immune checkpoint blockade is a novel treatment modality that has transformed the landscape of cancer treatments by showing remarkable efficacy in multiple types of cancer. However, further studies are required to improve the number and proportion of patients that can benefit from this treatment modality. Additionally, a deeper understanding of the mechanisms via which these therapies work will help in combating irAEs which strongly limit the applicability of these drugs. Further, combination therapy involving targeting two or more co-inhibitory receptors simultaneously or combining this check point blockade with other therapeutic modalities might provide better prospects.

## Glossary

**Checkpoint blockade** This is an immunotherapeutic approach where antibodies are used against co-inhibitory receptors to restore the function of T cells.

**Co-inhibitory receptors** These are receptors on T cells which are directly or indirectly involved in inhibiting the activity of T cells.

**CTLA4** Cytotoxic T-lymphocyte associated protein 4 (CTLA4) is expressed primarily in T cells where it regulates the amplitude of the early stages of T cell activation by competitively binding to B7 molecules on APCs and thereby inhibiting the interaction of B7 molecules with the co-stimulatory molecule CD28 on T cells. It is the first immune checkpoint receptor to be clinically targeted.

**Immune-related Adverse Events (irAE)** irAE refers to the immunological side effects associated with immunotherapeutic treatments like check point blockade.

**LAG-3** Lymphocyte Activation Gene-3 (LAG-3) is a CD4-related MHC class II binding molecule expressed on activated T cells and NK cells which maintains lymphocyte homeostatic balance via various inhibitory mechanisms.

**PD1** Programmed cell death protein 1 (PD1) is an inhibitory receptor that is expressed on activated T cells and it is known to limit the activity of T cells in the peripheral tissues at the time of an inflammatory response to infection and to limit autoimmunity.

**TIGIT** T cell immunoreceptor with IgG and ITIM domains(TIGIT) is an inhibitory receptor expressed on several types of lymphocytes. In multiple cancers, TIGIT is expressed on tumor-infiltrating cytotoxic T cells, helper T cells, regulatory T cells, and NK cells, and its main ligand CD155.

**TIM-3** T cell immunoglobulin and mucin domain-containing protein 3 (TIM3), a member of the TIM family, is a co-inhibitory receptor that is expressed on IFN-γ-producing T cells, FoxP3+ Treg cells, and innate immune cells (macrophages and dendritic cells) where it has been shown to suppress their responses upon interaction with their ligand(s).

## References

Acharya N, Anderson AC (2019) New clones on the block. Immunity. 51:606–608. https://doi.org/10.1016/j.immuni.2019.09.018

Acharya N, Acharya N, Sabatos-Peyton C, Anderson AC, Anderson AC (2020) Tim-3 finds its place in the cancer immunotherapy landscape. J Immunother Cancer. 8. https://doi.org/10.1136/jitc-2020-000911

Anderson AC, Joller N, Kuchroo VK (2016) Lag-3, Tim-3, and TIGIT: co-inhibitory receptors with specialized functions in immune regulation. Immunity. 44:989–1004. https://doi.org/10.1016/j.immuni.2016.05.001

Baumeister SH, Freeman GJ, Dranoff G, Sharpe AH (2016) Coinhibitory pathways in immunotherapy for cancer. Annu Rev Immunol. 34:539–573. https://doi.org/10.1146/annurev-immunol-032414-112049

Bettini M et al (2011) Cutting edge: accelerated autoimmune diabetes in the absence of LAG-3. J Immunol 187:3493–3498

Blackburn SD et al (2009) Coregulation of CD8+ T cell exhaustion by multiple inhibitory receptors during chronic viral infection. Nat Immunol 10:29–37

Callahan MK et al (2017) Nivolumab plus ipilimumab in patients with advanced melanoma: updated survival, response, and safety data in a Phase I dose-escalation study. J Clin Oncol 36:391–398

Chambers CA, Kuhns MS, Egen JG, Allison JP (2001) CTLA-4-mediated inhibition in regulation of T cell responses: mechanisms and manipulation in tumor immunotherapy. www.annualreviews.org

Daud AI et al (2016) Tumor immune profiling predicts response to anti-PD-1 therapy in human melanoma. J Clin Investig 126:3447–3452

de Mingo Pulido Á et al (2021) The inhibitory receptor TIM-3 limits activation of the cGAS-STING pathway in intra-tumoral dendritic cells by suppressing extracellular DNA uptake. Immunity 54:1154–1167.e7

Dixon KO et al (2021) TIM-3 restrains anti-tumour immunity by regulating inflammasome activation. Nature 595:101–106

Dong H et al (2002) Tumor-associated B7-H1 promotes T-cell apoptosis: a potential mechanism of immune evasion. Nat Med 8:793–800

Fife BT et al (2009) Interactions between PD-1 and PD-L1 promote tolerance by blocking the TCR-induced stop signal. Nat Immunol 10:1185–1192

Francisco LM et al (2009) PD-L1 regulates the development, maintenance, and function of induced regulatory T cells. J Exp Med 206:3015–3029

Freeman GJ et al (2000) Engagement of the PD-1 immunoinhibitory receptor by a novel B7 family member leads to negative regulation of lymphocyte activation. J Exp Med. 192:1027–1034. http://www.jem.org/cgi/content/full/192/7/1027

Fuhrman CA et al (2015) Divergent phenotypes of human regulatory T cells expressing the receptors TIGIT and CD226. J Immunol 195:145–155

Furness AJS, Vargas FA, Peggs KS, Quezada SA (2014) Impact of tumour microenvironment and Fc receptors on the activity of immunomodulatory antibodies. Trends Immunol. 35:290–298. https://doi.org/10.1016/j.it.2014.05.002

Gao J et al (2016) Loss of IFN-γ pathway genes in tumor cells as a mechanism of resistance to anti-CTLA-4 therapy. Cell 167:397–404.e9

Garcia-Diaz A et al (2017) Interferon receptor signaling pathways regulating PD-L1 and PD-L2 expression. Cell Rep 19:1189–1201

Garon EB et al (2015) Pembrolizumab for the treatment of non-small-cell lung cancer. N Engl J Med 372:2018–2028

Goldberg MV, Drake CG (2010) LAG-3 in cancer immunotherapy. Curr Top Microbiol Immunol. 344:269–278. https://doi.org/10.1007/82_2010_114

Huard B, Gaulard P, Faure F, Hercend T, Triebel E (1994) Cellular expression and tissue distribution of the human LAG-3-encoded protein, an MHC class II ligand. Immunogenetics. 39:213–217

Hung AL et al (2018) TIGIT and PD-1 dual checkpoint blockade enhances antitumor immunity and survival in GBM. Oncoimmunology 7:e1466769

Im SJ et al (2016) Defining CD8+ T cells that provide the proliferative burst after PD-1 therapy. Nature 537:417–421

Ishida Y, Agata Y, Shibahara K, Honjo, T. (1992) Induced expression of PD-1, a novel member of the immunoglobulin gene superfamily, upon programmed cell death. EMBO J. 1:3887–3895

Jha V et al (2014) Lymphocyte Activation Gene-3 (LAG-3) negatively regulates environmentally-induced autoimmunity. PLoS One 9:e104484

Johnston RJ et al (2014) The immunoreceptor TIGIT regulates antitumor and antiviral CD8+ T cell effector function. Cancer Cell 26:923–937

Joller N et al (2011) Cutting edge: TIGIT has T cell-intrinsic inhibitory functions. J Immunol 186:1338–1342
Kamada T et al (2019) PD-1+ regulatory T cells amplified by PD-1 blockade promote hyperprogression of cancer. Proc Natl Acad Sci U S A 116:9999–10008
Kisielow M, Kisielow J, Capoferri-Sollami G, Karjalainen K (2005) Expression of lymphocyte activation gene 3 (LAG-3) on B cells is induced by T cells. Eur J Immunol 35:2081–2088
Kurtulus S et al (2015) TIGIT predominantly regulates the immune response via regulatory T cells. J Clin Investig 125:4053–4062
Latchman Y et al (2001) PD-L2 is a second ligand for PD-1 and inhibits T cell activation. http://immunol.nature.com
Le DT et al (2015) PD-1 blockade in tumors with mismatch-repair deficiency. N Engl J Med 372:2509–2520
Levin SD et al (2011) Vstm3 is a member of the CD28 family and an important modulator of T-cell function. Eur J Immunol 41:902–915
Miller BC et al (2019) Subsets of exhausted CD8+ T cells differentially mediate tumor control and respond to checkpoint blockade. Nat Immunol 20:326–336
Mognol GP et al (2017) Exhaustion-associated regulatory regions in CD8+ tumor-infiltrating T cells. Proc Natl Acad Sci U S A 114:E2776–E2785
Myers G (2018) Immune-related adverse events of immune checkpoint inhibitors: a brief review. Curr Oncol 25:342–347
Nishimura H, Nose M, Hiai H, Minato N, Honjo T (1999) Development of lupus-like autoimmune diseases by disruption of the PD-1 gene encoding an ITIM motif-carrying immunoreceptor. Immunity. 11:141–151
Pauken KE et al (2016) Epigenetic stability of exhausted T cells limits durability of reinvigoration by PD-1 blockade. Science 354:1160–1165
Peggs KS, Quezada SA, Chambers CA, Korman AJ, Allison JP (2009) Blockade of CTLA-4 on both effector and regulatory T cell compartments contributes to the antitumor activity of anti-CTLA-4 antibodies. J Exp Med 206:1717–1725
Philip M et al (2017) Chromatin states define tumour-specific T cell dysfunction and reprogramming. Nature 545:452–456
Qureshi OS et al (2011) Trans-endocytosis of CD80 and CD86: a molecular basis for the cell-extrinsic function of CTLA-4. Science 332:600–603
Ribas A, Wolchok JD (2018) Cancer immunotherapy using checkpoint blockade. Science. 359:1350–1355. https://doi.org/10.1126/science.aar4060
Ribas A et al (2013) Phase III randomized clinical trial comparing tremelimumab with standard-of-care chemotherapy in patients with advanced melanoma. J Clin Oncol 31:616–622
Robert C et al (2015) Pembrolizumab versus ipilimumab in advanced melanoma. N Engl J Med 372:2521–2532
Rotte A (2019) Combination of CTLA-4 and PD-1 blockers for treatment of cancer. J Exp Clin Cancer Res. 38. https://doi.org/10.1186/s13046-019-1259-z
Rudd CE, Taylor A, Schneider H (2009) CD28 and CTLA-4 coreceptor expression and signal transduction. Immunol Rev. 229:12–26. https://doi.org/10.1111/j.1600-065X.2009.00770.x
Sade-Feldman M et al (2017) Resistance to checkpoint blockade therapy through inactivation of antigen presentation. Nat Commun 8:1136
Sade-Feldman M et al (2018) Defining T cell states associated with response to checkpoint immunotherapy in melanoma. Cell 175:998–1013.e20
Samstein RM et al (2019) Tumor mutational load predicts survival after immunotherapy across multiple cancer types. Nat Genet 51:202–206. https://doi.org/10.1038/s41588-018-0312-8
Scott AC et al (2019) TOX is a critical regulator of tumour-specific T cell differentiation. Nature 571:270–274
Sen DR et al (2016) The epigenetic landscape of T cell exhaustion. Science 354:1165–1169
Sharma P, Hu-Lieskovan S, Wargo JA, Ribas A (2017) Primary, adaptive, and acquired resistance to cancer immunotherapy. Cell. 168:707–723. https://doi.org/10.1016/j.cell.2017.01.017

Siddiqui I et al (2019) Intratumoral Tcf1 + PD-1 + CD8 + T cells with stem-like properties promote tumor control in response to vaccination and checkpoint blockade immunotherapy. Immunity 50:195–211.e10

Tan CL et al (2021) PD-1 restraint of regulatory T cell suppressive activity is critical for immune tolerance. J Exp Med 218:e20182232

Tivol EA et al (1995) Loss of CTLA-4 leads to massive lymphoproliferation and fatal multi-organ tissue destruction, revealing a critical negative regulatory role of CTLA-4. Immunity. 3:541–547

Topalian SL et al (2012) Safety, activity, and immune correlates of Anti–PD-1 antibody in cancer. N Engl J Med 366:2443–2454

Triebel F et al (1990) LAG-3, a novel lymphocyte activation gene closely related to CD4. 171:1393–1405

Velu V et al (2009) Enhancing SIV-specific immunity in vivo by PD-1 blockade. Nature 458:206–210

Waterhouse P, Penninger JM, Timms E, Wakeham A (1995) Lymphoproliferative disorders with early lethality in mice deficient in Ctla-4. Science. 270:985–988

Wolchok JD et al (2017) Overall survival with combined nivolumab and ipilimumab in advanced melanoma. N Engl J Med 377:1345–1356

Workman CJ et al (2009) LAG-3 regulates plasmacytoid dendritic cell homeostasis. J Immunol 182:1885–1891

Yost KE et al (2019) Clonal replacement of tumor-specific T cells following PD-1 blockade. Nat Med 25:1251–1259

Yu X et al (2009) The surface protein TIGIT suppresses T cell activation by promoting the generation of mature immunoregulatory dendritic cells. Nat Immunol 10:48–57

# Neoepitope-Based Cancer Vaccines: Principle and Applications

**Cory Alexander Brennick, Mariam Mathew George, and Marc Andrew Gillig**

**Abstract** It has been known that different cancers can elicit host immune responses that are highly specific for any given tumor. This high level of specificity stems from the fact that cancers arise due to mutations (both passenger and driver mutations) that ultimately give rise to mutated peptides that are seen as "non-self" by the adaptive immune compartment. Therefore, tumors are immunogenic, and the mutated peptides encoded by tumor-specific mutations and presented on surface of tumor cells, called *neoantigens* which are a source of their immunogenicity. Precise identification and prediction of neoantigens using advanced high-throughput gene sequencing and bioinformatic approaches have significantly amplified the probability of success of neo-antigen-based vaccines. Further, neoantigen-based vaccines are in many clinical trials showing promising results. However, many important concerns need to be addressed before neoantigen-based vaccines evolve as a major form of cancer immunotherapy. Here, we will discuss the prospects and major challenges that lie ahead.

**Keywords** Neoantigen · Cancer Immunotherapy · Tumor Rejection Mediating Neoantigens · T-Cell

Abbreviations

| | |
|---|---|
| APC | Antigen-presenting cell |
| CGA | Cancer-germline antigen |
| CTL | Cytotoxic T lymphocyte |

C. A. Brennick
Yale School of Medicine, Yale University, New Haven, CT, USA

M. M. George (✉)
Weill Cornell Medicine, New York, NY, USA
e-mail: mmg4005@med.cornell.edu

M. A. Gillig
Enable Life Sciences, Worcester, MA, USA


A. Das Mohapatra, P. S. Sahu (eds.), *Advances in Immunology and Immuno-techniques*, Interdisciplinary Biotechnological Advances,
https://doi.org/10.1007/978-981-97-5508-0_3

| | |
|---|---|
| DAI | Differential agretopicity index |
| DC | Dendritic cell |
| ER | Endoplasmic reticulum |
| ERAP | ER aminopeptidases associated with antigen processing |
| HSP | Heat shock protein |
| INDEL | Insertion and deletion |
| MHC | Major histocompatibility complex |
| MS | Mass spectrometry |
| pMHC | peptide MHC |
| TAA | Tumor-associated antigen |
| TAP | Transporter associated with antigen processing |
| TCR | T-cell receptor |
| TME | Tumor microenvironment |
| Treg | T-regulatory |
| TRMN | Tumor rejection mediating neoantigens |
| TSA | Tumor-specific antigen |

## 1 Introduction

In the early-mid twentieth century, it was elegantly demonstrated that if a mouse had a chemically induced tumor resected, it would be completely protected from subsequent implantation of that same tumor. If that same tumor was transplanted into a new mouse, even if the mouse was genetically identical, it would not be protected against the tumor. This early research work of Gross (Gross 1943), Prehn and Main (Prehn and Main 1957), and Klein et al. (1960) in murine models suggested that the tumor immunogenicity is dictated by the specificity of the individual hosts and tumor itself (Fig. 1). Their work laid the foundation for what would become known as tumor antigens. Tumor antigens encompass any antigen that is presented by the tumor in the context of the Major Histocompatibility Complexes (MHCs) and recognized by the adaptive immune system, specifically T-cell receptors (TCRs). The first two branches of tumor antigens that have been heavily researched, and are controversial, are tumor-associated antigens (TAAs) and cancer-germline antigens (CGAs). TAAs and CGAs are unmutated self-antigens that are presented on both the surface of a tumor cell and normal tissue, but at a significantly lower level in normal tissue. However, recently, it has been demonstrated that common TAAs and CGAs are not ideal vaccine targets in cancer (Haen et al. 2020).

With the improvement of sequencing and bioinformatic approaches, the second branch of tumor antigens was established and became known as tumor-specific antigens (TSAs). TSAs or more regularly referred to as neoantigens are antigens that are either altered or completely foreign to the host and recognized by the adaptive immune system as non-self. Neoantigens are either viral in origin in an open reading frame (Walboomers et al. 1999; Feng et al. 2008) or arise due to genetic alterations that result in a change in the amino acid sequence (Fig. 2). The most heavily

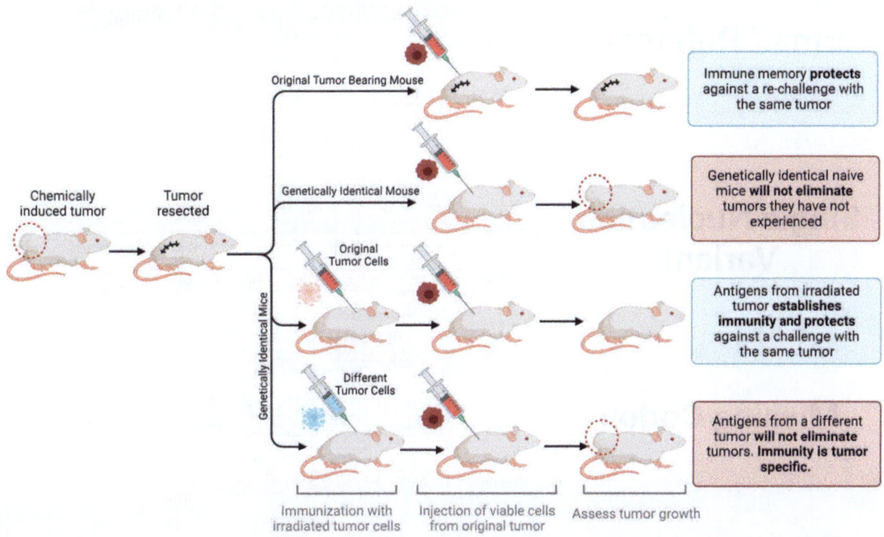


**Fig. 1** Immunity against tumors is specific. In genetically identical mice, a tumor was induced using a carcinogenic methylcholanthrene. That tumor was surgically resected, and viable tumor cells from the resected tumor were then implanted back into (top row) the original mouse (second row) into naive genetically identical mice. These experiments demonstrated that antitumor immunity is specific to individual tumors, which strongly suggested that tumors possess antigens uniquely found in those individual tumors and that anticancer immune responses target those tumor-specific antigens

studied genetic alterations for neoantigen selection are somatic mutations, insertion and deletions (INDELs), and frameshift mutations. These genomic alterations ultimately give rise to foreign or mutated proteins that are expressed specifically in the tumor. These proteins will then go through the normal antigen processing and presentation pathway and be presented on the surface of the cell in the context of MHC class I or II (MHC I or MHC II) and recognized by their cognate TCR on $CD8^+$ or $CD4^+$ T-cells, respectively (Monach et al. 1995). This immune recognition and specificity are what the early scientists unknowingly observed in their murine transplant studies, and it is only now that we have the tools to further investigate personalized cancer immunotherapy.

With advancements in high-throughput genomics and bioinformatic technologies, it has become possible to predict and identify tumor-specific neoantigens with modest accuracy (Brennick et al. 2017, 2020; Blass and Ott 2021). Therefore, the use of neoantigen-based cancer vaccines is a strong and unignorable candidate for personalized cancer immunotherapy. However, there are still major challenges to overcome before neoantigen-based vaccination is used as standard practice. Even today, there are no accurate means to predict which neoantigens will eventually elicit antitumor immune responses (Srivastava 2015). Major issues ranging from precise identification of all the tumor-specific somatic mutations to the prediction of

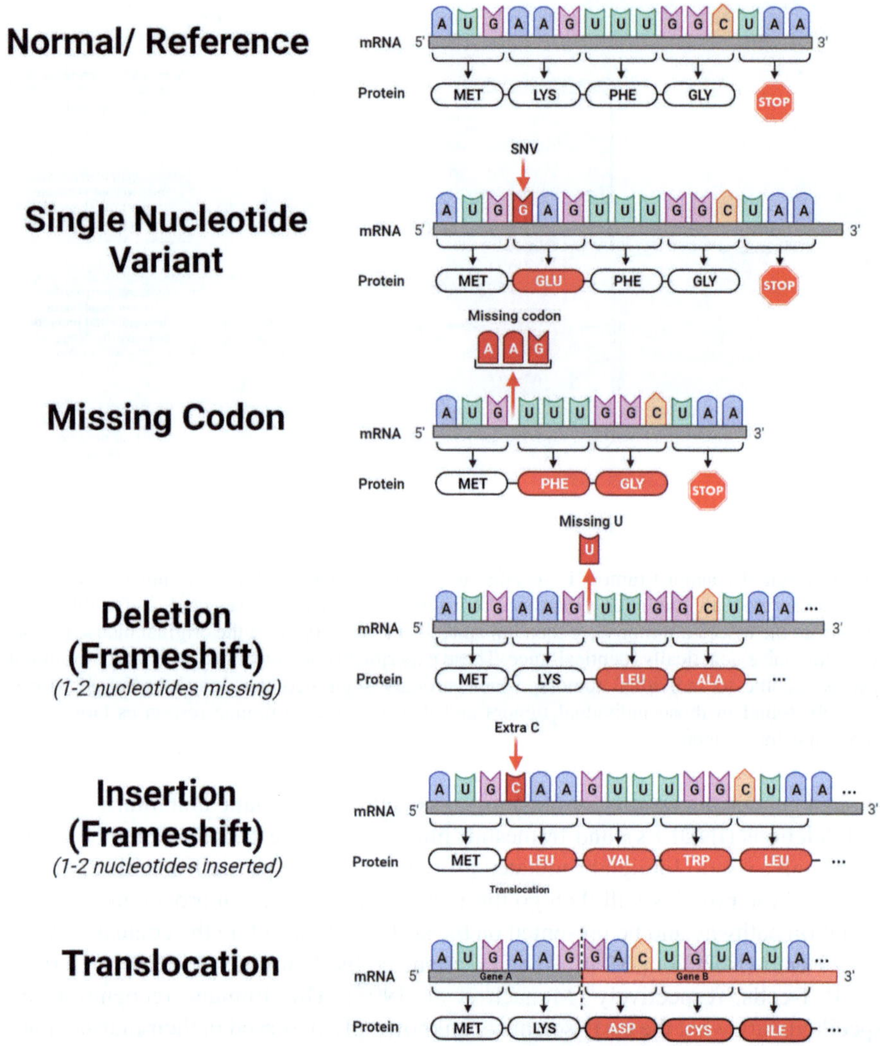


**Fig. 2** Common sources of neoantigens. Neoantigens arise from mutations in protein-coding genes that result in translation of non-self-amino acid sequences. A non-synonymous single-nucleotide variant (SNV) results in a single amino acid change. An insertion or deletion (InDel) of 1 or 2 nucleotides, or any number of nucleotides not a multiple of 3, results in translation of a non-self-sequence due to a shift in the reading frame. A translocation can give rise to neoantigens in two ways, depending on if the reading frame is maintained where the gene junction occurs. If the reading frame is maintained, neoantigens can arise from the non-self-sequence created by the junction point between the two genes; if the reading frame is disrupted, the potential for neoantigens is similar to an InDel mutation, where the non-self-sequence extends until a stop codon happens to occur

reliable neoantigens and translation of that information into the clinical setting signify the hurdles that we face.

Here we will discuss what is known in the context of neoantigens, the major obstacles in this field, and the prospects that lie ahead for neoantigen-based vaccines.

## 2 Immunological Basis for Neoantigen Recognition

The immune response against cancers is carried out by various cells in the innate and adaptive immune compartments. Macrophages, dendritic cells (DCs), and B-cells, to name a few, spearhead the presentation of tumor neoantigens to the adaptive immune compartment, while $CD4^+$ and $CD8^+$ T-cells recognize the neoantigens as "non-self" or "mutated self" and will specifically eliminate the cancer cells. However, it is known that $CD8^+$–T-cells are the major cell type which can identify tumor-specific antigens present on malignant cells. Upon recognizing the MHC class I restricted epitopes derived from tumor-specific antigens on malignant cells, the $CD8^+$ T cells get activated and then selectively kill these cancerous cells.. For this our understanding of the antigen processing and presentation machinery and the pathway is of paramount importance.

Neoantigens are specific peptides which are derived from somatic mutations and are seen as different from self and cross-presented by antigen-presenting cells (APCs), such as DCs and the tumor cells themselves (Srivastava 2015). The normal protein turnover in the cell and ~~also~~ the elimination of misfolded or defective proteins depends on its ubiquitination and degradation by the proteasome in the cytosol of the cell. The degraded peptides are then chaperoned across the cytoplasm to the endoplasmic reticulum (ER) by chaperone proteins such as Heat Shock Protein (HSP) 90 (Srivastava 1993). The active transport of the peptides by the transporter associated with antigen processing (TAP) and trimming to the appropriate length by ER aminopeptidases associated with antigen processing (ERAP), loading onto the peptide-binding cleft of the MHC molecules with the help of chaperone proteins, such as tapasin, and transport to the cell surface is essential for the eventual recognition by the TCR on T-cells. All tumor cells are inherently able to present endogenous peptides including neoantigens restricted to MHC class I, as long as there are no "loss of function" mutations in the genes involved in the antigen presentation machinery. However, in the context of neoantigen vaccines, where the antigen will be delivered exogenously, the ability of DCs to cross-present antigens plays an important role as the APCs are able to translocate exogenous antigens from the phagosome to be directly loaded onto MHC II within the endosome or into the cytosol for processing into MHC I epitopes by the proteasome and then transported into the presentation pathway within the ER (Kovacsovics-Bankowski and Rock 1995; Blum et al. 2013).

Two major events that need to happen for the successful elicitation of an antitumor immune response are as follows: (1) Not all peptides generated by the intracellular processing of antigens can bind and be presented by the MHC alleles expressed by the host and (2) Only those peptides that are presented in the context of MHC AND are recognized by T-cells of the host are bona fide epitopes. To address the first point, Sette and colleagues measured the binding affinity of a peptide to MHC I and used that as a surrogate marker to predict a T-cell response and arrived at the now widely accepted criteria that, when compared to a standard peptide, the majority of epitopes show a half maximal inhibitory concentration, $IC_{50}$ of <500 nM and preferably <50 nM (Sette et al. 1994; Assarsson et al. 2007). These conclusions have stood the test of time and have been extensively validated in viral and model antigen systems. Using binding affinity as a criterion for predicting immunogenicity, several computational algorithms that predict the affinity of any user-defined peptide to a well-characterized MHC I allele have been developed. Computational algorithms such as NetMHC (Nielsen et al. 2003) and IEDB (2019) are being extensively used to predict neoantigens based on their strong binding affinity to Class I MHC (Karandikar et al. 2019). However, thus far, there are no reliable methods to accurately verify that in silico-identified neoantigens can elicit tumor regression or an immune response. To partially address the second point, mass spectrometry (MS) may play a key role in prediction pipelines to account for the neoantigens that are presented on the tumor surface. Note that not all in silico predicted neoantigens will be presented by MHCs. However, this method has not nearly reached the level of being a high-throughput approach. More recently, improved sensitivity of MS in identifying tumor-derived neoantigens in patient's tumors validated by neoantigen-specific antitumor immune response has been recently reported (Bassani-Sternberg et al. 2016; Ebrahimi-Nik et al. 2019).

On a side note, it has been reported in some hematological malignancies that mutations occur in the genes that are involved in antigen processing and presentation (Leone et al. 2013). In such a case, the use of neoantigens as a cancer immunotherapeutic vaccine may not result in potent antitumor immune responses, because the presentation of neoantigens on the cell surface is abrogated, therefore impeding their recognition by the immune system.

The development of prediction algorithms or programs that incorporate all aspects of antigen processing, presentation, and TCR recognition remains an important step toward improving the accuracy of predicting successful neoantigens.

## 3 Peptide-MHC Binding Affinity—Not an Accurate Predictor of Bona Fide Neoantigens

Thus far, most preclinical models of neoantigen prediction and identification have relied on peptides having a strong binding affinity to MHC molecules (Gubin et al. 2014; Yadav et al. 2014). Further, the same criteria have also been applied to the identification and use of neoantigens in clinical trials (Keskin et al. 2019; Zaidi and Jaffee 2019). However, these clinical trials have consistently shown that "prominent $CD4^+$ T -cell responses against immunizing neoantigens were detected, despite the use of (MHC) class I binding prediction algorithms" (Keskin et al. 2019). The observation made by the work of van Rooij et al. reported that less than 1% of the high-affinity binders predicted from a melanoma patient were able to be recognized by the patient's T-cells (van Rooij et al. 2013). The thresholds used to determine which neoantigens are most likely to elicit a cytotoxic T lymphocyte (CTL) response and set the standard that an $IC_{50}$ ≤500 nM, preferably 50 nM or less, is considered a strong to an intermediate binder and determines an epitope's capacity to elicit a CTL response (Sette et al. 1994). Since neoantigens arise from altered self-peptides and are very close to self-peptides, neoantigen-specific T-cells potentially will face thymic deletion and tolerization. This issue was addressed by Duan et al. who created a differential agretopicity index (DAI) score, which compares the binding affinity of the mutated peptide with the reference or self-peptide (Duan et al. 2014). In this study, neoantigens with $IC_{50}$ >500 nM were identified that effectively inhibit tumor growth in two different BALB/c tumor models. By current standards, these neoantigens would have been eliminated and not predicted to be immunogenic based on their peptide-MHC (pMHC) binding affinity alone.

A recent and extensive study by Brennick and George et al. showed that our current methods of prediction discard the majority of true anticancer neoantigens and that the true tumor rejection mediating neoantigens "(TRMNs) are predicted to have a binding affinity, towards MHC class I molecules in a range, that would conventionally be regarded as poor binding affinity." These TRMNs have dramatically different properties from viral epitopes. Here, all mutations in any given tumor model were tested for tumor control in an unbiased fashion to eliminate any biases established by considering pMHC binding affinity. Since this study has the largest collection of true TRMNs (9 TRMNs), a new prediction algorithm was developed which is essentially a model to enrich for TRMNs from data in silico (Brennick et al. 2021). Ghorani et al. and Rech et al. examined mutational data and patient survival from several thousand patients and concluded that the presence of high-affinity MHC-binding neoantigens in tumors did not correlate with overall survival, progression-free survival, or T-cell infiltration in tumors (Ghorani et al. 2018; Rech et al. 2018). These studies provide convincing evidence that the use of pMHC binding affinity criteria, which was developed based on foreign antigen systems (viral and model antigens) is not sufficient to predict neoantigens that are far closer to self than foreign antigens.

Using binding affinity as a major predictor of neoantigens, while partially useful, may be restricting our ability to accurately identify all immunologically active neoantigens. A more comprehensive strategy that includes not just pMHC binding affinity but also other aspects of antigen processing and presentation and TCR recognition is warranted. Further, finetuning the prediction algorithms using larger data sets will provide a platform for better neoantigen identification.

# 4 Advances in Neoantigen Prediction

Since the number of potential neoantigens predicted in silico is large in most cases, it remains a challenge to accurately predict true neoantigens capable of eliciting antitumor activity. Several independent groups have developed bioinformatic algorithms for predicting neoantigens and scoring them based on factors that affect their likelihood of MHC presentation as well as their immunogenic potential.

As discussed in a previous section, the binding affinity between peptide and MHC may not be the most accurate way to predict neoantigens (Brennick et al. 2021). However, most existing neoantigen prediction methods still incorporate predicted binding affinity as a component of their neoantigen predictions (Albert et al. 2022). Most of these methods utilize the NetMHC algorithm, which was first introduced in 2003 as a free online tool for predicting the binding affinity of peptide sequences to a human MHC (HLA-A2) and a mouse MHC (H-2K$^k$) (Buus 2003). This algorithm is based on a neural network trained on various sources of real MHC-peptide binding affinity data. Notably, the training data was obtained using viral antigens and other model antigens (i.e., foreign antigens), not neoantigens (which are alterations of "self" antigens). Several updates to this algorithm over the last 20 years have expanded the database to include dozens of human HLA alleles as well as many common MHC alleles in other species; meanwhile, the algorithm was re-trained on expanded MHC-peptide binding datasets (Andreatta and Nielsen 2016). This same group also developed NetMHCpan, which can predict the binding of peptides to any HLA with a known sequence, and which also incorporates data about peptides that are naturally presented by cells (known as "eluted ligand" data). This combination of binding affinity data and eluted ligand data allows this more sophisticated algorithm to make more accurate predictions about what peptides will likely be presented (Reynisson et al. 2020).

In addition to the considerations of predicted binding affinity and likelihood of natural presentation, some groups have attempted to incorporate immunogenicity predictions into their algorithms. For example, the PRIME algorithm was trained by incorporating molecular properties of the actual TCR recognition of neoantigen peptides (Schmidt et al. 2021). Rather than focusing on the pMHC interactions, the binding of the TCR to that pMHC complex is also considered. This approach leans heavily on structural biology; for example, peptides have more immunogenic potential when aromatic side chains face outward from the MHC, pointing in the direction of the TCR.

Within the last 5 years, several new neoantigen prediction pipelines have been developed by various research groups, each claiming to have one or more advantages over previous and/or competing algorithms (Schenck et al. 2019; Rao et al. 2020; Wood et al. 2020; Albert et al. 2022). However, none of these pipelines have consistently demonstrated the ability to predict which neoantigens have the potential to elicit clinically relevant antitumor activity. To achieve the milestone of truly rational design of personalized neoantigen vaccines, more bioinformatic innovations will be necessary.

## 5 Neoantigen-Specific $CD8^+$ T-cell Response—An Inaccurate Surrogate for Antitumor Immune Response

It is a well-known fact that $CD8^+$ T-cells (among other immune elements) are essential for a successful antiviral and antitumor immune response. This fact has generated a code of belief that the antigen-specific $CD8^+$ T-cells as measured contemporarily (e.g., Flow cytometry and ELISpot) are true surrogate markers of antitumor $CD8^+$ activity in vivo. Many studies have identified neoantigens based on their ability to stimulate PBMCs obtained from patient blood, in vitro (Linnemann et al. 2015; Fehlings et al. 2019). However, the conventional assays used to measure antitumor immune responses, for the most part, fail to detect tumor-specific $CD8^+$ T-cells because they are essentially anti-altered-self responses. This dogma persists despite the observations in mice (Ramarathinam et al. 1995; Sarma et al. 2003; Rosato et al. 2003) and in humans (Rosenberg et al. 1998; Marchand et al. 1999; Anichini et al. 1999) that there is little connection between measurable $CD8^+$ T-cell responses and clinical activity. Further, the study by Brennick and George et al. failed to detect $CD8^+$ T-cells responses against any of the 9 TRMNs shown to elicit potent tumor control by the conventional ELISpot assay (Brennick et al. 2021). This is despite the fact that the antitumor immunity elicited by these 9 TRMNs was indeed $CD8^+$ T-cell dependent, as observed by the abrogation of immunity by depletion of $CD8^+$ cells as well as by successful adoptive transfer of $CD8^+$ T-cells from TRMN-immunized mice to naive mice. Several reasons could contribute to this discrepancy including stricter regulation of anti-neoantigen-specific $CD8^+$ T-cells and lower precursor frequencies.

Such discrepancies call for the development of more sensitive assays such as those based on the quantitation of TCR clones. This may also address this disparity between $CD8^+$ T-cell dependent responses in vivo and the lack of a $CD8^+$ T-cell response in vitro.

Another interesting development in the identification of tumor-specific $CD8^+$ T-cells is embodied in recent studies showing that the antigen-specific $CD8^+$ T-cells adopt a more stem-like phenotype upon chronic antigen stimulation. These stem-like $CD8^+$ T-cells, defined by higher levels of Tcf1 and lower levels of PD1, have emerged as a significant factor in their function, in vivo in viral and tumor models (Utzschneider et al. 2016; Philip et al. 2017; Kurtulus et al. 2019; Siddiqui et al. 2019). This population of Tcf1 expressing tumor-specific $CD8^+$ T-cells express markers of exhaustion like PD1, CD38, CD101, CD39, and TIGIT and would have otherwise been characterized as terminally exhausted. However, they remain functional (as shown by IFNg and other cytotoxic markers) and are essential for antitumor immune responses. Therefore, these cells have been named stem-like early dysfunctional $CD8^+$ T-cells. Of note, the treatment of mice and humans with Immune Checkpoint Blockade (ICB) induced an increase in the proliferation of these Tcf1-expressing $CD8^+$ T-cells. Apart from ICB, immunization of mice with TRMNs also elicited an increase in early dysfunctional stem-like $CD8^+$ tumor infiltrating lymphocytes (TILs), thereby improving overall tumor control, despite the fact that neoantigen-specific $CD8^+$ T-cell responses could not be detected by a conventional ELISpot assay (Brennick et al. 2021). These studies, therefore, provide a strong case for combining neoantigen vaccinations with ICB in order to obtain the most potent antitumor immune responses.

## 6 Bringing in the Help with Neoantigen-Specific $CD4^+$ T-Cell Response

A majority of neoantigen studies have been in the context of $CD8^+$ T-cells, as they are the cells involved with ~~in~~ the direct killing of tumor cells and that all tumors express MHC class I; unless the MHC allele has been mutated or loses expression within the tumor. $CD4^+$ T-cells recognize peptide-MHC II complexes, which is only normally expressed on professional APCs (Axelrod et al. 2019). Unless there are cancer cells that express MHC II, the $CD4^+$ cells will not directly target the cells. Even though the cancer cells do not directly present MHC II, neoantigen-specific $CD4^+$ T-cells are found within naïve tumors (Linnemann et al. 2015), and increase in patients that have been treated with cancer immunotherapies (Aarntzen et al. 2013; Alspach et al. 2019). Professional APCs in and around the tumor will cross-present tumor-specific MHC II antigens and recruit $CD4^+$ T -cells to its location and begin the orchestration of $CD4^+$ help within the tumor microenvironment (TME). It has been well established that $CD4^+$ T-cells are important in licensing DCs which is critical in priming a strong memory $CD8^+$ T-cell response (Laidlaw et al. 2016). In murine models, the inclusion of $CD4^+$ specific neoantigens in a vaccine generated strong effector memory $CD8^+$ T-cells with a downregulation in co-inhibitory receptors (Alspach et al. 2019). Not only are $CD4^+$ T-cells important for $CD8^+$ T-cell development, but also for the recruitment and stimulation of immune populations, such as macrophages and natural killer cells, in the TME (Marzo et al. 2000).

One major consideration when targeting MHC II neoantigens is which subset of $CD4^+$ T-cells are being induced. A recent study observed neoantigen-specific T-regulatory cells (Treg) in patients with melanoma, gastrointestinal and ovarian cancers (Ahmadzadeh et al. 2019). If the $CD4^+$ positive cells within the tumor are pushed toward a Treg phenotype that could be very detrimental to the immune system clearing the tumor. Tregs are known for their strong ability to inhibit the immune system and their negative role in the TME. With the identification of these "inhibitory" antigens, we can now include them in our selection platforms to ensure that these inhibitory antigens are not used for treatment (Nogueira et al. 2018; Cohen et al. 2020). These observations further stress the importance of immunophenotyping of the neoantigen-specific T-cells when considering the ideal TRMN.

## 7 Neoantigen-Based Immunization Strategies

While it was long understood that unique antigenic signatures arose from unique mutations in the cancer cell exome, it was costly and laborious to identify these cancer-specific mutations—and quite impossible to do so in a high-throughput manner—until the technological breakthroughs made in the new millennium. Next-generation sequencing platforms have made it possible to quickly and inexpensively sequence the exons within the protein-coding genes of the DNA genome (i.e., the exome) as well as the RNA transcriptome of cancerous tissue. This data can then be used to identify mutations in a particular tumor sample that result in changes to the amino acid sequences of proteins that are expressed by the tumor cells (Srivastava 2015). Thus, next-generation sequencing has transformed the concept of personalized therapeutic cancer vaccines from impossibility into a promising modality rapidly progressing toward widespread clinical implementation (Fig. 3).

Of course, the identification of a tumor sample's non-synonymous mutations is only the first step in the development of a personalized therapeutic vaccine. Two major questions immediately follow: (1) in tumors with numerous non-synonymous mutations, which mutations should be chosen as targets for the induced T-cell response, and (2) how will the immune response be generated against the mutated amino acid sequences? Question 1 remains an open question in the field; although advances have been made in predicting (a) the affinity of neoantigen peptides for MHC molecules, (b) the likelihood of the natural presentation of those peptides, and (c) the potential for their immunogenicity, it nonetheless remains extremely challenging to predict which neoantigens will mediate real antitumor activity (see section "Advances in Neoantigen Prediction"). Concerning Question 2, neoantigen-specific T-cell responses have been generated by immunizing with (a) synthetic peptides matching the neoantigen sequence and (b) messenger RNA (mRNA) or DNA encoding the neoantigen.

Unlike full-length proteins, which need to be expressed in living cells using nucleic acid vectors, peptides of up to 80 amino acids in length can be chemically synthesized with high sequence accuracy and high purity. Peptides presented by MHC class I molecules, including neoantigens, tend to be between 8 and 11 amino acids in length, with longer peptides (up to 15-mers) seen occasionally (Trolle et al. 2016). Once mutations have been identified and neoantigen peptides have been predicted (using the aforementioned prediction methods), synthetic versions of the predicted peptides can be combined and formulated into vaccines (Calvo Tardón et al. 2019). However, rather than using neoantigen peptides of the precise length that they appear in the context of MHC presentation, many groups use longer peptides that are extended in both the N-terminal and C-terminal directions. These "long" peptides are sometimes used for immunization for several reasons. Most importantly, the N-terminal and C-terminal bounds of the precise neoantigen are often unknown. After immunizing with a "long" peptide that includes all possibilities of the precise neoantigen, antigen-presenting cells trim it down to the precise neoantigen as part of the normal antigen processing pathway. Secondly, injected peptides

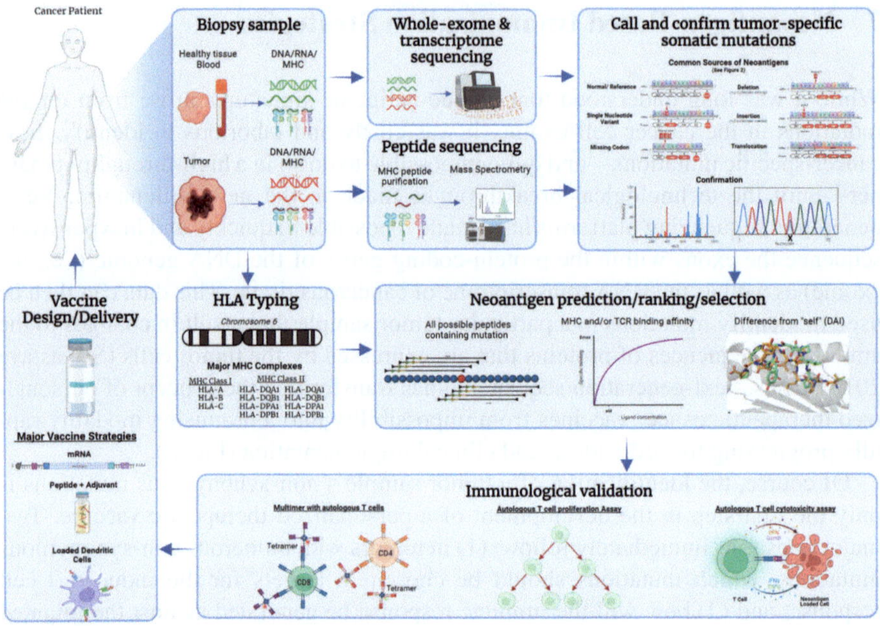


**Fig. 3** Personalized neoantigen-based therapeutic vaccine development pipeline. First, exome sequencing is performed on tumor tissue DNA. Tumor-specific mutations are identified (i.e., "called") by comparing this sequencing data against the patient's normal exome sequence. RNA sequencing is also performed to identify which mutated genes are expressed by the tumor tissue. Meanwhile, the patient's MHC (i.e., HLA) alleles are identified from the sequencing data. In some cases, presented neoantigen peptides can be identified directly using mass spectrometry-based sequencing of peptides eluted from the MHC molecules of tumor cells. Otherwise, potential neoantigens are predicted using bioinformatic algorithms and ranked based on their likelihood of being presented by tumor cells. Sometimes, predicted neoantigens can be validated using T cell-based assays. Whether validated or not, several neoantigens are combined into a vaccine that is administered to the patient using either synthetic peptides or mRNA or DNA vaccines

are often degraded by peptidases in the extracellular and intracellular environments; an extended peptide thus allows for a greater likelihood of the precise neoantigen survival from injection to presentation on the surface MHC molecules of APCs. Thirdly, immunizing with a longer sequence increases the likelihood that additional epitopes—perhaps even MHC class II epitopes that will activate $CD4^+$ helper T-cells—will be incorporated into the resulting immune response. In some cases, the amino acid sequences of several different neoantigens have been combined sequentially into one multi-epitope long peptide (Stephens et al. 2021). In addition to the relative ease of peptide synthesis compared to recombinant expression of whole antigen, and the advantage of combining several neoantigens into one peptide, it has been demonstrated that synthetic long peptides are more immunogenic than the corresponding whole antigens (Rosalia et al. 2013).

Within the past 15 years, rapid advancements have been made in mRNA-based immunization strategies. mRNA immunization involves the injection of antigen-encoding mRNA rather than the injection of the antigen itself. APCs and other cells then translate the mRNA, which is produced in such a way that it fully resembles naturally processed mRNA molecules, into protein antigens. These proteins then follow the antigen presentation pathway, ultimately resulting in T-cell responses specific to the MHC-presented peptides derived from those proteins. Because "naked" mRNA would be rapidly degraded by nucleases following injection, the antigen-encoding mRNA has been delivered into the body using various types of nanoparticles and liposomes. Prior to the COVID-19 pandemic, progress was already being made toward employing this experimental immunization approach for neoantigen-based anticancer therapy development (Pardi et al. 2018). With the success of the mRNA-based vaccines developed in 2020 to combat the pandemic, large-scale proof of the efficacy of this vaccination modality became firmly established. Neoantigen-encoding mRNA vaccines are now being vigorously pursued as possible future cancer immunotherapies.

# 8 Checkpoint Blockade Therapy and Its Impact on Neoantigen-Based Vaccines

Established tumors, much like chronic infections, result in subpopulations of T-cells with an altered phenotype and reduced functionality (Fig. 4). These dysfunctional T-cells have been called "exhausted," owing to the fact that they can arise as a result

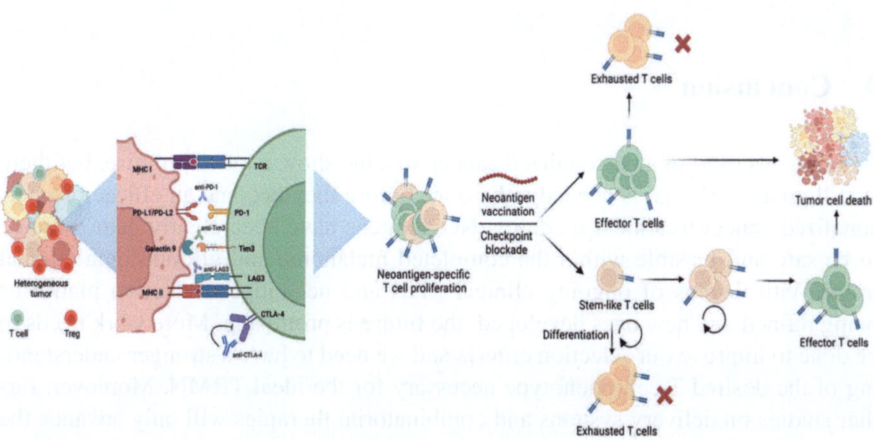


**Fig. 4** In established tumors, the TIL often expresses inhibitory cell surface receptors, such as PD-1, Tim-3, and Lag-3, that are associated with dysfunctional T-cells and a poor prognosis. While neoantigen vaccines can increase an existing T-cell or induce a de novo response, they are sometimes unable to overcome this inhibition. The development of ICB therapies successfully capitalized on the concept of blocking such inhibitory interactions. By removing inhibitory signals using blocking antibodies, the TRMN response is often rejuvenated, and/or a new clonally unique T-cell response develops

of chronic antigen stimulation. Exhausted T-cells have been identified based on the expression of inhibitory cell surface receptors such as PD-1, Tim-3, and Lag-3 (Wherry and Kurachi 2015). In established tumors, the cancer cells often express ligands for these receptors (e.g., PD-L1 for the PD-1 receptor), thereby inhibiting the T-cells' cytotoxic or immune stimulatory functions. The development of "checkpoint blockade" monoclonal antibody therapies (e.g., anti-PD-1 and anti-CTLA-4) successfully capitalized on the concept of blocking such inhibitory interactions. By removing inhibitory signals using blocking antibodies, the anticancer T-cell response is often rejuvenated, and/or a new clonally unique T-cell response develops (McLane et al. 2019). In recent years, it has become clear that T-cell responses developing as a result of these antibody treatments often consist of neoantigen-specific T-cells. In other words, cancer patients' immune systems have in many cases mounted a natural neoantigen-targeted anticancer T-cell response, but the T-cells have been prevented from performing their function due to inhibitory signals such as PD-1 engagement with PD-L1. This phenomenon was first observed in mice when checkpoint blockade was used to identify neoantigen-specific T-cell responses that were not otherwise observed (Gubin et al. 2014); it has since been confirmed in human cancers (Rizvi et al. 2015). This realization has been groundbreaking from a clinical perspective: combination approaches have emerged, where neoantigen vaccination is delivered along with checkpoint blockade therapy like anti-PD-1. This combination therapy has seen early successes (Ott et al. 2017, 2020; Sahin et al. 2017), and (as of early 2023) over 40 clinical trials are ongoing when searching "neoantigen" and "checkpoint" at ClinicalTrials.gov (NIH: U.S. National Library of Medicine n.d.).

## 9 Conclusion

Neoantigens used in a personalized cancer vaccine show a lot of promise, but there is still great work that is needed to be done before they become an efficacious personalized cancer treatment modality. Neoantigens have been clearly demonstrated to be safe and feasible within the completed melanoma and glioblastoma clinical trials. With dozens of ongoing clinical trials and neoantigen selection platforms being refined and new ones developed, the future is promising. More work needs to be done to improve our selection criteria and we need to have a stronger understanding of the desired T-cell phenotype necessary for the ideal TRMN. Moreover, further studies on delivery systems and combinatorial therapies will only advance the field in the development of an effective, precise, and durable immune response against cancer.

## Glossary

**Antigen-presenting cells (APCs)** Innate immune cells which professionally process and present antigens to activate antigen-specific T cells.

**Cancer-germline antigens (CGAs)** Antigens whose expression is limited to cancer cells as well as germline tissues like testis, ovary, or placenta.

**Checkpoint blockade** This is an immunotherapeutic approach where antibodies are used against co-inhibitory receptors to restore the function of T cells.

**Differential Agretopicity index (DAI)** The difference in predicted class I major histocompatibility complex (MHC-I) binding affinity between wild-type and corresponding mutant peptides.

**Neoantigens** Mutated peptides that are uniquely presented on tumor cells and derived from tumor-specific mutations.

**Tumor microenvironment (TME)** The ecosystem that surrounds a tumor inside the body and includes immune cells, the extracellular matrix, blood vessels, and other cells like fibroblasts, etc.

**Tumor-associated antigens (TAAs)** Antigens which are expressed at a higher level on tumor cells but are also expressed at lower levels on healthy cells.

**Tumor-specific antigens (TSAs)** Antigens that are uniquely presented on tumor cells but absent on normal cells.

## References

Aarntzen EHJG, Vries IJMD, Lesterhuis WJ et al (2013) Targeting CD4+ T-Helper cells improves the induction of antitumor responses in dendritic cell-based vaccination. Cancer Res 73:19–29. https://doi.org/10.1158/0008-5472.CAN-12-1127

Ahmadzadeh M, Pasetto A, Jia L et al (2019) Tumor-infiltrating human CD4+ regulatory T cells display a distinct TCR repertoire and exhibit tumor and neoantigen reactivity. Sci Immunol 4. https://doi.org/10.1126/sciimmunol.aao4310

Albert BA, Yang Y, Shao XM et al (2022) Deep neural networks predict MHC-I epitope presentation and transfer learn neoepitope immunogenicity. https://doi.org/10.1101/2022.08.29.505690

Alspach E, Lussier DM, Miceli AP et al (2019) MHC-II neoantigens shape tumor immunity and response to immunotherapy. Nature 574:696–701. https://doi.org/10.1038/s41586-019-1671-8

Andreatta M, Nielsen M (2016) Gapped sequence alignment using artificial neural networks: application to the MHC class I system. Bioinformatics 32:511–517. https://doi.org/10.1093/bioinformatics/btv639

Anichini A, Molla A, Mortarini R et al (1999) An expanded peripheral T cell population to a cytotoxic t lymphocyte (Ctl)-defined, melanocyte-specific antigen in metastatic melanoma patients impacts on generation of peptide-specific Ctls but does not overcome tumor escape from immune surveillance in metastatic lesions. J Exp Med 190:651–668. https://doi.org/10.1084/jem.190.5.651

Assarsson E, Sidney J, Oseroff C et al (2007) A quantitative analysis of the variables affecting the repertoire of T cell specificities recognized after vaccinia virus infection. J Immunol 178:7890–7901. https://doi.org/10.4049/jimmunol.178.12.7890

Axelrod ML, Cook RS, Johnson DB, Balko JM (2019) Biological consequences of MHC-II expression by tumor cells in cancer. Clin Cancer Res 25:2392–2402. https://doi.org/10.1158/1078-0432.CCR-18-3200

Bassani-Sternberg M, Bräunlein E, Klar R et al (2016) Direct identification of clinically relevant neoepitopes presented on native human melanoma tissue by mass spectrometry. Nat Commun 7:13404. https://doi.org/10.1038/ncomms13404

Blass E, Ott PA (2021) Advances in the development of personalized neoantigen-based therapeutic cancer vaccines. Nat Rev Clin Oncol 18:215–229. https://doi.org/10.1038/s41571-020-00460-2

Blum JS, Wearsch PA, Cresswell P (2013) Pathways of antigen processing. Annu Rev Immunol 31:443–473. https://doi.org/10.1146/annurev-immunol-032712-095910

Brennick CA, George MM, Corwin WL et al (2017) Neoepitopes as cancer immunotherapy targets: key challenges and opportunities. Immunotherapy 9:361–371. https://doi.org/10.2217/imt-2016-0146

Brennick CA, George MM, Srivastava PK, Karandikar SH (2020) Prediction of cancer neoepitopes needs new rules. Semin Immunol 101387. https://doi.org/10.1016/j.smim.2020.101387

Brennick CA, George MM, Moussa MM et al (2021) An unbiased approach to defining bona fide cancer neoepitopes that elicit immune-mediated cancer rejection. J Clin Invest:131. https://doi.org/10.1172/JCI142823

Buus S, Lauemøller SL, Worning P et al (2003) Sensitive quantitative predictions of peptide-MHC binding by a 'Query by Committee' artificial neural network approach. Tissue Antigens 62:378–384. https://doi.org/10.1034/j.1399-0039.2003.00112.x

Calvo Tardón M, Allard M, Dutoit V et al (2019) Peptides as cancer vaccines. Curr Opin Pharmacol 47:20–26. https://doi.org/10.1016/j.coph.2019.01.007

Cohen RB, Twardowski P, Johnson ML et al (2020) GEN-009, a neoantigen vaccine containing ATLAS selected neoantigens, to generate broad sustained immunity against immunogenic tumor mutations and avoid inhibitory peptides. JCO 38:3107–3107. https://doi.org/10.1200/JCO.2020.38.15_suppl.3107

Duan F, Duitama J, Seesi SA et al (2014) Genomic and bioinformatic profiling of mutational neoepitopes reveals new rules to predict anticancer immunogenicity. J Exp Med 211:2231–2248. https://doi.org/10.1084/jem.20141308

Ebrahimi-Nik H, Michaux J, Corwin WL et al (2019) Mass spectrometry–driven exploration reveals nuances of neoepitope-driven tumor rejection. JCI Insight 4. https://doi.org/10.1172/jci.insight.129152

Fehlings M, Jhunjhunwala S, Kowanetz M et al (2019) Late-differentiated effector neoantigen-specific CD8+ T cells are enriched in peripheral blood of non-small cell lung carcinoma patients responding to atezolizumab treatment. J Immunother Cancer 7:249. https://doi.org/10.1186/s40425-019-0695-9

Feng H, Shuda M, Chang Y, Moore PS (2008) Clonal integration of a polyomavirus in human Merkel cell carcinoma. Science 319:1096–1100. https://doi.org/10.1126/science.1152586

Ghorani E, Rosenthal R, McGranahan N et al (2018) Differential binding affinity of mutated peptides for MHC class I is a predictor of survival in advanced lung cancer and melanoma. Ann Oncol 29:271–279. https://doi.org/10.1093/annonc/mdx687

Gross L (1943) Intradermal immunization of C3h mice against a sarcoma that originated in an animal of the same line. Cancer Res 3:326–333

Gubin MM, Zhang X, Schuster H et al (2014) Checkpoint blockade cancer immunotherapy targets tumour-specific mutant antigens. Nature 515:577–581. https://doi.org/10.1038/nature13988

Haen SP, Löffler MW, Rammensee H-G, Brossart P (2020) Towards new horizons: characterization, classification and implications of the tumour antigenic repertoire. Nat Rev Clin Oncol 17:595–610. https://doi.org/10.1038/s41571-020-0387-x

IEDB IEDB Analysis Resource (n.d.) In: Epitope Prediction and Analysis Tools. http://tools.iedb.org/main/references/

Karandikar SH, Sidney J, Sette A et al (2019) Identification of epitopes in ovalbumin that provide insights for cancer neoepitopes. JCI Insight 4. https://doi.org/10.1172/jci.insight.127882

Keskin DB, Anandappa AJ, Sun J et al (2019) Neoantigen vaccine generates intratumoral T cell responses in phase Ib glioblastoma trial. Nature 565:234. https://doi.org/10.1038/s41586-018-0792-9

Klein G, Sjögren HO, Klein E, Hellström KE (1960) Demonstration of resistance against methylcholanthrene induced sarcomas in the primary autochthonous host. Cancer Res 20:1561–1572

Kovacsovics-Bankowski M, Rock KL (1995) A phagosome-to-cytosol pathway for exogenous antigens presented on MHC class I molecules. Science 267:243–246. https://doi.org/10.1126/science.7809629

Kurtulus S, Madi A, Escobar G et al (2019) Checkpoint blockade immunotherapy induces dynamic changes in PD-1−CD8+ tumor-infiltrating T cells. Immunity 50:181–194.e6. https://doi.org/10.1016/j.immuni.2018.11.014

Laidlaw BJ, Craft J, Kaech SM (2016) The multifaceted role of CD4+ T cells in the regulation of CD8+ T cell memory maturation. Nat Rev Immunol 16:102–111. https://doi.org/10.1038/nri.2015.10

Leone P, Shin E-C, Perosa F et al (2013) MHC Class I Antigen processing and presenting machinery: organization, function, and defects in tumor cells. J Natl Cancer Inst 105:1172–1187. https://doi.org/10.1093/jnci/djt184

Linnemann C, van Buuren MM, Bies L et al (2015) High-throughput epitope discovery reveals frequent recognition of neo-antigens by CD4+ T cells in human melanoma. Nat Med 21:81–85. https://doi.org/10.1038/nm.3773

Marchand M, van Baren N, Weynants P et al (1999) Tumor regressions observed in patients with metastatic melanoma treated with an antigenic peptide encoded by gene MAGE-3 and presented by HLA-A1. Int J Cancer 80:219–230. https://doi.org/10.1002/(SICI)1097-0215(19990118)80:2<219::AID-IJC10>3.0.CO;2-S

Marzo AL, Kinnear BF, Lake RA et al (2000) Tumor-specific CD4+ T cells have a major "Post-Licensing" role in CTL mediated anti-tumor immunity. J Immunol 165:6047–6055. https://doi.org/10.4049/jimmunol.165.11.6047

McLane LM, Abdel-Hakeem MS, Wherry EJ (2019) CD8 T cell exhaustion during chronic viral infection and cancer. Annu Rev Immunol 37:457–495. https://doi.org/10.1146/annurev-immunol-041015-055318

Monach PA, Meredith SC, Siegel CT, Schreiber H (1995) A unique tumor antigen produced by a single amino acid substitution. Immunity 2:45–59

Nielsen M, Lundegaard C, Worning P et al (2003) Reliable prediction of T-cell epitopes using neural networks with novel sequence representations. Protein Sci 12:1007–1017

NIH: U.S. (n.d.) National Library of Medicine ClinicalTrials.gov. https://clinicaltrials.gov. Accessed 15 Feb 2023

Nogueira C, Kaufmann JK, Lam H, Flechtner JB (2018) Improving cancer immunotherapies through empirical neoantigen selection. Trends Cancer 4:97–100. https://doi.org/10.1016/j.trecan.2017.12.003

Ott PA, Hu Z, Keskin DB et al (2017) An immunogenic personal neoantigen vaccine for patients with melanoma. Nature 547:217–221. https://doi.org/10.1038/nature22991

Ott PA, Hu-Lieskovan S, Chmielowski B et al (2020) A Phase Ib trial of personalized neoantigen therapy plus anti-PD-1 in patients with advanced melanoma, non-small cell lung cancer, or bladder cancer. Cell 183:347–362.e24. https://doi.org/10.1016/j.cell.2020.08.053

Pardi N, Hogan MJ, Porter FW, Weissman D (2018) mRNA vaccines—a new era in vaccinology. Nat Rev Drug Discov 17:261–279. https://doi.org/10.1038/nrd.2017.243

Philip M, Fairchild L, Sun L et al (2017) Chromatin states define tumour-specific T cell dysfunction and reprogramming. Nature 545:452–456. https://doi.org/10.1038/nature22367

Prehn RT, Main JM (1957) Immunity to methylcholanthrene-induced sarcomas. J Natl Cancer Inst 18:769–778

Ramarathinam L, Sarma S, Maric M et al (1995) Multiple lineages of tumors express a common tumor antigen, P1A, but they are not cross-protected. J Immunol 155:5323–5329

Rao AA, Madejska AA, Pfeil J et al (2020) ProTECT—prediction of T-cell epitopes for cancer therapy. Front Immunol 11:483296

Rech AJ, Balli D, Mantero A et al (2018) Tumor immunity and survival as a function of alternative neopeptides in human cancer. Cancer Immunol Res 6:276–287. https://doi.org/10.1158/2326-6066.CIR-17-0559

Reynisson B, Alvarez B, Paul S et al (2020) NetMHCpan-4.1 and NetMHCIIpan-4.0: improved predictions of MHC antigen presentation by concurrent motif deconvolution and integration of MS MHC eluted ligand data. Nucleic Acids Res 48:W449–W454. https://doi.org/10.1093/nar/gkaa379

Rizvi NA, Hellmann MD, Snyder A et al (2015) Cancer immunology. Mutational landscape determines sensitivity to PD-1 blockade in non-small cell lung cancer. Science 348:124–128. https://doi.org/10.1126/science.aaa1348

Rosalia RA, Quakkelaar ED, Redeker A et al (2013) Dendritic cells process synthetic long peptides better than whole protein, improving antigen presentation and T-cell activation. Eur J Immunol 43:2554–2565. https://doi.org/10.1002/eji.201343324

Rosato A, Zoso A, Milan G et al (2003) Individual analysis of mice vaccinated against a weakly immunogenic self tumor-specific antigen reveals a correlation between CD8 t cell response and antitumor efficacy. J Immunol 171:5172–5179. https://doi.org/10.4049/jimmunol.171.10.5172

Rosenberg SA, Yang JC, Schwartzentruber DJ et al (1998) Immunologic and therapeutic evaluation of a synthetic peptide vaccine for the treatment of patients with metastatic melanoma. Nat Med 4:321–327. https://doi.org/10.1038/nm0398-321

Sahin U, Derhovanessian E, Miller M et al (2017) Personalized RNA mutanome vaccines mobilize poly-specific therapeutic immunity against cancer. Nature 547:222–226. https://doi.org/10.1038/nature23003

Sarma S, Bai X-F, Liu J et al (2003) On the role of unmutated antigens in tumor rejection in mice with unperturbed T-cell repertoires. Cancer Res 63:6051–6055

Schenck RO, Lakatos E, Gatenbee C et al (2019) NeoPredPipe: high-throughput neoantigen prediction and recognition potential pipeline. BMC Bioinformatics 20:264. https://doi.org/10.1186/s12859-019-2876-4

Schmidt J, Smith AR, Magnin M et al (2021) Prediction of neo-epitope immunogenicity reveals TCR recognition determinants and provides insight into immunoediting. Cell Rep Med 2:100194. https://doi.org/10.1016/j.xcrm.2021.100194

Sette A, Vitiello A, Reherman B et al (1994) The relationship between class I binding affinity and immunogenicity of potential cytotoxic T cell epitopes. J Immunol 153:5586–5592

Siddiqui I, Schaeuble K, Chennupati V et al (2019) Intratumoral Tcf1+PD-1+CD8+ T cells with stem-like properties promote tumor control in response to vaccination and checkpoint blockade immunotherapy. Immunity 50:195–211.e10. https://doi.org/10.1016/j.immuni.2018.12.021

Srivastava PK (1993) Peptide-binding heat shock proteins in the endoplasmic reticulum: role in immune response to cancer and in antigen presentation. In: Klein GFVW and G (ed) Advances in cancer research. Academic Press, Cambridge, MA, pp 153–177

Srivastava PK (2015) Neoepitopes of cancers: looking back, looking ahead. Cancer Immunol Res 3:969–977. https://doi.org/10.1158/2326-6066.CIR-15-0134

Stephens AJ, Burgess-Brown NA, Jiang S (2021) Beyond just peptide antigens: the complex world of peptide-based cancer vaccines. Front Immunol 12:696791. https://doi.org/10.3389/fimmu.2021.696791

Trolle T, McMurtrey CP, Sidney J et al (2016) The length distribution of class I-restricted T cell epitopes is determined by both peptide supply and MHC allele-specific binding preference. J Immunol 196:1480–1487. https://doi.org/10.4049/jimmunol.1501721

Utzschneider DT, Charmoy M, Chennupati V et al (2016) T cell factor 1-expressing memory-like CD8+ T cells sustain the immune response to chronic viral infections. Immunity 45:415–427. https://doi.org/10.1016/j.immuni.2016.07.021

van Rooij N, van Buuren MM, Philips D et al (2013) Tumor exome analysis reveals neoantigen-specific T-cell reactivity in an ipilimumab-responsive melanoma. J Clin Oncol 31. https://doi.org/10.1200/JCO.2012.47.7521

Walboomers JM, Jacobs MV, Manos MM et al (1999) Human papillomavirus is a necessary cause of invasive cervical cancer worldwide. J Pathol 189:12–19. https://doi.org/10.1002/(SICI)1096-9896(199909)189:1<12::AID-PATH431>3.0.CO;2-F

Wherry EJ, Kurachi M (2015) Molecular and cellular insights into T cell exhaustion. Nat Rev Immunol 15:486–499. https://doi.org/10.1038/nri3862

Wood MA, Nguyen A, Struck AJ et al (2020) Neoepiscope improves neoepitope prediction with multivariant phasing. Bioinformatics 36:713–720. https://doi.org/10.1093/bioinformatics/btz653

Yadav M, Jhunjhunwala S, Phung QT et al (2014) Predicting immunogenic tumour mutations by combining mass spectrometry and exome sequencing. Nature 515:572–576. https://doi.org/10.1038/nature14001

Zaidi N, Jaffee EM (2019) Immune cells track hard-to-target brain tumours. Nature 565:170–171. https://doi.org/10.1038/d41586-018-07728-9

# Bispecific Antibodies: An Emerging Concept in Antibody-Based Cancer Therapies

**Rafael Oliveira Brandão, Xin Jiang, Sudhagar Selvaraju, and Purusottam Mohapatra**

**Abstract** Bispecific antibody-based treatment options have emerged as an impending therapeutic strategy for various deadly cancers including several hematological malignancies and solid tumors. A bispecific antibody is a synthetic antibody construct that contains binding sites directed against different cellular and molecular targets. Its function and activity can be dependent on this bispecific linkage that would not normally be obtained by the presence of monospecific antibodies with the same specificities, or it can be an alternative for a monospecific antibody mixture. Bispecific antibodies have emerged as modern promising targeted therapy options that could combine different modes of actions in the same molecule potentially facilitating its development, approval, manufacturing, and clinical usage comparing to monospecific antibodies. Here, we have shed light on the basics of bispecific antibodies' mode of action and their manufacturing. Next, we have provided current bispecific antibody treatment strategies in various preclinical and clinical models of solid and hematological cancers. Furthermore, we have included an updated clinical cancer trials details that employing various formats of bispecific antibodies.

**Keywords** Cancer · Bispecific antibody · Precision cancer medicine · Clinical Trials

---

Rafael Oliveira Brandão and Xin Jiang contributed equally to this work.

---

R. O. Brandão · X. Jiang
Biotech Research & Innovation Centre (BRIC), Biomedical Institute, University of Copenhagen, Copenhagen, Denmark

S. Selvaraju · P. Mohapatra (✉)
Department of Biotechnology, National Institute of Pharmaceutical Education and Research (NIPER) Guwahati, Kamrup, Assam, India
e-mail: purusottam@niperguwahati.in





A. Das Mohapatra, P. S. Sahu (eds.), *Advances in Immunology and Immuno-techniques*, Interdisciplinary Biotechnological Advances,
https://doi.org/10.1007/978-981-97-5508-0_4

# 1 Introduction to Bispecific Antibodies

Harvesting the power of antibodies to treat human diseases is now possible due to the development of the hybridoma technique in 1975 that gave rise to a multitude of monoclonal antibodies for the treatment of diverse types of cancer (Köhler and Milstein 1975). Although the therapeutic importance of monospecific antibodies in the cancer field is no longer questionable, these antibodies still do not allow the engagement of the cytotoxic T lymphocytes in the action against cancer cells.

A bispecific antibody is a synthetic antibody construct that contains binding sites directed against different targets. Its function and activity can be dependent on this bispecific linkage that would not normally be obtained by the presence of monospecific antibodies with the same specificities, or it can be an alternative for a monospecific antibody mixture (Labrijn et al. 2019; Suurs et al. 2019). The idea of a bispecific antibody was envisioned and developed as a potential cancer treatment by having one binding affinity to a specific tumor antigen and another to the CD3 complex of the T-cells. By binding to CD3, the antibody can successfully activate the T-cells by bypassing the peptide-major histocompatibility complex restriction. This activated T-cell can therefore be brought in close proximity to its target to form an extracellular synapse allowing the T-cell to release perforin and other cytotoxic components killing the cancer cell (Suurs et al. 2019; Wu and Cheung 2018; Offner et al. 2006; Haas et al. 2009). Over the years, other conditions were also considered as treatment targets for bispecific antibodies, such as fighting transplant autoimmunity, chronic inflammatory diseases, bleeding disorders, and neurodegeneration (Brinkmann and Kontermann 2017) (Fig. 1).

The design of bispecific antibodies is highly customizable depending on the intended therapeutic applications. Different kinds of bispecific antibody architectures show differences in their binding affinity or stability in vivo making this technology highly versatile for exploiting and targeting different cellular mechanisms. The list of designs developed, and that are currently undergoing clinical trials, are quite extensive; however, the majority can be grouped into three distinct groups based on their architecture: (1) Fc-less antibody, contains the antigen-binding sites of two antibodies; (2) Asymmetric, that retain the native architecture of the antibody potentially preserving its functional characteristics and lowering its immunogenicity; (3) Symmetric, where both specificities are kept in a single polypeptide chain or heavy-light chain pair while retaining the Fc region (Labrijn et al. 2019; Brinkmann and Kontermann 2017).

The production of bispecific antibodies can be achieved using a quadroma—a cell resulting the fusion of two hybridoma cells. This cell produces a mixture of antibody chains that associate randomly resulting in few antibody molecules that are functional while others will not be functional—also known as the chain association issue (Köhler and Milstein 1975; Kontermann 2012). To circumvent this efficiency issue, other methods were developed such as chimeric quadromas which take advantage of species restricted pairing, and recombinant proteins that drive the correct chain association (Merchant et al. 1998; Lindhofer et al. 1995; Ridgway et al. 1996).

Although the most common use of bispecific antibodies is bridging cells by having affinity to membrane proteins in both cells, other types of linkage are also being explored, such as: (1) receptor by binding in (in-cis) to two specific cell surface receptors that could result in its activation or inhibition, (2) bringing an enzyme and its substrate in close proximity which facilitates enzymatic activity in a way that mimics the presence of a cofactor, and (3) "piggybacking" which facilitates the internalization of proteins through binding to cell surface receptors (Labrijn et al. 2019). A particularly interesting study in brain drug uptake shows effective passage through the blood–brain barrier of a bispecific antibody with affinity to both transferrin receptor and β-secretase 1 (BACE1). In this design, the affinity to the transferrin receptor provides a way to overcome the barrier by exploiting the receptor internalization pathway while the second affinity to the BACE1 inhibited the protein which is of great interest in the Alzheimer's treatment (Yu et al. 2011).

The number of antibodies that are approved to treat diverse diseases has been increasing in the last few years while the majority deals with cancer treatment. As of today, 130 monoclonal antibodies were approved or currently in review in the European Union or United States (TheAntibodySociety 2021) while only four studies involving bispecific antibodies have so far reached Phase 3 clinical trials (ClinicalTrials 2021).

Throughout the years of development of the bispecific antibody technology, there were several different designs reported, each with their advantages and caveats. Smaller versions such as Fc-less antibodies have a minimalistic design making it easier to express in lower eukaryotic and prokaryotic organisms. However, the absence of Fc region decreases the antibody plasmatic half-life which require continuous administration to the patient. On the other hand, antibodies that retain the Fc region stay longer in the patient body but are more difficult and costlier to produce.

The design of the bispecific antibody does not only influence the specific binding ability to its targets but could have repercussions on its off-target binding affinity which could result in severe conditions. A common side effect of the T-cell dependent bispecific antibody treatment is the uncontrolled release of cytokines at the systemic level in a process named CRS (cytokine release syndrome). Although yet not totally understood, this systemic inflammatory response was observed in some patients with various degrees of severity (Shimabukuro-Vornhagen et al. 2018; Xu et al. 2000). An example of a bispecific antibody that has shown safety concerns is catumaxomab which induced fatal acute liver failure in some patients resulting in the termination of the study (Borlak et al. 2016). In vivo studies in mice shown that CRS is dependent on the TNF-α release by the T-cells which function as a signal that trigger monocyte activation where blocking IL-6, IL-1β, or TNF-α shows promising results in the treatment of CRS (Li et al. 2019). In the clinic, CRS is now treated with administration of corticosteroids or tocilizumab, an antibody against IL-6 receptor (Shimabukuro-Vornhagen et al. 2018).

Bispecific antibodies have emerged as a targeted therapy that could combine different modes of action in the same molecule potentially facilitating its development,

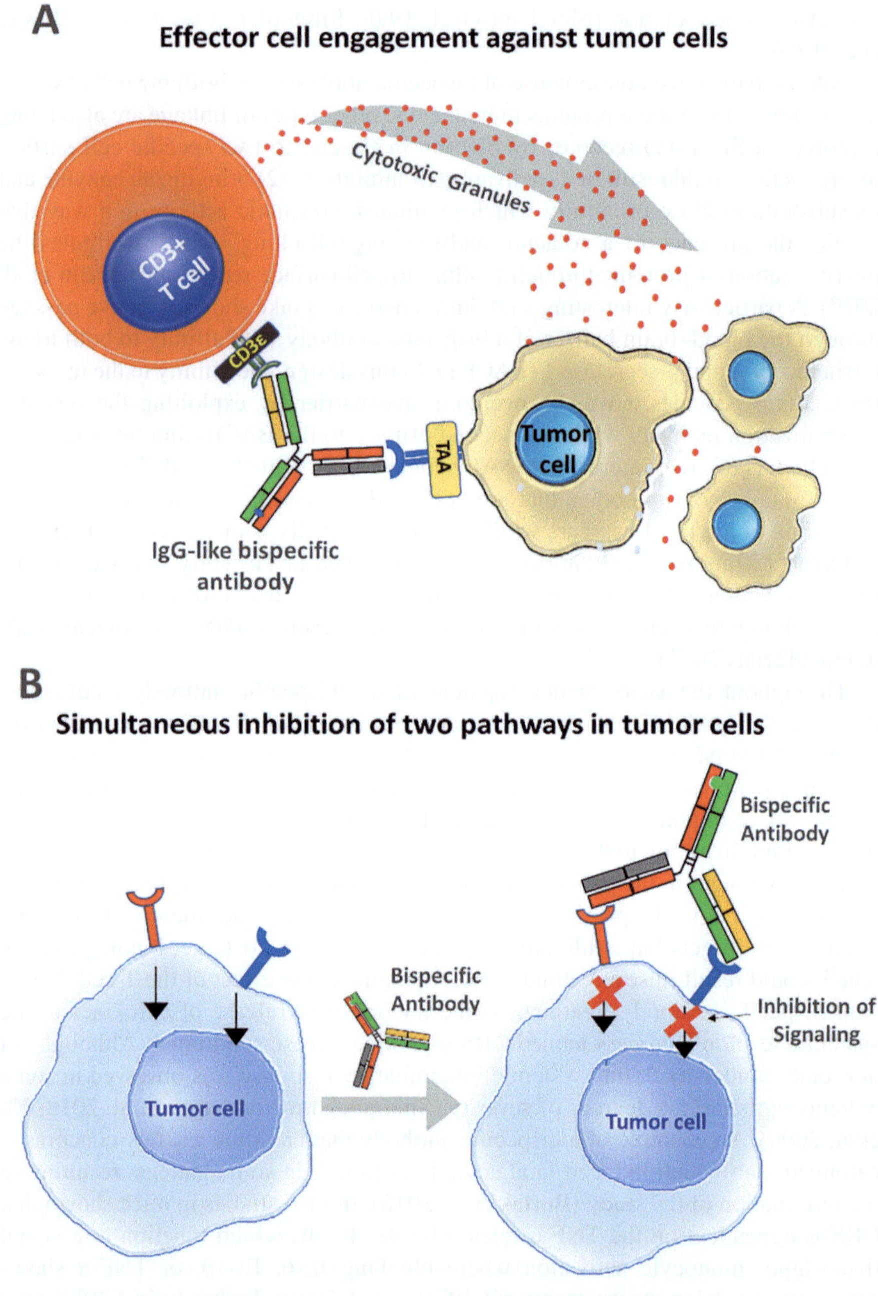


**Fig. 1** Schematic showing mechanisms of the action of IgG-based bispecific antibody against tumor cells

approval, manufacturing, and clinical usage comparing to monospecific antibodies (Kontermann 2012). Most bispecific antibodies undergoing clinical trials are directed against cancer; however, many other usages of bispecific antibodies are currently being explored which could lead to exciting novel therapy methods. It is therefore anticipated that the bispecific antibody technology will be further developed and its use in the clinic would be more widespread in the upcoming years potentially replacing lesser effective treatments for cancer and other diseases.

## 2 Key Technologies in Bispecific Antibody Generation

The size, stability, binding affinity, and immunogenicity of any bispecific antibody profoundly influence their safety, effectiveness, and clinical success. In general, the antigen-binding sites of any bispecific antibody are manufactured by the variable domains of two different light (VL) and heavy (VH) chains which account to one of the key difficulties in the generation of bispecific antibodies. There are several technologies developed during the past years to generate bispecific antibodies, the most common methods are briefly described here.

### 2.1 *Chemical Engineering Approach*

Oxidative recombination is a method employed to chemically conjugate two different purified monoclonal antibodies in order to generate bispecific antibodies. For example: Yankelevich and co-worker prepared Anti-CD3 × Anti-GD2 bispecific antibody and Anti-CD3 × Anti-Her2/neu bispecific antibodies by chemical heteroconjugation against GD2-positive neuroblastomas (Yankelevich et al. 2012). Purification of chemically crosslinked bispecific antibody is a challenging task, and the purification process might affect the activity of the antibodies by disturbing the stability of chemical-crosslinking.

### 2.2 *Quadroma Technology*

Quadroma technology is one of the common technologies employed for bispecific antibody production where two different hybridoma cells producing monoclonal antibodies are fused to produce an array of antibodies including bispecific antibodies. For example, catumaxomab was the first approved bispecific antibody for metastatic tumors with EpCAM expression. Bispecific antibodies produced through the quadroma methods usually similar to the typical antibodies which exert complement-mediated cytotoxicity, cell-mediated cytotoxicity, and cellular phagocytosis (Spiess et al. 2015). Purification is the major challenge in quadroma technology since arrays

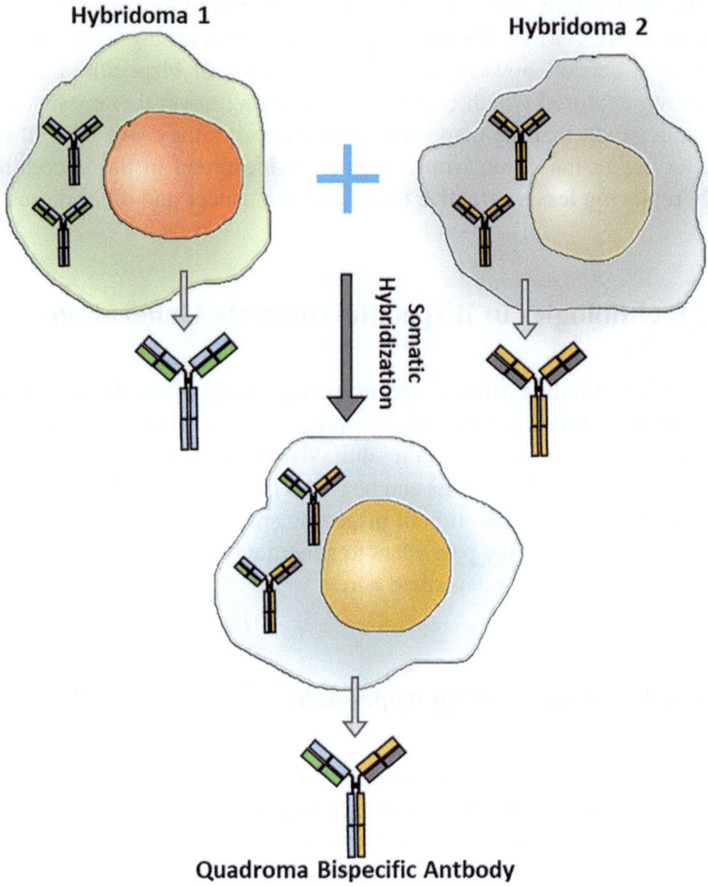


**Fig. 2** Schematic showing quadroma technology for bispecific antibody generation

of non-functional antibodies are also produced during the hybridoma cell fusion process. To reduce the non-functional antibody production, fusion of two different hybridoma from different species is currently employed in the bispecific antibody generation. This approach is highly effective which reduces the mismatch non-functional antibody production to less than 10%. For example, catumaxomab bispecific antibody produced by the fusion of mouse and rat quadroma used for the treatment of solid tumors (Fig. 2).

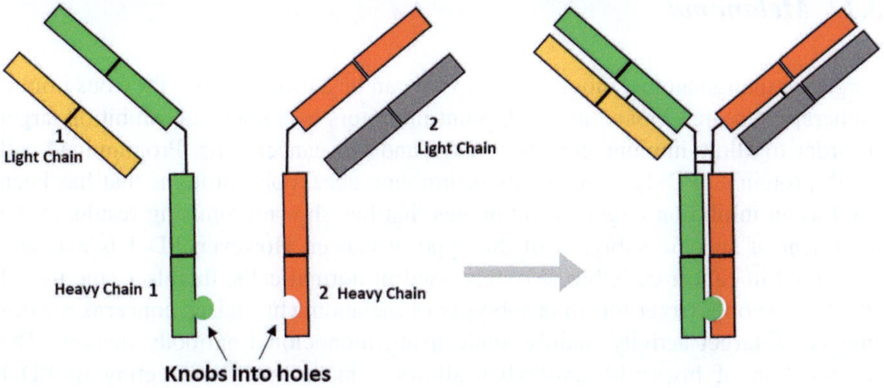


**Fig. 3** Schematic showing overview of Knobs into Holes bispecific antibody format

## 2.3 *Knobs into Holes (KIH) Technology*

Knobs into holes (KIH) technology of bispecific antibody production is a modern technique that yields more heterodimers of two heavy chains via introduction of various mutation into two CH3 domains. One heavy chain part of the antibody is synthesized with a mutation at T366W that adds a Knob like structure to the CH3 domain, whereas the second part is synthesized with mutations T366S, L368, or Y407V that replaces a smaller amino acid to make a "hole" (Fig. 3). The resulting bispecific antibody synthesized through the KIH technology is considered to be more thermodynamically stable with an extended half-life.

# 3 Bispecific Antibody Therapy for Various Cancers

Bispecific antibodies have emerged as a promising therapy for fighting cancer. Its mode of action is dependent on the binding to two antigens at the same time. Depending on the intended outcome, these antibodies can bind receptors in the same cell which can induce receptor degradation or inhibition by blocking the ligand interaction. Furthermore, a very common use of bispecific antibodies is to bring T-cells in close proximity to tumor cells providing a way for the formation of a cytotoxic synapse in order to induce tumor cell death (Krishnamurthy and Jimeno 2018). In this subchapter, we will explore the current status of the research involving bispecific antibodies and their promising application in the treatment of diverse types of cancer.

## 3.1 Melanoma

Targeting malignant melanoma cells in vivo can be a difficult task. Previous immunotherapy research looked at check point inhibitors as a potential inhibition target in order to allow immune cells to engage and kill cancer cells. Programmed cell death protein 1 (PD-1) is one of those immune checkpoint proteins that has been used as an inhibition target by antibodies that has shown promising results in the treatment of specific subtypes of this type of cancer. However, PD-1 is not only expressed in cancer cells but also expressed in normal cells, therefore, the use of PD-1 as the only target for some subtypes of melanoma has raised concerns regarding its off-target activity mainly while using monoclonal antibody therapy. The introduction of bispecific antibodies allows a more specific targeting of PD-1 because this type of antibodies allow co-targeting of the cell surface proteoglycan CSPG4 that is known to be overexpressed in some types of cancer including melanoma. Use of this bispecific antibody (bAbPD-1xCSPG4) was shown in in vitro studies to allow a more precise targeting of melanoma cells while not affecting normal cells that also express PD-1. Importantly, this bispecific antibody retained its ability to activate T-cells and direct them to the cancer cells in a CSPG4-dependent manner (Koopmans et al. 2019).

Other bispecific antibodies that have a different mode of action have also emerged as potential inhibitors of tumor cell immune checkpoints. B7 Homolog 3 (B7-H3) is an important immune checkpoint protein that is present in diverse type of cancer resulting in blocking of cancer cell targeting by the T-cells. Its overexpression has been associated in and increased cell proliferation and poor survival of melanoma cancer patients. Bispecific antibodies co-targeting B7-H3 and the CD3 receptor in $CD3^+$ T-cells were shown to effectively inhibit this immune check point while at the same time redirect the $CD3^+$ T-cell to the cancer cells. This method allowed the activation and recruitment of the T-cells to the tumor sites and effectively induce the killing of tumor cells which was shown in both in vitro and in vivo preclinical xenograft models (Ma et al. 2019, 2016).

The quest to target immune checkpoints in cancer cells has been going on for years, which in some cases lead to the development and approval of antibody-based drugs such as ipilimumab—an anti-CTLA-4 monoclonal antibody that was successfully used in the treatment of patients with metastatic melanoma. However, the use of this antibody was also associated with severe immune reactions which greatly limit their clinical use. Ensuring that a T-cell activating antibody is restricted to the tumor microenvironment is of major importance to direct its effect to the target cells while reducing potential fatal off-target effects. A targeted combination of two proteins that are overexpressed in the tumor microenvironment has the potential to better direct the effect to its intended location. Thus, researchers have devised a bispecific antibody against both CTLA-4 and OX40 protein which are highly overexpressed in the tumor microenvironment. Studies with this bispecific antibody in cancer mouse models have shown tumor site specific depletion of T-regulatory cells and effective activation of T-cells correlating with an increase in animal survival

(Kvarnhammar et al. 2019). Others have also used a similar rational for the development of a bispecific antibody that simultaneously targets RANKL and PD-1 which have shown a beneficial treatment efficiency in comparison to the use of monospecific antibodies targeting these two proteins (Dougall et al. 2019).

## 3.2 Breast Cancer

B7 homolog 4 (B7-H4) is an immune checkpoint protein that is often found overexpressed in breast cancer independently of the HER2 or PD-L1 expression. In vivo studies in xenograft models for breast cancer using a bispecific antibody against the B7-H4 immune checkpoint and CD3 T-cell receptor shown a specific reduction of tumor cells in the $B7\text{-}H4^{+}$ MDA-MB-468 tumors which was not observed in the $B7\text{-}H4^{-}$ NCI-H2170 tumors used as control (Iizuka et al. 2019). Cell surface receptors such as prolactin receptor (PRLR) were also considered as targets for bispecific antibody therapy. PRLR is found overexpressed in some types of breast cancer, and its inhibition was shown to be of therapeutic importance. In fact, monospecific antibodies against this receptor were previously developed that shown promising results by blocking the prolactin binding to the receptor in in vitro and in vivo studies prompting the start of a phase 1 clinical trial (Agarwal et al. 2016). However, the lack of encouraging clinical trial results leads to further development of a bispecific antibody against this same tumor receptor and the CD3 T-cell receptor which shown a significant increase in cancer cell cytotoxicity that is positively correlated with the expression of PRLR in cancer cells. Further in vivo studies aimed at comparing these two antibodies have shown that this bispecific antibody carried a better tumor inhibition and animal survival compared to the monospecific antibody therapy (Zhou et al. 2020).

HER2 is a very well-studied cell surface receptor present in 25% of the breast cancers (Slamon et al. 1987). Targeting of this protein using bispecific antibodies against this receptor and CD3 receptor of the T-cells was previously attempted; however, it was shown to induce non-specific toxicity due to the presence of HER2 in normal cells. A carboxyl-terminal fragment of HER2 named p95HER2 was shown to be present only in 40% of overexpressed HER2 cancer cells (Arribas et al. 2011). Researchers developed a bispecific T-cell engaging antibody against this p95HER2 fragment which shown promising in vivo results in breast cancer while not affecting cells with normal expression of HER2 (Rius Ruiz et al. 2018).

Triple-negative breast cancer represents 10–20% of all the invasive breast cancers showing a high recurrence and a worse survival rate than other breast cancers (Paradiso and Singer 2017). Due to the lack of effective treatments, immune therapy became an attractive treatment option for patients with this type of breast cancer. Mesothelin (MSLN), a membrane protein whose expression is only present in the normal peritoneum, pericardium, and pleura is often found overexpressed in this type of breast cancer where it is associated with an increased tumor cell survival, proliferation, migration, and invasion (Chang et al. 1992; Tang et al. 2013). A

Fab-like Anti-MSLN x CD15 bispecific antibody was recently developed showing promising in vitro and in vivo results. In this study, NK cells successfully infiltrated triple-negative breast cancer spheroids and orthotopic xenograft models in the presence of this antibody. However, some model-to-model variation in the results measured by tumor volume was observed which could limit the clinical application of this antibody (Del Bano et al. 2019).

Targeting tumor microenvironment, such as tumor vasculature instead of the directly targeting tumor cells was previously regarded as a potential strategy for developing new cancer treatments. Inhibiting vascular endothelial growth factor A (VEGFA) was previously attempted as a potential cancer treatment; however, it has shown limited effectiveness due to a mechanism of tumor adaptive response often resulting in cancer relapse (Bergers and Hanahan 2008; Apte et al. 2019). A recent elegant study reports the development of a bispecific antibody against TGFBR2 x CD4 showing dramatic effects in in vivo cancer models leading to an increase in PYMT animal survival. This antibody is able to specifically inhibit TGF-β in $CD4^+$ T-cells resulting in cancer suppression by inducing vascular reorganization. Combination treatment of this bispecific antibody with a monoclonal antibody against VEGF further increases the survival of the animals to double the time compared to the treatment with anti-VEGF alone (Li et al. 2020).

### *3.3 Prostate Cancer*

Treatment of prostate cancer still heavily relies on surgery, chemotherapy, and radiotherapy. Therefore, there is a very high incentive for the development of novel therapies for this type of cancer such as immunotherapy that allows specific targeting of prostate cancer cells using tumor-specific markers while taking advantage of the host immune cells (Cha et al. 2020). A bispecific antibody for prostate-specific membrane antigen (PSMA) and CD3 receptor in the T-cells has shown efficient lysis of diverse human prostate cancer cell lines complemented with in vivo results showing a significant delay of tumor growth in mouse xenograft models (Friedrich et al. 2012).

Targeting metastatic microenvironment of the cancer cells is also an important aspect of cancer therapy. Prostate cancer cells show a preference for metastatic colonization of the bone marrow which support their lethality (Dinatale and Fatatis 2019). Mesenchymal stromal cells from bone marrow have been implicated in the chemotactic and adhesive response in prostate cancer cells via the secretion of cleaved fibronectin fragments which induce $\alpha5\beta1$ signaling in the prostate cancer cells. The interaction of fibronectin fragments with $\alpha5\beta1$ in these cancer cells has been suggested as a mechanism for cancer cell metastasis in the bone marrow microenvironment (Shiozawa et al. 2011; Méndez-Ferrer et al. 2010). αv integrin was proposed as a cooperator of $\alpha5\beta1$ integrin in the binding to the fibronectin fragments aiding in the process of metastatic niche interaction with the prostate cancer cells. A bispecific antibody anti-$\alpha5\beta1$ x αv (BsAba5b1/av) was developed in order to simultaneously target both integrins in the same cell showing an increased efficiency in the blocking of the fibronectin fragment induced signaling and effectively diminishing the cancer cells migration, adhesion, and endothelial migration in a co-culture in vitro setting. This same study showed that, mechanistically, this bispecific antibody induces the loss of cellular integrin expression through a process of induced lysosomal degradation (Joshi et al. 2020).

### 3.4 *Lung Cancer*

Immunotherapy for lung cancer has been looked at with increased interest by the scientific community. However, only 34 clinical trial studies are currently being carried while none has yet reached Phase 3 (ClinicalTrials 2021). The need to inhibit key effector proteins in the lung cancer cells has driven the development of multiple kinase inhibitors over the last few years. However, cancer cells quickly develop resistance—in less than a year on average—by acquiring mutations in the EGFR or the activation of alternative pathways such as cMet (Liu et al. 2020; Moores et al. 2016; Zucali et al. 2008). Bispecific antibodies carry the promise to better overcome tumor resistance compared to previously developed small molecule inhibitors which can be used in combination in order to further increase their treatment efficiency. A bispecific antibody against EGFR and cMet has shown promising results in the inhibition of these two pathways in lung adenocarcinoma cell lines with EGFR T790M and cMet amplification. Inhibition of these two receptors was successfully demonstrated by a clear reduction in the levels of phosphorylation of known downstream effectors where small molecule inhibitors have previously failed. Moreover, this bispecific antibody carried a modified Fc region resulting in better binding to the human FcgRIIIa on natural killer cells leading to a more effective antibody-dependent cellular cytotoxicity (Moores et al. 2016).

Another study using a T-cell engaging bispecific antibody with affinity for the Delta-like ligand 3 (DLL3) was shown to effectively activate T-cells and induce tumor cell lysis in in vitro and in vivo xenograft studies with small cell lung cancer cells (Hipp et al. 2020). A similar designed antibody has also targeted T-cells to the tumor cells via DLL3 binding where its combination with a PD-1 inhibitor further increased the killing efficiency of small cell lung cancer cells (Chen et al. 2020).

### 3.5 *Pancreatic Cancer*

Pancreatic cancer such as adenocarcinoma is one of the deadliest types of cancer where most of the patients do not survive more than 1 year after the diagnosis. When detected, only a small group of the patients is eligible for tumor surgical removal because, when detected, it has already metastasized to other organs. Thus, the lack of early diagnostic tools that allow proper staging of the tumors led to the development novel early detection methods that could devise further treatment (Vincent et al. 2011).

A bispecific antibody against MUC1 (a glycoprotein that is specifically overexpressed in the pancreatic adenocarcinoma cells) and In-DTPA (In-diethylenetriamine penta-acetic acid, a chelating agent used for binding of radiopharmaceuticals) was used in the pretreatment of xenograft mouse models of pancreatic adenocarcinoma. These mice were later treated with a radiolabeled peptide that binds to In-DTPA allowing the imaging of the xenograft tumor cells with a relatively high signal-to-noise ratio compared to previously used monospecific antibodies (Cardillo et al. 2004). Others have further improved the technology by developing an antibody divalent for MUC1 and monovalent for HSG (histamine-succinyl-glycine). Similar to the approach mentioned before, this technology allowed the in vivo visualization of the pancreatic adenocarcinoma cells with high specificity showing an increased signal-to-noise ratio compared to the previous design (Gold et al. 2008). To the present date, very few studies are describing the use of bispecific antibodies for the treatment of pancreatic cancer. However, some ongoing clinical trials are currently testing the efficiency of some bispecific antibodies, such as bAb anti-MUC1 x CD3 or bAb anti-EGFR armed activated T-cells for the treatment of high-grade pancreatic cancer (ClinicalTrials 2021).

## *3.6 Colorectal Cancer*

Colorectal cancer, the forth deadliest cancer has been a topic of great interest in the immunotherapy field given that it represents 10% of all diagnosed types of cancer worldwide (Dekker et al. 2019; Balibegloo and Rezaei 2020). Bispecific antibody therapy for this type of cancer arisen as a promising immunotherapy given the presence of some specific colorectal antigens. One of those, TAG-72 (tumor associated glycoprotein 72) is found in the surface of some types of cancer including colon cancer. Early studies using a bispecific antibody designed to co-target this glycoprotein and CD3 in the T-cells shown a significant decrease in tumor growth in an in vivo xenograft mouse model (Shpitz et al. 1996).

Immune checkpoint proteins, such as CTLA-4 – a protein expressed in regulatory T-cells ($T_{regs}$) and other activated T-cells, are commonly upregulated in the tumor microenvironment (Śledzińska et al. 2015). Previous attempts to target CTLA-4 using a monospecific antibody resulted in a systemic activation of the T-cells which greatly limited its clinical use. A bispecific antibody against CTLA-4 and OX40—a costimulatory immune checkpoint molecule also found upregulated in cancer—shown effective activation of $CD8^+$ T-cells while decreasing the number of $T_{regs}$ in an M38 mouse model for colon cancer (Kvarnhammar et al. 2019). GpA33, also known as cell surface A33 antigen, is a cell surface antigen that is expressed consistently in almost all cases of colon cancer (Ackerman et al. 2008; Sakamoto et al. 2000). Bispecific antibodies against this protein and CD3 shown a potent activity at eliminating colorectal cancer in xenograft mouse models (Moore et al. 2018; Wu et al. 2018).

CEA (carcinoembryonic antigen) is another attractive target for cancer therapy due to its low prevalent expression in normal cells. Under normal conditions, CEA is found expressed in the apical surface and the luminal portion of the adult colon. However, in colorectal cancer, CEA is overexpressed and shows loss of polarized distribution within the cancer cells (Hammarström 1999). Since CEA localization in the normal cells of the colon is limited to a few structures in the cell compared to the global distribution in the cancer cells, it became a potential candidate for targeted immunotherapy. The interest in this protein leads to the design of several bispecific antibodies that co-target CEA and CD3 leading to encouraging preclinical results. Among them, MEDI-565, a BiTE CEAxCD3 antibody is able to specifically bind to tissues where CEA is expressed. This antibody was able to mediate the killing of $CEA^{+}$ cancer cells with its cytotoxic activity correlating with CEA expression. Importantly, the presence of common oncogenic mutations or the loss of function mutations in tumor suppressor genes did not affect the in vitro activity of this antibody (Oberst et al. 2014). Although bispecific antibodies targeting CEA-expressing cells have shown remarkable therapeutic outcomes in the treatment of colorectal cancer, it is important to consider the heterogenic expression of this protein in the tumors. This heterogenicity, captured in patient derived colorectal organoid studies, shown a lower sensitivity of these bispecific antibodies. These results draw attention to possible mechanism of cancer immunotherapy resistance by changing expression levels of target antigens (Gonzalez-Exposito et al. 2019).

In addition to the ones mentioned here, several other bispecific antibody designs were generated, some of which have shown a remarkable effect in colorectal cancer studies. However, the effectiveness of these antibodies in clinical trials might face further challenges given the genetic variability of the patients. A study involving 129 healthy volunteers aimed at evaluating the impact of polymorphisms in the activation and cytotoxicity of peripheral blood mononuclear cells revealed that T-cells from homozygous patients for a CD247 polymorphism (rs2949655) showed significant lower toxicity toward colorectal cancer cells compared to heterozygotes when incubated with a bispecific antibody CD3xEpCAM (Zhang et al. 2018).

### *3.7 Hematological Malignancies*

Several research groups have extensively studied the use of bispecific T-cell engagers (BiTEs) for the treatment of non-Hodgkin's lymphoma and acute lymphoblastic leukemia. An example of an antibody that shown promising results in the clinical trials was the T-cell engager bispecific antibody blinatumomab. Blinatumomab binds both to the CD3 receptor on the T-cells and CD19 receptor in the tumor cells. Upon binding to the tumor cells and the host T-cells, this bispecific antibody allows the formation of a cytotoxic synapse where enzymes like perforin and granzymes are released (Rousalova and Krepela 2010). Early clinical studies allowed the assessment of the performance of this antibody in relapsed or refractory non-Hodgkin's lymphoma and chronic lymphocytic leukemia. By administrating the drug twice or three times weekly, it has shown a low overall benefit/risk ratio which resulting in the early termination of the study. Further changes in the drug administration strategy such as continuous infusion of blinatumomab and the combination with the administration of glucocorticoids have significantly increased the ratio of benefit/risk for the patients receiving this treatment compared to the short-term infusion (Nagorsen et al. 2012; Kantarjian et al. 2017). In the following clinical trial studies (Phase II), a larger number of patients were tested for the efficacy and safety of the drug. While the data from this phase corroborates the results from the Phase I in terms of efficacy, the results from the Phase II highlight the importance of the drug side effects particularly those at the neurological level (Viardot et al. 2016).

Other slightly different antibody design includes the Anti-CD20/CD3 bispecific antibody. This design retains the biochemical and pharmacological characteristics of the full-length IgG1 antibodies and, although not yet clinically tested, it promises on a weekly to monthly schedule of patient dosing compared to the continuous administration in the example above (Sun et al. 2015). More recent strategies like SMITE (simultaneous multiple interaction T-cell engaging) use two bispecific antibodies with affinity to the cancer cell and to either CD3 and CD28 in the T-cells. In here, researchers have shown that the combination of these two antibodies leads to a more pronounced T-cell response due to the receptor co-stimulation. Thus, the use of this combination of antibodies requires a much lower antibody concentration down to levels where either of the two antibodies are inactive which could potentially reduce the off-target risks for the patient (Correnti et al. 2018).

## *3.8 Other Solid Tumors*

Studies on the development of bispecific antibodies to treat other types of solid tumors such as head and neck cancer are still very scarce. However, some cell receptors are currently being considered as attractive targets for the use of bispecific antibodies in the treatment of the disease. Preliminary studies have analyzed the expression pattern of B7-H3 in multiple samples from cancer patients showing overexpression in 66.1% of the samples. Treatment of the cells with a bispecific antibody anti-B7H3 x CD3 shown promising results such as increased in vivo cancer cell lysis and a dramatic reduction of xenograft tumor volume over the course of an 18-day treatment (Hu et al. 2019).

To our knowledge, other types of cancer such as cervical, ovarian, and intestinal have at the present moment have not yet shown studies showing the development of bispecific antibodies. Targeted therapies for these types of cancer have so far relied mostly in small molecule inhibitors and monoclonal antibody therapy. However, it is believed that in the future, we will witness an increased growth in the development of bispecific antibody-based immunotherapy.

**Table 1** Bispecific antibodies already approved for marketing in various cancer treatment

| Name of the bispecific antibody | Name of product | Company | Target | Mechanism | Approved time | Indication |
|---|---|---|---|---|---|---|
| Catumaxomab (2017 delist) | Removab | Trion Parma | CD3 and EpCAM | Simultaneously recognizes the surface antigen CD3 of T-cell and EpCAM, a marker of cancer progesterone and eliminates cancer cells | 2009.4 (FMA) (2017 delist) | Malignant ascites caused by EpCAM+ tumor |
| Blinatumomab | Blincyto | Amgen | CD3 and CD19 | Simultaneously targets B-cell surface antigen CD19 and T-cell surface antigen CD3 Mediates the targeted lysis of tumor B-cells by T-cells | 2014.12(FDA) 2015.11(EMA) | Relapsed or refractory B-cells Acute lymphocytic leukemia |
| Emicizumab | Hemlibra | Roche | FIX and FX | Simulates coagulation factor IX (FIX) and to factor X (FX), which leads to a significant increase in the production of thrombin to achieve hemostasis | 2017.11(FDA) 2018.02(EMA) | Hemophilia type A |

## 4 Current Clinical Trials Using Bispecific Antibodies

So far, there are nearly 176 clinical trials with more than 80 bispecific antibodies, registered on http://www.clinicaltrials.gov and actively recruiting patients for evaluation in different cancer types. Among those, there are three bispecific antibody formats which are already in clinics (Table 1) and more than 80 bispecific antibodies are under clinical trials for different cancers type, some of which are shown in Tables 2, 3, 4, 5, 6, 7, and 8 based on their applications in various cancer types. As shown in Table 1, three bispecific antibodies are now in the market, namely Trion's catumaxomab, Amgen's blinatumomab, and Roche's emicizumab. Catumaxomab was the world's first commercialized bispecific antibody, but it was delisted in 2017. Catumaxomab recognizes the surface antigen CD3 of T-cell and human epithelial

**Table 2** Clinical trials involving the use of bispecific antibodies in melanoma treatment

| Bispecific antibodies | Format | Target | Cancer stage/ type | Phases | NCT Number |
|---|---|---|---|---|---|
| BCA101 | IgG-fusion proteins | EGFR and TGFβ | Uveal melanoma | 1 | NCT04429542 |
| INBRX-105 | (sdAb)2 + (sdAb)2 + Fc, 2 + 2 | PD-L1 and 41BB | Melanoma | 1 | NCT03809624 |
| rM28 | Tandem scFv | CD28 and MAPG | Malignant melanoma | 1\| 2 | NCT00204594 |
| RO7121661 | Fab + Fab + Fc, 1 + 1, CrossMAb | PD-1 and TIM-3 | Metastatic melanoma | 1 | NCT03708328 |
| RO7247669 | Fab + Fab + Fc, 1 + 1, CrossMAb | PD1 and LAG3 | Metastatic melanoma | 1 | NCT04140500 |
| XmAb20717 | Fab + scFv + Fc, 1 + 1, XmAb | PD-1 and CTLA-4 | Melanoma | 1 | NCT03517488 |
| XmAbÂ®22841 | Fab + scFv + Fc, 1 + 1, XmAb | CTLA-4 and LAG-3 | Melanoma | 1 | NCT03849469 |
| XmAbÂ®23104 | Fab + scFv + Fc, 1 + 1, XmAb | PD-1 and ICOS | Melanoma (excluding uveal melanoma) | 1 | NCT03752398 |

cell adhesion molecule (EpCAM), expressed on the majority of the various epithelial tumors that cause metastatic ascites. Studies reveals that it can destroy EpCAM expressing cancer cells in the peritoneal cavity (Linke et al. 2010). Moreover, blinatumomab targets B-cell surface antigen CD19 and T-cell surface antigen CD3. In 2014, the FDA granted breakthrough therapy designation to blinatumomab in adults with r/r Philadelphia chromosome-negative acute lymphoblastic leukemia (ALL). Five-year follow-up data verified that receiving blinatumomab treatment could achieve a complete remission with negative minimal residual disease (an important predictor of recurrence of acute lymphoblastic leukemia) and prolong survival, with a median overall survival of 36.5 months (Topp et al. 2015). Lastly, clinical trials confirmed that emicizumab can simulate coagulation factor IX (FIX) and to factor X (FX). Emicizumab therapy led to a decreased bleeding rate compared to a previous factor VIII prophylaxis (Mahlangu et al. 2018).

As depicted above, there have been numerous clinical studies involving the use of bispecific antibodies in the treatment of melanoma, breast cancer, colorectal cancer, hematologic cancers (i.e., leukemia, lymphoma), and lung cancer, respectively. The majority of these bispecific antibodies are in clinical phase I/II and preclinical research phases. A series of recent clinical trials have confirmed that blocking molecules such as HER2, EGFR, FBP, and VEGF receptors specifically expressed on tumor cells, and secreted cytokines can be a good option that is frequently applied in tumor therapy, especially in solid tumors. Additionally, bispecific antibodies can

**Table 3** Clinical trials involving the use of bispecific antibodies in breast cancer treatment

| Bispecific antibodies | Format | Target | Cancer stage/type | Phases | NCT Number |
|---|---|---|---|---|---|
| BCA101 | IgG-fusion proteins | EGFR and TGFβ | Triple-negative breast cancer(TNBC) | 1 | NCT04429542 |
| Catumaxomab (Removab) | TrioMab | EpCAM and CD3 | EpCam positive tumor (breast) | 2\| 3 | NCT00836654 |
| CD3-MUC1 BsAb | undisclosed | CD3 and MUC1 | Advanced breast cancer | 2 | NCT03524261 |
| CDX-527 | IgG + (scFv)2, 2 + 2 | PD-L1 and CD27 | Breast cancer | 1 | NCT04440943 |
| ertumaxomab | TrioMab | Her-2/neu and CD3 | Metastatic breast cancer \|Advanced breast cancer | 2 | NCT00452140 |
| ertumaxomab | TrioMab | Her-2/neu and CD3 | Metastatic breast cancer \|Advanced breast cancer | 2 | NCT00522457 |
| GEM3PSCA | scFv + scFv, 1 + 1 | PSCA and CD3 | Breast cancer | 1 | NCT03927573 |
| GEN1044 | Fab + Fab + Fc, 1 + 1,DuoBody | CD3 and 5 T4 | TNBC | 1\| 2 | NCT04424641 |
| HER2 BAT | HER2 BAT | Her-2/neu and CD3 | Breast cancer Female | 1 | NCT03661424 |
| HER2 BATs | HER2 BAT | HER2Bi armed activated T-cells | Metastatic breast cancer | 1\| 2 | NCT03272334 |
| ISB 1302 | Fab + scFv + Fc, 1 + 1, BEAT | CD3 and HER2 | Breast cancer | 1\| 2 | NCT03983395 |
| MCLA-128 | Fab + Fab + Fc, 1 + 1, Biclonics | HER2 and HER3 | Breast cancer metastatic | 2 | NCT03321981 |
| MM-111 | scFv-HSA | HER2 and HER3 | Her2 amplified solid tumors \|Metastatic breast cancer | 1 | NCT00911898 |
| MM-111 | scFv-HSA | HER2 and HER3 | Breast neoplasms | 1 | NCT01097460 |
| TF2 | D&L Fab3 | CEA and HSG | HER2 negative breast carcinoma expressing CEA | 1\| 2 | NCT01730612 |
| XmAb20717 | Fab + scFv + Fc, 1 + 1XmAb | PD-1 and CTLA-4 | Breast carcinoma | 1 | NCT03517488 |
| XmAbÂ®22841 | Fab + scFv + Fc, 1 + 1, XmAb | CTLA-4 and LAG-3 | TNBC | 1 | NCT03849469 |

(continued)

**Table 3** (continued)

| Bispecific antibodies | Format | Target | Cancer stage/type | Phases | NCT Number |
|---|---|---|---|---|---|
| XmAbÂ®23104 | Fab + scFv + Fc, 1 + 1, XmAb | PD-1 and ICOS | Breast carcinoma, i.e., estrogen receptor, progesterone receptor, and Her2 negative | 1 | NCT03752398 |
| ZW25 (Zanidatamab) | Fab + scFv + Fc, 1 + 1 | Target two epitopes of HER2 | Breast cancer | 1\| 2 | NCT04276493 |
| ZW25 (Zanidatamab) | Fab + scFv + Fc, 1 + 1 | Target two epitopes of HER2 | HER2+/HR+ breast cancer | 2 | NCT04224272 |

execute their therapeutic effects by binding to receptors, without requiring receptor internalization (Yang et al. 2021). For hematological tumors, antigen molecules such as CD19, CD20 are utilized to target against B-cell malignancies (May et al. 2012).

Here we have taken example of various bispecific antibodies application on non-small cell lung cancer (NSCLC) and colorectal cancer. JNJ-61186372, a bispecific antibody against EGFR and cMet developed by Johnson & Johnson and Genmab, is shown to have effective anti-tumor activity in both wild type and EGFR mutated NSCLC in in vivo tumor models (Grugan et al. 2017). Ongoing studies have shown promising preliminary clinical activity of JNJ-61186372 in NSCLC patients with diverse EGFR mutations (Haura et al. 2019).

GpA33, frequently expressed in primary and metastatic colorectal cancers is a potential target in colorectal cancer immunotherapy. MGD007 developed by MacroGenics is a new type of gpA33×CD3 bispecific antibody that is in phase I clinical trials for colorectal cancer patients. In vitro experiments have shown that MGD007 mediates the redirected killing of colorectal cancer cell lines expressing gpA33 by T-cells. Clinical investigations revealed good recognition and elimination of gpA33-positive colorectal cancer cells (Murer et al. 2020).

Altogether, bispecific antibodies have been under investigation in the preclinical literature for about 30 years. The success of blinatumomab and ongoing clinical and preclinical studies show the significant potential of bispecific antibodies in the treatment of cancer. More clinical trials should be performed to effectively confirm the safety and efficacy of bispecific antibodies in clinical practice in the coming future.

**Table 4** Clinical trials involving the use of bispecific antibodies in colorectal cancer treatment

| Bispecific antibodies | Format | Target | Cancer stage/ type | Phases | NCT number |
|---|---|---|---|---|---|
| BCA101 | IgG-fusion proteins | EGFR and TGFβ | Colorectal cancer | 1 | NCT04429542 |
| CD3-MUC1 BsAb | undisclosed | CD3 and MUC1 | Advanced colorectal cancer | 2 | NCT03524274 |
| CDX-527 | IgG + (scFv)2, 2 + 2 | PD-L1 and CD27 | MSI-H colorectal cancer | 1 | NCT04440943 |
| MCLA-158 | Fab + Fab + Fc, 1 + 1Biclonics | EGFR and LGR5 | Colorectal cancer | 1 | NCT03526835 |
| MGD007 | Fv + Fv + Fc, 1 + 1DART-Fc | gpA33 and CD3 | Colorectal carcinoma | 1 | NCT02248805 |
| MGD007 | Fv + Fv + Fc, 1 + 1DART-Fc | gpA33 and CD3 | Colorectal cancer metastatic | 1\| 2 | NCT03531632 |
| OMP-305B83 | Fab + Fab + Fc, 1 + 1common LC | DLL4 and VEGF | Metastatic colorectal cancer | 1 | NCT03035253 |
| TF2 | D&L Fab3 | CEA and HSG | Colorectal cancer | 1 | NCT00860860 |
| TF2 | D&L Fab3 | CEA and HSG | Colorectal cancer | 1 | NCT00895323 |
| TF2 | D&L Fab3 | CEA and HSG | Metastatic colorectal cancer | 2 | NCT02587247 |
| TF2 | D&L Fab3 | CEA and HSG | Metastatic colorectal cancer | 1 | NCT01273402 |
| XmAb20717 | Fab + scFv + Fc, 1 + 1, XmAb | PD-1 and CTLA-4 | Colorectal carcinoma | 1 | NCT03517488 |
| XmAbÂ®23104 | Fab + scFv + Fc, 1 + 1, XmAb | PD-1 and ICOS | Colorectal carcinoma | 1 | NCT03752398 |
| ZW25 (Zanidatamab) | Fab + scFv + Fc, 1 + 1 | target two epitopes of HER2 | Colorectal cancer | 2 | NCT03929666 |

## 5 Current Challenges in Bispecific Antibody Therapy

The challenges in bispecific antibody therapy lies in various level ranging from selection of target antigen, design, production, and purification. Currently, all available antibody or bispecific antibody therapy targets cell surface antigens or secreted proteins by malignant cells that are accessible by the therapeutic antibodies. However, most of the known oncogenic proteins such as transcription factors, non-receptor kinases, and phosphatases are presented within the cytosol of the cancer

**Table 5** Clinical trials involving the use of bispecific antibodies in hematological cancer treatment

| Bispecific anibodies | Format | Target | Cancer stage/type | Phases | NCT number |
|---|---|---|---|---|---|
| IBI322 | sdAb + Fab + Fc, 1 + 1 | CD47 and PD-L1 | Hematologic malignancy | 1 | NCT04795128 |
| MGD006 | Fv + Fv, 1 + 1, DART | CD123 and CD3 | Refractory hematologic malignancy | 2 | NCT03739606 |
| MGD006 | Fv + Fv, 1 + 1, DART | CD123 and CD3 | Recurrent hematologic malignancy | 1 | NCT04681105 |
| Talquetamab | Fab + Fab + Fc, 1 + 1,DuoBody | GPRC5D and CD3 | Hematological malignancies | 2 | NCT04634552 |
| Talquetamab | Fab + Fab + Fc, 1 + 1, DuoBody | GPRC5D and CD3 | Hematological malignancies | 1 | NCT03399799 |
| Teclistamab | Fab + Fab + Fc, 1 + 1, DuoBody | BCMA and CD3 | Hematological malignancies | 1 | NCT04696809 |
| Teclistamab | Fab + Fab + Fc, 1 + 1, DuoBody | BCMA and CD3 | Hematological malignancies | 2 | NCT04557098 |

cells. Designing bispecific antibodies that target intracellular oncogenic or mutated protein molecules responsible for the malignancy remains a major challenge due to antibodies complex structure and size. Intracellular oncogenic protein that are processed and presented by cancer cells on their surface is worth considering as targetable antigens by bispecific antibodies to induce immune response against the tumor. Furthermore, designing bispecific antibodies with one Fab region targeting cancer surface antigen and the other targeting intracellular protein may increase tumor cell capability to internalize the antibodies leading to neutralization of the target intracellular protein and maximizing antibody efficacy. Exploring strategies that make bispecific antibodies access to intracellular tumor target would widen the scope of immunotherapy and can possibly overcome the immune evasion and greater clinical outcome.

Unfortunately, the majority of the current bispecific antibodies development and trials focus on hematological malignancies compared to solid tumor. Owing to the tumor microenvironment, targeting solid tumors with bispecific antibodies has its own challenge because of poor vascular architecture of solid tumor inner core hampers the penetration of bispecific antibodies, therefore affecting their clinical efficacy. Focus need to be given for the best route of administration and dosage form to concentrate the bispecific antibodies level in target tissues and that may help to decrease the systemic side effects. Further, maintaining constant circulatory bispecific antibodies level remains a critical parameter for successful immune therapy. The half-life of bispecific antibodies can be extended by adding Fc region in their structure. The addition of Fc region on bispecific antibodies structure remains a crucial component of antibody development. The Fc receptors (FcRs) play an important role in antibody-mediated immune therapy. Reports suggest that engagement of both activating FcRs and inhibitory FcRs or either alone by bispecific

**Table 6** Clinical trials involving the use of bispecific antibodies in leukemia treatment

| Bispecific antibodies | Format | Target | Cancer stage/type | Phases | NCT number |
|---|---|---|---|---|---|
| 4G7xH22 | Fab2 | CD19 and CD64 | Leukemia | 1 | NCT00014560 |
| Blinatumomab | BiTE | CD19 and CD3 | Acute lymphoblastic leukemia | 3 | NCT03643276 |
| Blinatumomab | BiTE | CD19 and CD3 | B-cell acute lymphoblastic leukemia | 2 | NCT01207388 |
| Blinatumomab | BiTE | CD19 and CD3 | Acute lymphoblastic leukemia | 2 | NCT01466179 |
| Blinatumomab | BiTE | CD19 and CD3 | Acute lymphoblastic leukemia | 2 | NCT02744768 |
| Blinatumomab (MT103) | BiTE | CD19 and CD3 | Acute lymphoblastic leukemia | 2 | NCT00560794 |
| Blinatumomab (MT103) | BiTE | CD19 and CD3 | Acute lymphoblastic leukemia | 1\| 2 | NCT01471782 |
| DT2219ARL | (scFv)2 | CD19 and CD22 | Refractory B-lineage leukemia\| Relapsed B-lineage leukemia | 1\| 2 | NCT02370160 |
| FBTA05 | TrioMab | CD20 and CD3 | Leukemia | 1\| 2 | NCT01138579 |
| Flotetuzumab | Fv + Fv, 1 + 1DART | CD123 and CD3 | Recurrent acute myeloid leukemia(AML) \|Refractory AML | 1 | NCT04158739 |
| GEM333 | scFv + scFv, 1 + 1 | CD3 and CD33 | \|Relapsed AML\|Refractory AML | 1 | NCT03516760 |
| JNJ-67571244 | Fab + Fab + Fc, 1 + 1DuoBody | CD33 and CD3 | AML | 1 | NCT03915379 |
| JNJ-75348780 | Fab + Fab + Fc, 1 + 1DuoBody | CD3 and CD22 | Lymphoma, non-Hodgkin \|leukemia, lymphocytic, chronic, B-Cell | 1 | NCT04540796 |
| MCLA-117 | Fab + Fab + Fc, 1 + 1Biclonics | CLEC12A and CD3 | Acute myelogenous leukemia\| AML | 1 | NCT03038230 |
| MGD006 | Fv + Fv, 1 + 1DART | CD123 and CD3 | Acute biphenotypic leukemia\|Acute leukemia\|Chronic myelogenous leukemia, BCR-ABL1 positive\|Hairy cell leukemia\|Recurrent acute lymphoblastic leukemia \|Refractory acute lymphoblastic leukemia | 2 | NCT03739606 |

(continued)

**Table 6** (continued)

| Bispecific antibodies | Format | Target | Cancer stage/type | Phases | NCT number |
|---|---|---|---|---|---|
| MGD006 | Fv + Fv, 1 + 1DART | CD123 and CD3 | Recurrent acute leukemia\|Recurrent B acute lymphoblastic leukemia\|Recurrent chronic myelogenous leukemia, BCR-ABL1 positive\|Recurrent hairy cell leukemia\|Recurrent T acute lymphoblastic leukemia\|Refractory acute leukemia\|Refractory B acute lymphoblastic leukemia\|Refractory chronic myelogenous leukemia, BCR-ABL1 positive\|Refractory hairy cell leukemia\|Refractory T acute lymphoblastic leukemia | 1 | NCT04681105 |
| REGN 1979 | Fab + Fab + Fc, 1 + 1 | CD20 and CD3 | Chronic lymphocytic leukemia | 1 | NCT02290951 |

antibodies can drive the immune response against the tumor. Hence, the choice of Fc region needs to be taken wisely to balance engagement between the activating and inhibitory FcRs. Like small molecule anticancer therapy, bispecific antibodies also have their own side effects caused by the constant release of cytokine similar to any other T-cell activating therapy. Cytokine release syndrome can be controlled by using corticosteroids therapy as well as administrating CD20 and CD3 antibodies treatment and opting stepwise dosing of bispecific antibodies.

Despite several advancement in the design format of bispecific antibody strategies, the process optimization remains a major hurdle for commercial scale production. Like any other immunoglobulin or biosimilar, bispecific antibodies manufacturing requires sophisticated production line with purification methodology to ensure appropriate safety and stability of final products resulting the entire process time consuming and costly. Development of cost-effective cell line models and new technological breakthrough in purification techniques may lower the production cost. Hence translating laboratory tested bispecific antibodies against solid tumor to clinically viable drug possesses a greater challenge and opportunities.

**Table 7** Clinical trials involving the use of bispecific antibodies in lymphoma treatment

| Bispecific antibodies | Format | Target | Cancer stage/type | Phases | NCT number |
|---|---|---|---|---|---|
| 4G7xH22 | Fab2 | CD19 and CD64 | Lymphoma | 1 | NCT00014560 |
| AFM13 | TandAb | CD30 and CD16A | Lymphoma | 1\| 2 | NCT03192202 |
| AFM13 | TandAb | CD30 and CD16A | Recurrent anaplastic large cell lymphoma\|Recurrent B-Cell Non-Hodgkin lymphoma (NHL)\|Recurrent classic Hodgkin lymphoma\|Recurrent peripheral T-Cell lymphoma, not otherwise specified\|Refractory anaplastic large cell lymphoma\|Refractory B-Cell NHL\|Refractory classic Hodgkin lymphoma\|Refractory peripheral T-cell lymphoma | 1 | NCT04074746 |
| AK104 | IgG +(scFv)2, 2 + 2 | PD-1 and CTLA-4 | Peripheral T-cell lymphoma | 1\| 2 | NCT04444141 |
| anti-CD3 x anti-CD20 BAT | CD20-BAT | CD3 and CD20 | Lymphoma | 1 | NCT00521261 |
| anti-CD3 x anti-CD20 BAT | CD20-BAT | CD3 and CD20 | Lymphoma | 1 | NCT00244946 |
| Blinatumomab | BiTE | CD19 and CD3 | Diffuse large B-cell lymphoma | 2 | NCT01741792 |
| Blinatumomab (MT103) | BiTE | CD19 and CD3 | NHL, Relapsed | 1 | NCT00274742 |
| DT2219ARL | (scFv)2 | CD19 and CD22 | Refractory B-lineage lymphoma \|Relapsed B-lineage lymphoma | 1\| 2 | NCT02370160 |
| Glofitamab | CrossMAb | CD20 and CD3 | B-cell lymphoma | 1 | NCT04313608 |

(continued)

**Table 7** (continued)

| Bispecific antibodies | Format | Target | Cancer stage/type | Phases | NCT number |
|---|---|---|---|---|---|
| Glofitamab | CrossMAb | CD3 and CD20 | Diffuse large B-cell lymphoma refractory\|Refractory indolent adult NHL \|Refractory transformed B-cell NHL \|Refractory primary mediastinal large B-cell lymphoma\|Refractory mantle cell lymphoma | 2 | NCT04703686 |
| IBI318 | Fab + Fab + Fc, 1 + 1 | PD1 and PD-L1 | Extranodal NK/T-cell lymphoma, nasal type | 1\| 2 | NCT04602065 |
| JNJ-75348780 | Fab + Fab + Fc, 1 + 1DuoBody | CD3 and CD22 | NHL | 1 | NCT04540796 |
| MCLA-145 | Fab + Fab + Fc, 1 + 1, Biclonics | PD-L1 and CD137 | B-cell lymphoma, adult | 1 | NCT03922204 |
| MGD006 | Fv + Fv, 1 + 1, DART | CD123 and CD3 | Recurrent Hodgkin lymphoma \|Refractory Hodgkin lymphoma | 2 | NCT03739606 |
| MGD006 | Fv + Fv, 1 + 1, DART | CD123 and CD3 | Recurrent Hodgkin lymphoma \|Refractory Hodgkin lymphoma | 1 | NCT04681105 |
| NVG-111 | tandem scFv | ROR1 and CD3 | Small lymphocytic lymphoma \|Mantle cell lymphoma | 1\| 2 | NCT04763083 |
| REGN 1979 | Fab + Fab + Fc, 1 + 1 | CD20 and CD3 | NHL | 2 | NCT03888105 |
| REGN 1979 | Fab + Fab + Fc, 1 + 1 | CD20 and CD3 | NHL | 1 | NCT02290951 |
| REGN1979 | Fab + Fab + Fc, 1 + 1 | CD20 and CD3 | NHL | 1 | NCT02651662 |
| TG-1801 | Fab + Fab + Fc, 1 + 1κλ body | CD47 and CD19 | B-Cell lymphoma | 1 | NCT03804996 |
| TNB-486 | sdAb + Fab + Fc, 1 + 1 | CD19 and CD3 | B-cell NHL \|Diffuse large B-cell lymphoma \|High-grade B-cell lymphoma\|Follicular lymphoma | 1 | NCT04594642 |

**Table 8** Clinical trials involving the use of bispecific antibody in lung cancer treatment

| Bispecific antibodies | Format | Target | Cancer stage/type | Phases | NCT number |
|---|---|---|---|---|---|
| AK112 | IgG + (scFv)2, 2 + 2 | PD-1 and VEGF | Non-small cell lung cancer (NSCLC) | 1\| 2 | NCT04900363 |
| Amivantamab | Fab + Fab + Fc, 1 + 1, DuoBody | EGFR and cMet | NSCLC | 1 | NCT02609776 |
| BCA101 | IgG-fusion proteins | EGFR and TGFβ | Carcinoma of the lung | 1 | NCT04429542 |
| CD3-MUC1 | Undisclosed | CD3 and MUC1 | Advanced lung cancer | 2 | NCT03501056 |
| CDX-527 | IgG + (scFv)2, 2 + 2 | PD-L1 and CD27 | NSCLC | 1 | NCT04440943 |
| EMB-01 | (Fab)2 + IgG, 2 + 2, FIT-Ig | EGFR and cMET | NSCLC | 1\| 2 | NCT03797391 |
| GEM3PSCA | scFv + scFv, 1 + 1 | PSCA and CD3 | NSCLC | 1 | NCT03927573 |
| IBI318 | Fab + Fab + Fc, 1 + 1 | PD-L1 and PD-1 | Small cell lung cancer (SCLC) | 1 | NCT04672928 |
| IBI318 | Fab + Fab + Fc, 1 + 1 | PD-1 and PD-L1 | NSCLC | 2 | NCT04777084 |
| INBRX-105 | (sdAb)2 + (sdAb)2 + Fc, 2 + 2 | PD-L1 and 41BB | NSCLC | 1 | NCT03809624 |
| JNJ-61186372 | Fab + Fab + Fc, 1 + 1DuoBody | EGFR and cMet | NSCLC | 1 | NCT04077463 |
| M1231 | scFv-Fab | MUC1 and EGFR | NSCLC | 1 | NCT04695847 |
| MCLA-129 | Fab + Fab + Fc, 1 + 1Biclonics | EGFR and cMET | NSCLC Metastatic | 1\| 2 | NCT04868877 |
| RO7121661 | Fab + Fab + Fc, 1 + 1CrossMAb | PD-1 and TIM-3 | NSCLC\| SCLC | 1 | NCT03708328 |
| RO7247669 | Fab + Fab + Fc, 1 + 1CrossMAb | PD1 and LAG3 | NSCLC | 1 | NCT04140500 |
| TargomiRs | Undisclosed | EGFR and EDV | NSCLC | 1 | NCT02369198 |
| TF2 | D&L Fab3 | CEA and HSG | SCLC \|CEA-expressing NSCLC | 1\| 2 | NCT01221675 |

(continued)

**Table 8** (continued)

| Bispecific antibodies | Format | Target | Cancer stage/ type | Phases | NCT number |
|---|---|---|---|---|---|
| XmAb18087 | Fab + scFv + Fc, 1 + 1, XmAb | SSTR2 and CD3 | SCLC | 1I 2 | NCT04590781 |
| XmAb20717 | Fab + scFv + Fc, 1 + 1, XmAb | PD-1 and CTLA-4 | NSCLC | 1 | NCT03517488 |
| XmAbÂ®22841 | Fab + scFv + Fc, 1 + 1, XmAb | CTLA-4 and LAG-3 | SCLC INSCLC | 1 | NCT03849469 |
| XmAbÂ®23104 | Fab + scFv + Fc, 1 + 1, XmAb | PD-1 and ICOS | SCLC INSCLC | 1 | NCT03752398 |

Note: Information from ClinicalTrials.gov (http://clinicaltrials.gov). Criteria: Exclusion of drugs with not yet recruiting 1, phase 1 clinical trials; 2, phase 2 clinical trials. Bispecific T-cell Engager (BiTE); Dual affinity retargeting (DART); Single domain antibody (sdAb); Single chain variable fragment (scFv); IgG subclasses (IgGSc); kappa lambda body, κλ body; anti-CD3 and CD20 bispecific antibody-armed activated T-cells (CD20 BAT)

## 6 Future Prospective in Bispecific Antibody in Cancer Therapy

The past advancement in the development of new designing methods for bispecific antibodies generation has greatly helped to increase the therapeutic properties of these therapeutic molecules. The new generation of bispecific antibodies is more stable showing remarkable serum half-life and improved pharmacokinetics. Currently, more than hundreds of bispecific antibodies are in clinical trials for various cancers which indicates that the promising therapeutic potential of these agents as a precision future cancer medicine against deadly cancer forms. The production methods of bispecific antibodies are complex and expensive and therefore new technologies to generate affordable bispecific antibodies are emerging rapidly and are a prime objective for various pharma industries. Since these heavily engineered bispecific antibodies have a higher immunogenic potential, the improved safety is a major concern that still need to be addressed. The future of bispecific antibody therapy looks very promising for advanced cancer treatments since bispecific antibodies can also be combined with routine anticancer agents improving the patient responses. In summary, bispecific antibodies have emerged as impending therapeutic options for various deadly cancers showing immense potential to endure the frontiers of modern antibody therapeutics.

**Acknowledgments** The authors like to thank the Ramalingaswami Fellowship Fund, DBT, Govt. of India (to Purusottam Mohapatra) for supporting this work.

## Glossary

**Bispecific antibodies** Antibodies with two binding sites directed at two different antigens or two different epitopes on the same antigen; their clinical therapeutic effects are superior to those of monoclonal antibodies, with broad applications for tumor immunotherapy as well as for the treatment of other diseases.

**Immune checkpoints** Regulators of the immune system; immune checkpoint proteins refer to molecules that act as gatekeepers of immune responses (including inhibitory and stimulatory immune checkpoint molecules).

**Leukemia** A broad term for cancers of the blood cells.

**Lymphoma** A broad term for cancer that begins in cells of the lymph system.

**Melanoma** A kind of skin cancer that starts in the melanocytes (melanocytes are cells that make melanin pigment that gives skin its color); melanoma typically starts on skin that's often exposed to the sun.

**Quadroma** A cell resulting from the fusion of two hybridoma cells; used in the production of bispecific antibodies.

**Solid tumor** Heterotypic aggregates of many cell types, including cancer cells, cancer stem cells, connective tissue cells, and immune cells. That grows anywhere in the body especially affecting organs.

## References

Ackerman ME, Chalouni C, Schmidt MM, Raman VV, Ritter G, Old LJ et al (2008) A33 antigen displays persistent surface expression. Cancer Immunol Immunother 57(7):1017–1027

Agarwal N, Machiels JP, Suárez C, Lewis N, Higgins M, Wisinski K et al (2016) Phase I study of the prolactin receptor antagonist LFA102 in metastatic breast and castration-resistant prostate cancer. Oncologist 21(5):535

Apte RS, Chen DS, Ferrara N (2019) VEGF in signaling and disease: beyond discovery and development. Cell 176(6):1248–1264

Arribas J, Baselga J, Pedersen K, Parra-Palau JL (2011) p95HER2 and breast cancer: figure 1. Cancer Res 71(5):1515–1519

Balibegloo M, Rezaei N (2020) Development and clinical application of bispecific antibody in the treatment of colorectal cancer. Expert Rev Clin Immunol 16(7):689–709

Bergers G, Hanahan D (2008) Modes of resistance to anti-angiogenic therapy. Nat Rev Cancer 8(8):592–603

Borlak J, Länger F, Spanel R, Schöndorfer G, Dittrich C (2016) Immune-mediated liver injury of the cancer therapeutic antibody catumaxomab targeting EpCAM, CD3 and Fcγ receptors. Oncotarget 7(19):28059–28074

Brinkmann U, Kontermann RE (2017) The making of bispecific antibodies. mAbs 9(2):182–212

Cardillo TM, Karacay H, Goldenberg DM, Yeldell D, Chang CH, Modrak DE et al (2004) Improved targeting of pancreatic cancer: experimental studies of a new bispecific antibody, pretargeting enhancement system for immunoscintigraphy. Clin Cancer Res 10(10):3552–3561

Cha H-R, Lee JH, Ponnazhagan S (2020) Revisiting immunotherapy: a focus on prostate cancer. Cancer Res 80(8):1615–1623

Chang K, Pastan I, Willingham MC (1992) Frequent expression of the tumor antigen CAK1 in squamous-cell carcinomas. Int J Cancer 51(4):548–554

Chen X, Amar N, Zhu Y, Wang C, Xia C, Yang X et al (2020) Combined DLL3-targeted bispecific antibody with PD-1 inhibition is efficient to suppress small cell lung cancer growth. J Immunother Cancer 8(1):e000785

ClinicalTrials (2021). Available from: https://clinicaltrials.gov

Correnti CE, Laszlo GS, De Van Der Schueren WJ, Godwin CD, Bandaranayake A, Busch MA et al (2018) Simultaneous multiple interaction T-cell engaging (SMITE) bispecific antibodies overcome bispecific T-cell engager (BiTE) resistance via CD28 co-stimulation. Leukemia 32(5):1239–1243

Dekker E, Tanis PJ, Vleugels JLA, Kasi PM, Wallace MB (2019) Colorectal cancer. Lancet 394(10207):1467–1480

Del Bano J, Florès-Florès R, Josselin E, Goubard A, Ganier L, Castellano R et al (2019) A bispecific antibody-based approach for targeting mesothelin in triple negative breast cancer. Front Immunol 10:1593

Dinatale A, Fatatis A (2019) The bone microenvironment in prostate cancer metastasis. Springer, Cham, pp 171–184

Dougall WC, Roman Aguilera A, Smyth MJ (2019) Dual targeting ofRANKLandPD-1 with a bispecific antibody improves anti-tumor immunity. Clin Transl Immunol. 8(10):e01081

Friedrich M, Raum T, Lutterbuese R, Voelkel M, Deegen P, Rau D et al (2012) Regression of human prostate cancer xenografts in mice by AMG 212/BAY2010112, a novel PSMA/CD3-bispecific BiTE antibody cross-reactive with non-human primate antigens. Mol Cancer Ther 11(12):2664–2673

Gold DV, Goldenberg DM, Karacay H, Rossi EA, Chang CH, Cardillo TM et al (2008) A novel bispecific, trivalent antibody construct for targeting pancreatic carcinoma. Cancer Res 68(12):4819–4826

Gonzalez-Exposito R, Semiannikova M, Griffiths B, Khan K, Barber LJ, Woolston A et al (2019) CEA expression heterogeneity and plasticity confer resistance to the CEA-targeting bispecific immunotherapy antibody cibisatamab (CEA-TCB) in patient-derived colorectal cancer organoids. J Immunother Cancer. 7(1):101

Grugan K, Dorn K, Jarantow S, Bushey B, Pardinas J, Laquerre S et al (2017) Fc-mediated activity of EGFR x c-Met bispecific antibody JNJ-61186372 enhanced killing of lung cancer cells. MAbs 9(1):114–126

Haas C, Krinner E, Brischwein K, Hoffmann P, Lutterbüse R, Schlereth B et al (2009) Mode of cytotoxic action of T cell-engaging BiTE antibody MT110. Immunobiology 214(6):441–453

Hammarström S (1999) The carcinoembryonic antigen (CEA) family: structures, suggested functions and expression in normal and malignant tissues. Semin Cancer Biol 9(2):67–81

Haura EB, Cho BC, Lee JS, Han J-Y, Lee KH, Sanborn RE et al (2019) JNJ-61186372 (JNJ-372), an EGFR-cMet bispecific antibody, in EGFR-driven advanced non-small cell lung cancer (NSCLC). American Society of Clinical Oncology, Alexandria, VA

Hipp S, Voynov V, Drobits-Handl B, Giragossian C, Trapani F, Nixon AE et al (2020) A Bispecific DLL3/CD3 IgG-Like T-cell engaging antibody induces antitumor responses in small cell lung cancer. Clin Cancer Res 26(19):5258–5268

Hu J, Jiang C, Zheng M, Guo Y, Tang X, Ren J et al (2019) Overexpression of B7-H3 as an opportunity for targeted therapy in head and neck cancers. Am J Transl Res 11(8):5183–5196

Iizuka A, Nonomura C, Ashizawa T, Kondou R, Ohshima K, Sugino T et al (2019) A T-cell-engaging B7-H4/CD3-bispecific Fab-scFv antibody targets human breast cancer. Clin Cancer Res 25(9):2925–2934

Joshi R, Ren W, Mathew P (2020) A bispecific antibody targeting the αv and α5β1 integrins induces integrin degradation in prostate cancer cells and is superior to monospecific antibodies. Mol Cancer Res 18(1):27–32

Kantarjian H, Stein A, Gökbuget N, Fielding AK, Schuh AC, Ribera J-M et al (2017) Blinatumomab versus chemotherapy for advanced acute lymphoblastic leukemia. N Engl J Med 376(9):836–847

Köhler G, Milstein C (1975) Continuous cultures of fused cells secreting antibody of predefined specificity. Nature 256(5517):495–497

Kontermann R (2012) Dual targeting strategies with bispecific antibodies. mAbs. 4(2):182–197

Koopmans I, Hendriks MAJM, Van Ginkel RJ, Samplonius DF, Bremer E, Helfrich W (2019) Bispecific antibody approach for improved melanoma-selective PD-L1 immune checkpoint blockade. J Investig Dermatol 139(11):2343–51.e3

Krishnamurthy A, Jimeno A (2018) Bispecific antibodies for cancer therapy: a review. Pharmacol Ther 185:122–134

Kvarnhammar AM, Veitonmäki N, Hägerbrand K, Dahlman A, Smith KE, Fritzell S et al (2019) The CTLA-4 x OX40 bispecific antibody ATOR-1015 induces anti-tumor effects through tumor-directed immune activation. J Immunother Cancer. 7(1):103

Labrijn AF, Janmaat ML, Reichert JM, Parren PWHI (2019) Bispecific antibodies: a mechanistic review of the pipeline. Nat Rev Drug Discov 18(8):585–608

Li J, Piskol R, Ybarra R, Chen Y-JJ, Li J, Slaga D et al (2019) CD3 bispecific antibody-induced cytokine release is dispensable for cytotoxic T cell activity. Sci Transl Med 11(508):eaax8861

Li S, Liu M, Do MH, Chou C, Stamatiades EG, Nixon BG et al (2020) Cancer immunotherapy via targeted TGF-β signalling blockade in TH cells. Nature 587(7832):121–125

Lindhofer H, Mocikat R, Steipe B, Thierfelder S (1995) Preferential species-restricted heavy/light chain pairing in rat/mouse quadromas. Implications for a single-step purification of bispecific antibodies. J Immunol 155(1):219–225

Linke R, Klein A, Seimetz D (2010) Catumaxomab: clinical development and future directions. MAbs 2(2):129–136

Liu WJ, Du Y, Wen R, Yang M, Xu J (2020) Drug resistance to targeted therapeutic strategies in non-small cell lung cancer. Pharmacol Ther 206:107438

Ma J, Ma P, Zhao C, Xue X, Han H, Liu C et al (2016) B7-H3 as a promising target for cytotoxicity T cell in human cancer therapy. Oncotarget 7(20):29480–29491

Ma J, Shang T, Ma P, Sun X, Zhao J, Sun X et al (2019) Bispecific anti-CD3 x anti-B7-H3 antibody mediates T cell cytotoxic ability to human melanoma in vitro and in vivo. Investig New Drugs 37(5):1036–1043

Mahlangu J, Oldenburg J, Paz-Priel I, Negrier C, Niggli M, Mancuso ME et al (2018) Emicizumab prophylaxis in patients who have hemophilia A without inhibitors. N Engl J Med 379(9):811–822

May C, Sapra P, Gerber H-P (2012) Advances in bispecific biotherapeutics for the treatment of cancer. Biochem Pharmacol 84(9):1105–1112

Méndez-Ferrer S, Michurina TV, Ferraro F, Mazloom AR, Macarthur BD, Lira SA et al (2010) Mesenchymal and haematopoietic stem cells form a unique bone marrow niche. Nature 466(7308):829–834

Merchant AM, Zhu Z, Yuan JQ, Goddard A, Adams CW, Presta LG et al (1998) An efficient route to human bispecific IgG. Nat Biotechnol 16(7):677–681

Moore PA, Shah K, Yang Y, Alderson R, Roberts P, Long V et al (2018) Development of MGD007, a gpA33 x CD3-bispecific DART protein for T-cell immunotherapy of metastatic colorectal cancer. Mol Cancer Ther 17(8):1761–1772

Moores SL, Chiu ML, Bushey BS, Chevalier K, Luistro L, Dorn K et al (2016) A novel bispecific antibody targeting EGFR and cMet is effective against EGFR inhibitor-resistant lung tumors. Cancer Res 76(13):3942–3953

Murer P, Plüss L, Neri D (2020) A novel human monoclonal antibody specific to the A33 glycoprotein recognizes colorectal cancer and inhibits metastasis. MAbs 12(1):1714371

Nagorsen D, Kufer P, Baeuerle PA, Bargou R (2012) Blinatumomab: a historical perspective. Pharmacol Ther 136(3):334–342

Oberst MD, Fuhrmann S, Mulgrew K, Amann M, Cheng L, Lutterbuese P et al (2014) CEA/CD3 bispecific antibody MEDI-565/AMG 211 activation of T cells and subsequent killing of human tumors is independent of mutations commonly found in colorectal adenocarcinomas. mAbs. 6(6):1571–1584

Offner S, Hofmeister R, Romaniuk A, Kufer P, Baeuerle PA (2006) Induction of regular cytolytic T cell synapses by bispecific single-chain antibody constructs on MHC class I-negative tumor cells. Mol Immunol 43(6):763–771

Paradiso A, Singer CF (2017) Therapeutic strategies in triple-negative breast cancer. Breast Care 12(1):6–7

Ridgway JB, Presta LG, Carter P (1996) 'Knobs-into-holes' engineering of antibody CH3 domains for heavy chain heterodimerization. Protein Eng 9(7):617–621

Rius Ruiz I, Vicario R, Morancho B, Morales CB, Arenas EJ, Herter S et al (2018) p95HER2–T cell bispecific antibody for breast cancer treatment. Sci Transl Med. 10(461):eaat1445

Rousalova I, Krepela E (2010) Granzyme B-induced apoptosis in cancer cells and its regulation (review). Int J Oncol 37(6):1361–1378

Sakamoto J, Kojima H, Kato J, Hamashima H, Suzuki H (2000) Organ-specific expression of the intestinal epithelium-related antigen A33, a cell surface target for antibody-based imaging and treatment in gastrointestinal cancer. Cancer Chemother Pharmacol 46(S1):S27–S32

Shimabukuro-Vornhagen A, Gödel P, Subklewe M, Stemmler HJ, Schlößer HA, Schlaak M et al (2018) Cytokine release syndrome. J Immunother Cancer. 6(1):56

Shiozawa Y, Pedersen EA, Havens AM, Jung Y, Mishra A, Joseph J et al (2011) Human prostate cancer metastases target the hematopoietic stem cell niche to establish footholds in mouse bone marrow. J Clin Investig 121(4):1298–1312

Shpitz B, Stern H, Sandhu JS, Reilley RM, Tropak MB, Jansz G et al (1996) Cytotoxicity of B72.3XOKT3 bispecific antibody recognizing human colon cancer. J Surg Res 61(1):134–138

Slamon DJ, Clark GM, Wong SG, Levin WJ, Ullrich A, McGuire WL (1987) Human breast cancer: correlation of relapse and survival with amplification of the HER-2/neu oncogene. Science 235(4785):177–182

Śledzińska A, Menger L, Bergerhoff K, Peggs KS, Quezada SA (2015) Negative immune checkpoints on T lymphocytes and their relevance to cancer immunotherapy. Mol Oncol 9(10):1936–1965

Spiess C, Zhai Q, Carter PJ (2015) Alternative molecular formats and therapeutic applications for bispecific antibodies. Mol Immunol. 67(2 Pt A):95–106

Sun LL, Ellerman D, Mathieu M, Hristopoulos M, Chen X, Li Y et al (2015) Anti-CD20/CD3 T cell-dependent bispecific antibody for the treatment of B cell malignancies. Sci Transl Med 7(287):287ra70–287ra70

Suurs FV, Lub-De Hooge MN, De Vries EGE, De Groot DJA (2019) A review of bispecific antibodies and antibody constructs in oncology and clinical challenges. Pharmacol Ther 201:103–119

Tang Z, Qian M, Ho M (2013) The role of mesothelin in tumor progression and targeted therapy. Anti Cancer Agents Med Chem 13(2):276–280

TheAntibodySociety. Antibody therapeutics approved or in regulatory review in the EU or US 2021. Available from: https://www.antibodysociety.org/resources/approved-antibodies/

Topp MS, Gökbuget N, Stein AS, Zugmaier G, O'Brien S, Bargou RC et al (2015) Safety and activity of blinatumomab for adult patients with relapsed or refractory B-precursor acute lymphoblastic leukaemia: a multicentre, single-arm, phase 2 study. Lancet Oncol 16(1):57–66

Viardot A, Goebeler M-E, Hess G, Neumann S, Pfreundschuh M, Adrian N et al (2016) Phase 2 study of the bispecific T-cell engager (BiTE) antibody blinatumomab in relapsed/refractory diffuse large B-cell lymphoma. Blood 127(11):1410–1416

Vincent A, Herman J, Schulick R, Hruban RH, Goggins M (2011) Pancreatic cancer. Lancet 378(9791):607–620

Wu Z, Cheung NV (2018) T cell engaging bispecific antibody (T-BsAb): from technology to therapeutics. Pharmacol Ther 182:161–175

Wu Z, Guo H-F, Xu H, Cheung N-KV (2018) Development of a tetravalent Anti-GPA33/Anti-CD3 bispecific antibody for colorectal cancers. Mol Cancer Ther 17(10):2164–2175

Xu D, Alegre M-L, Varga SS, Rothermel AL, Collins AM, Pulito VL et al (2000) In vitro characterization of five humanized OKT3 effector function variant antibodies. Cell Immunol 200(1):16–26

Yang Y, Yang Z, Yang Y (2021) Potential role of CD47-directed bispecific antibodies in cancer immunotherapy. Front Immunol 12(2773):686031

Yankelevich M, Kondadasula SV, Thakur A, Buck S, Cheung NK, Lum LG (2012) Anti-CD3 × anti-GD2 bispecific antibody redirects T-cell cytolytic activity to neuroblastoma targets. Pediatr Blood Cancer 59(7):1198–1205

Yu YJ, Zhang Y, Kenrick M, Hoyte K, Luk W, Lu Y et al (2011) Boosting brain uptake of a therapeutic antibody by reducing its affinity for a transcytosis target. Sci Transl Med. 3(84):84ra44

Zhang CM, Yu LY, Lv JF, Gong L, Zhou HH, Chen XP et al (2018) Effects of immuno-related gene polymorphisms on a bispecific antibody targeting colorectal cancer cell. Per Med 15(3):167–179

Zhou Y, Zong H, Han L, Xie Y, Jiang H, Gilly J et al (2020) A novel bispecific antibody targeting CD3 and prolactin receptor (PRLR) against PRLR-expression breast cancer. J Exp Clin Cancer Res. 39(1):87

Zucali PA, Ruiz MG, Giovannetti E, Destro A, Varella-Garcia M, Floor K et al (2008) Role of cMET expression in non-small-cell lung cancer patients treated with EGFR tyrosine kinase inhibitors. Ann Oncol 19(9):1605–1612

# Progresses and Challenges in HIV Vaccine

**Subrat Kumar, Sunil Kumar Jada, Subhra Subhadra, and Priyadarshi S. Sahu**

**Abstract** Human immunodeficiency virus (HIV) that causes AIDS has become a worldwide epidemic. Currently, there are about 38.4 million people around the planet with HIV, which includes 36.7 million adults and 1.7 million children. In order to stop HIV progression, a vaccine is urgently needed. Despite advances in our scientific understanding of biology, replication, and pathogenesis of HIV, we are yet to develop a trustworthy vaccine to control HIV. Development of an effective HIV vaccine is decelerated by various factors like genetic heterogeneity of the virus, mutations acquired during viral replication, and lack of immune corelates for the virus. Still we have made tremendous progress in HIV vaccine development and in this review, we will emphasize these advancements and the future trends in developing an effective, affordable, and safe HIV vaccine.

**Keywords** Vaccine · Retrovirus · HIV · AIDS · Clinical trials

S. Kumar (✉)
Infection Biology Laboratory, School of Biotechnology, KIIT University, Bhubaneswar, Odisha, India
e-mail: skumar@kiitbiotech.ac.in

S. K. Jada
Department of Microbiology, Sri Venkata Sai Medical College and Hospital, Affiliated to KNR University of Health Sciences, Warangal, Telangana, India

S. Subhadra
Regional Medical Research Center, Bhubaneswar, Odisha, India

P. S. Sahu
Department of Microbiology & Immunology, School of Medicine, Medical University of the Americas, Nevis, St. Kitts and Nevis



A. Das Mohapatra, P. S. Sahu (eds.), *Advances in Immunology and Immuno-techniques*, Interdisciplinary Biotechnological Advances,
https://doi.org/10.1007/978-981-97-5508-0_5

# 1 Introduction

The retrovirus known as the Human Immunodeficiency Virus (HIV) is a member of the retroviridae family. Retroviruses have the remarkable capacity to create DNA from viral RNA with the aid of an enzyme called reverse transcriptase. The family of retroviridae includes many animal viruses that cause illnesses and are clinically significant to humans. The Retroviridae family is divided into two sub-families: Orthoretrovirinae and Spumaretrovirinae. The Orthoretrovirinae is further divided into two genuses: Deltaretrovirus which includes two viruses, human T-cell lymphotropic virus type 1 (oncogenic virus) and human T-cell lymphotropic virus type 2. While the second genus, Lentivirus, includes Human Immunodeficiency Virus (HIV). The first human case of AIDS was reported in USA in 1981 among homosexuals and drug abusers in whom the incidence of Kaposi sarcoma and *Pneumocystis carinii* pneumonia were alarmingly high. Francoise Baree-Sinoussi and Luc Montagnier discovered the etiological agent for AIDS in 1983 in patients with enlarged lymph nodes at the Pasteur Institute in Paris. Lymphadenopathy-associated virus was the name given to it (LAV). Later, a retrovirus known as human T-cell lymphotropic virus-III(HLTV-III) was isolated and described by Robert Gallo and colleagues at the National Institute of Health in the United States. In 1986, the International Committee on Taxonomy of viruses (ICTV) named them human immunodeficiency virus (HIV). HIV exists in two types: HIV-1, which is prevalent worldwide and is responsible for AIDS epidemic and HIV-2, which is low virulent and less prevalent. HIV-1 virus has been isolated from USA, Europe, and Central Africa, while HIV-2 is majorly restricted to West Africa. In 2009, Luc Montagnier was awarded Nobel Prize for discovering HIV (Vahlne 2009).

It is generally accepted that several interspecies transmissions of the SIV (Simian Immunodeficiency Virus) from nonhuman primates to humans are about how HIV was first transmitted to humans. This most likely happened when hunters butchered and ate monkey meat in Africa. HIV infection is managed by administering a cocktail of antiretroviral (ARV) drugs. There are six classes of ARV drugs which are used in combination. Two nucleoside reverse transcriptase inhibitors (NRTI) are frequently used with one protease inhibitor (PI), integrase inhibitor, or non-nucleoside reverse transcriptase inhibitor (NNRTI). Various studies have shown that adherence to antiretroviral therapy (ART) can help in reducing the viral titer and increase the CD4 cell counts. But non-adherence and inconsistency in HIV medication have led to the rise of drug resistance strains. Later in 1996, highly active antiretroviral therapy (HAART) was introduced and since then the mortality associated with AIDS-related opportunistic infections has significantly declined. ART alone cannot control the HIV epidemic, and we need a safe and effective vaccine. Till date we don't have a vaccine for HIV, in this review, we have tried to highlight the present challenges faced by vaccine developers and how to tackle them.

## 2 Structure and Replication of HIV

HIV virus is about 100 nm in diameter. HIV core comprises of two identical copies of ss RNA (9200–9800 nucleotides) together with protease, integrase, and reverse transcriptase enzymes. Apart from these proteins, several others proteins with regulatory and immune-modulatory function are also coded by both HIV-1 and HIV-2. The outer lipid envelope consists of two glycoproteins, gp41 and gp120 (Fig. 1). Glycoprotein 41 is a transmembrane glycoprotein and glycoprotein 120 is called as docking glycoprotein. Both these glycoproteins form a heterodimer which is responsible for the entry of virus into the host cell.

HIV genome contains structural genes, which encodes for proteins and are helpful in formation of functional structure of the virus. Gag gene encodes for core and shell of the virus. HIV shell is produced as p55, a precursor protein, that is broken into p18, (an antigen that makes up the matrix or shell), p24, and p15 (constitute the core antigens). A precursor protein that is encoded by the pol gene is cleaved into the proteins p31 (integrase), p51 (reverse transcriptase), and p66. The envelope gene produces glycoprotein (gp160) that is broken into gp120 and gp41 (Cleghorn et al. 2005).

Apart from these structural genes, HIV also encodes for many non-structural and regulatory genes in order to perform various function. Some of which includes vif

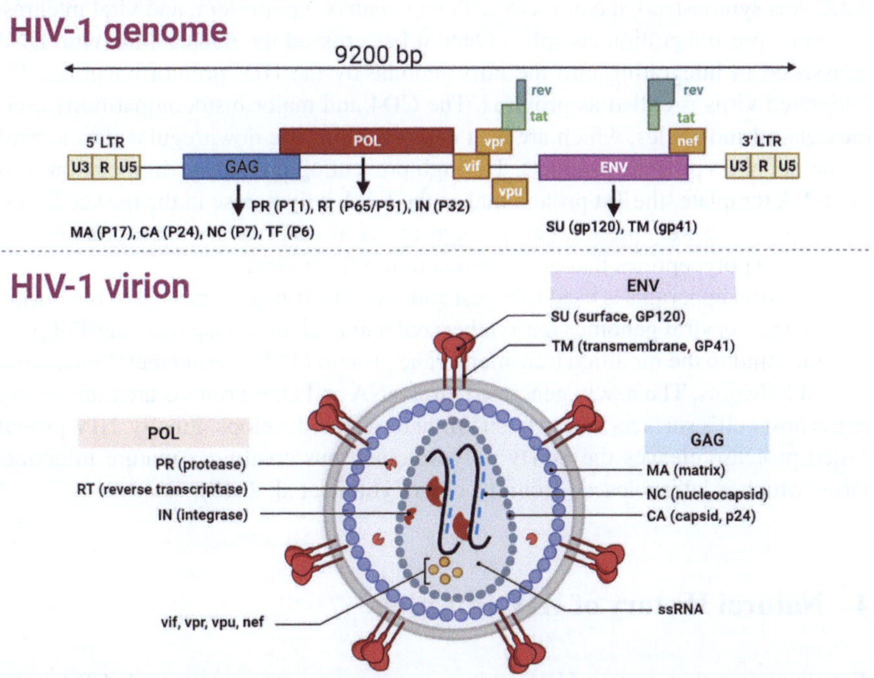


**Fig. 1** Genome and Viral structure of HIV-1

(viral infectivity factor gene), vpr (viral promoter region), vpu (in HIV-1) and vpx (in HIV-2), tat (transcriptional transactivator gene), rev (regulator of viral genes), nef (negative factor gene), and LTR (long terminal repeat) (Ratner et al. 1987).

## 3 Replication of HIV

Virus-encoded proteins carry out the earliest stages of HIV replication, which include viral attachment, entrance, and reverse transcription of the virus into the host genome. Production of mRNAs, processing of viral genomes, and the synthesis of structural proteins all take place during the later stages of replication. For transcription and translation purposes, the virus is fully dependent on the host cell machinery. Most cell types experience cell death as a result of HIV replication.

Internalization of HIV starts with the attachment of viral gp120 to the CD4 receptors on host cell. Chemokine co-receptors like CXCR4 or CCR5 facilitate a strong binding of CD4 cells with viral gp120 glycoprotein. After binding to the host cell, viral gp41 protein fuses with the cell wall, and the HIV nucleocapsid is released inside the host cell cytoplasm. Once inside the host cell, viral capsid breaks up, and two copies of single-stranded RNA and viral enzymes are released. Viral RNA is reverse transcribed into ssDNA using reverse transcriptase enzyme. The RNA is degraded by viral endonuclease, and ssDNA replicates to form dsDNA. Once the dsDNA is synthesized, it combines with gag, matrix, vpr protein, and viral integrase to form a pre-integration complex. Once it has crossed the nucleus, the viral DNA is assisted in integrating into the host genome by the HIV protein integrase. The integrated virus is called as provirus. The CD4 and major histocompatibility complex class I molecules, which are host cell receptors, are downregulated as a result of the nef and vpu gene products. Through preventing premature dissociation from the DNA template, the Tat protein makes the RNA polymerase in the host cell more processive. The Rev. protein ensures that the viral mRNAs are appropriately converted into polypeptides that get assembled into fresh virions.

The Golgi apparatus is used to digest and move the Env polyprotein to the plasma membrane. As viral genomes are synthesized and uncleaved Gag and Gag-Pol polyproteins bind to the modified transmembrane protein (TM) plasma membrane, virus assembly begins. The newly generated viral RNA and HIV proteins are translocated to the host cell's surface, where fresh, immature HIV develops. Finally, HIV protein called protease cleaves the newly polyproteins, thus creating a mature infectious virus, which is later released from the cell (Cynthia et al. 2019).

## 4 Natural History of HIV Infection

The discovery that certain HIV-positive persons can maintain a high CD4 count and/or suppressed viral load without antiretroviral therapy indicates that the natural course of untreated HIV disease differs from person to person. About 80–90% of

HIV infected individuals are typically confined to a survival time of approximately 10 years. In most of the people, infection progresses from continued viral multiplication to progressive CD4 depletion, leading to AIDS and death. The following five stages make up the usual course of HIV infection.

### 4.1 Acute HIV Disease or Acute Retroviral Syndrome

During this stage, HIV kills the infected T cells and spreads to the bloodstream, resulting in primary viremia (or acute mononucleosis like syndrome). There will be a large decrease in the quantity of circulating CD4 T cells after 3–6 weeks following the first infection.

### 4.2 In Asymptomatic Conditions (Clinical Latency)

Adequate immune response develops within 1 month in most of the patients. Both effective cell-mediated immune response and humoral immune response help in maintaining an normal CD4 T cell count via effective immune response toward the virus; this is called clinical latency condition. The period of clinical latency is variable; it may last for 10 years but ranges from few months to 30 years.

### 4.3 Persistent Generalized Lymphadenopathy (PGL)

PGL is seen in about 25–30% of asymptomatic infected people. PGL is characterized as enlarging lymph nodes (>1 cm) in two or more non-contiguous places that last for at least 3 months. PGL must be distinguished from other causes of lymphadenopathies such as lymphoma.

### 4.4 Symptomatic HIV Infection (AIDS-Related Complex—ARC)

Once the clinical latency period is over, the CD4 cell count started dropping and the patients show the following symptoms like unexplained diarrhea, lasting for more than 1 month, weight loss (>10%), fatigue, malaise, night sweat, and mild opportunistic infections such as oral thrush. The end stage of HIV infection is called AIDS. ARC is the most serious stage of HIV infection. At this stage, CD4 count drops to 200 cells/μL or even below 200 cells/μL, high viral load and fibrous tissues formation in lymphoid organs are also observed.

### 4.5 Progression to HIV Infection to AIDS

Due to their weakened immune systems, people with AIDS are more vulnerable to serious illnesses brought on by opportunistic infections. Depending on the CD4 T cell count, various infections occur like development of neoplasia (CNS lymphoma) and development of direct HIV-induced manifestations such as HIV encephalopathy (Kumarasamy et al. 2003).

## 5 Immune Responses During HIV Infection

Understanding intricate immune mechanisms during HIV-1 infections is essential, as it would probably open up avenues to find and develop interventions for its cure and/or prevention (Hartana and Yu 2021). However, information about the mechanisms of protective immunity to HIV is limited; a lack of information is therefore a major obstacle for developing a good vaccine (Johnson 2002). Besides that, HIV infection causes a wide array of immunological anomalies which ultimately causing severe immunosuppression, opportunistic infections, as well as death (Rosenberg and Fauci 1989). Numerous of these immunological flaws can be related to the virus-induced quantitative and qualitative inadequacies of T4 cells. Other problems might be caused by long-term HIV exposure to or infection of immune cell, like B cells, monocytes, or macrophages. Recently understood mediators in an immune response and/or immunomodulation factors are discussed in the following sections (Items 4.1–4.6).

### 5.1 Natural Killer (NK) Cell-Mediated Response

During viral infections, NK cells play a vital role in innate responses, thus representing the first line of defence. Recently, the NK cell-mediated innate immune response has captivated more focus for its role in controlling HIV infection and the disease succession. In a report by Chang et al. (2012), the immunogenic potential of HIV-1 Gag-virus-like particles (Gag-VLPs) was investigated in order to induce NK cell responses. It was observed that Gag-VLPs were able to activate DCs, which in turn induce innate and Gag-specific responses in NK cells.

### 5.2 HIV/Autophagy Interplay

In addition to regulating cell homeostasis, the degradative process of autophagy is crucial for both the innate and adaptive responses against viruses. Studies show that early HIV-1 infection features, such as virus adherence to CD4 or membrane fusion, allow the virus to change autophagy pathways, priming cells for infection (Cabrera-Rodríguez et al. 2021). Virus entry is regulated when it interacts with CD4, its primary receptor, and a co-receptor, primarily CCR5 or CXCR4. This interaction is crucial for HIV-1 disease progression. Along with eliminating viruses that are invasive, autophagy plays a pivotal role in the immunological response by presenting antigens via MHC class I and class II molecules, which serve as the innate immunological barrier against these invaders. The interaction between autophagy and viruses is complicated since viruses have developed a variety of ways to either hinder or take advantage of autophagy. Therefore, this mechanism may either promote or prevent viral replication (Chen et al. 2020). Since it has been demonstrated that autophagy is involved in final phages of the viral cycle (viral multiplication and exit), in preventing viral infection in CD4+ T lymphocytes and favoring its multiplication inside macrophages, the role of autophagy has become quite pertinent in HIV-1 infection (Espert et al. 2009).

## 5.3 *Role of CD8+ T Cells HIV-1 Control*

Recent investigations have supported the importance of virus -specific CD8+ T cells which are key driver of immunological control of HIV-1. However, growing evidence indicates that a sophisticated contact with innate immune cell improves their ability to perform their function (Hartana and Yu 2021). Because of this, mounting an efficient antiviral defence or immunity may be possible through the combination of various immunological effector pathways.

## 5.4 *HIV-Nef*

The negative factor (Nef) protein of HIV is an accessory protein that is a primary determinant of viral pathogenesis. Nef is abundantly expressed during infection and reroutes a variety of cell surface proteins to disrupt host immunity and thereby it can promote viral replication. This viral accessory protein counteracts host defences by sequestering and/or degrading its targets via the endocytic and secretory pathways. Nef does this by physically engaging several host trafficking proteins (Buffalo et al. 2019).

## 5.5 *Role of RIG-I-like Receptors Signalling*

The RNA helicase family of retinoic acid-inducible gene I (RIG-I)-like receptors (RLRs) is an important component of host defence mechanisms. It detects the virus and initiates a cascade of innate immune signals which controls viral replication and dissemination (Onomoto et al. 2021). RLRs act as cytoplasm-based sensors and recognize foreign RNA in the cell and activates cascade of antiviral reactions like, induction of type I interferons, activation of inflammasome, and manifestation of pro-inflammatory cytokines and chemokines (Ireton and Gale Jr. 2011). However, a protease-mediated sequester of RIG-I inhibits RIG-I-mediated antiviral signalling in HIV-1 infection (Solis et al. 2011).

There are various ways by which HIV-1 mediates defects in the IFN pathway, these include disablement of receptors involved in virus detection, signalling for IFN1 upregulation, and articulation of IFN1-inducible proteins. In spite of this, in vitro studies have shown killing of latently HIV infected cells (Sandstrom et al. 2017). Therefore, more studies are to be done in physiologically relevant models to understand the blocking of IFN1 signalling and can this be reversed. Such a strategy will be helpful for the development of novel therapeutics which can eradicate HIV-1 reservoir (Sandstrom et al. 2017).

### *5.6 AIDS—The End Stage*

In the final stage of AIDS, virus attacks the immune system similar to other autoimmune diseases (ADs), leaving the body prone to opportunistic infections. HIV is also responsible for an immune dysfunction (with more or less clinical symptoms). The overall pathogenic process is facilitated by this immune dysregulation, and it can lead to the development of autoimmune and systemic diseases (Zandman-Goddard and Shoenfeld 2002). Although ADs are uncommon during HIV infection; however, some ADs (viz. immune thrombocytopenic purpura, sarcoidosis, inflammatory myositis, and Guillain–Barre Syndrome) have been more frequently reported among HIV affected cases than in the overall population (Virot et al. 2017).

## 6 Types of HIV Vaccines

One of the most promising community health measures to mitigate HIV infection is vaccination. Thus, progress on a safe, universally effective and accessible vaccine is central to stopping the HIV/AIDS pandemic, and it would be a great public health tool. An HIV vaccine should be able to prevent the onset of disease or its further progression to AIDS (Nath 2010).

Scientists are still working very hard to find a vaccine or treatment for HIV infection; however, we do have antiretrovirals that can control the viral load in patients (Vanangamudi et al. 2021). The need of the hour is reduction or prevention of viral transmission by measures like condoms or male circumcision, but this is often limited by social and behavioral influences.

Antiretroviral therapies have their own concerns like non-adherence, side-effects, and occasional development of drug resistance. Thus, the advancement of a cheap, effective, and readily available preventive vaccine still remains a key goal in controlling the HIV pandemic. In spite of this urgent need, we have shown a very modest progress toward an effective HIV vaccine. The main reason for this slow progress is the capability of the virus to evolve rapidly and its ability to evade host immune responses (Hokello et al. 2021).

HIV is a uniquely complex virus, and it has posed a wide range of challenges to the vaccine developers (Barouch 2008). Major challenges in developing successful HIV vaccines are discussed under Item-8 in this chapter.

Various vaccine concepts and multiple vaccination strategies have been tested so far for HIV. These include DNA vaccines, subunit vaccines, live vectored recombinant vaccines, and various prime-boost vaccine combinations (Girard et al. 2006).

An ideal strategy to design vaccine immunogens includes their ability to activate both parts of adaptive response (antibodies and cell mediated), thereby inducing a sustainable and durable protection against HIV. World over, developers are now advancing and testing engineered vaccine immunogens that mimic actual structure of HIV envelope—envelope is the target of all broadly neutralizing antibodies against the virus (https://www.iavi.org/our-science/hiv-vaccines). The prospective HIV vaccines may be categorized as per the following types:

### 6.1 Live Attenuated Vaccines

Although live attenuated HIV vaccines have been studied in lab animals and have demonstrated significant protection against HIV, they are currently not being evaluated in people due to safety concerns.

Live attenuated HIV-1 vaccine has provided protection against challenge with simian immunodeficiency virus (SIV) in rhesus macaque model, still they may never be evaluated clinically in human subjects (Miller and Abel 2005). In a study using attenuated SHIV 89.6, immunized rhesus macaques showed antiviral CD8+ T cells in the genital tract, which was able to control viral replication and delayed the progression to AIDS after vaginal SIV challenge (Genescà et al. 2009, 2012). SIV with nef gene deletion was found to be attenuated in macaques and provided protection against wild-type SIV challenge (Whitney and Ruprecht 2004). In reality, it was observed that Nef-deletion mutants still exhibited virulence; in contrast to wild-type virus, disease progression was reduced but not completely eliminated (Whitney and Ruprecht 2004). Due to this reason, there is no hope left that mere deletion of nef would actually transform a virulent virus into a live attenuated vaccine.

### 6.2 Subunit Vaccines

Majority of the initial HIV vaccine candidates developed were recombinant/subunit vaccines. As the name suggests, these vaccines carry part of the pathogen or its protein which acts as a foreign antigen and stimulates B cells to produce antibodies. Envelope of HIV is made up of glycoprotein gp160 that gets broken into gp120 and gp41. The mature envelope forms a trimer composing three gp120/gp41 complexes. During the early clinical trials, researchers have analyzed both recombinant gp160 (rgp160) and gp120 (rgp120) as immunogens. In fact, many initial clinical trials focussed on a recombinant gp120 as vaccine (Girard et al. 2011; Cohen and Dolin 2013). It was later evident that rgp160 vaccine was able to induce neutralizing antibodies against homologous vaccine strains only, and it also stimulated limited antibody response (Dolin et al. 1991; Cohen and Dolin 2013). This was the first HIV vaccine which has undergone complete testing in humans (Girard et al. 2011). However, this vaccine failed efficacy trial and due to this reason researchers are still working to develop a better vaccine strategy.

### *6.3 DNA Vaccines*

DNA vaccines contain copies of one or many genes from the pathogen. These pathogenic genes get integrated with the host gene leading to a protein. This protein acts as a foreign antigen and stimulates the immune system to produce antibodies. This is a very ubiquitous strategy and is currently, being practiced to develop vaccine, and for this reason, many of the current AIDS vaccine candidates are DNA vaccines. They are regarded as safe because they are unable to cause HIV infection, as they lack all the genes required for the live pathogen.

The currents trends in HIV DNA vaccine efforts are majorly aimed at using HIV-based lentiviral vectors for antigen delivery. In spite of safety issues belonging to the use of HIV-based lentiviral vectors for vaccine release, it has its own vantages. Along with their ability to transduce CD4 cells and for in vivo gene therapy in AIDS disease, these vectors can also be pseudo typed into a recombinant form which may infect various cells, like neuronal and cancer cells. Various pre-clinical studies have shown that lentiviral vectors have the capability to induce strong and specific adaptive immune responses (Hokello et al. 2021). Upon testing in both mouse models and in humans, lentiviral vectors expressing viral surface proteins were able to induce strong HIV-specific B cells and cytotoxic T-lymphocytes (CTLs) responses (Hokello et al. 2021). Interestingly, in a study, it was observed that a lentiviral vector-based vaccine containing HIV Gag, Pol, and Rev. (VRX1023) was able to induce more powerful and lasting mucosal and systemic immune responses in comparison to adenovirus-based vectors (Lemiale et al. 2010).

Each viral vector used for delivery of DNA vaccine has distinct advantages and disadvantages (Table 1).

**Table 1** Major advantages and disadvantages of viral vectors for DNA vaccine delivery

| Virus | Advantages | Disadvantages |
|---|---|---|
| Adenovirus | High immunogenicity, safely used in many clinical trials, high titer production | Pre-existing immunity |
| Adeno-associated virus | Non-pathogenic virus, long term gene expression, induces CTL response | Low titer production<br>Pre-existing immunity |
| Cytomegalovirus | Protects against SIV infection in animal models | Pre-existing immunity<br>Risk of pathogenesis in specific individuals |
| Lentivirus | Infects both dividing and non-dividing cells, long term expression, can generate high immunogenicity | Generation of replication competent virus, Potential for tumorigenesis |
| Retrovirus | Long term expression | Generation of replication competent virus, Potential for tumorigenesis |
| Sendai virus | High immunogenicity | Pre-existing immunity |
| Vaccinia virus | High immunogenicity, safety, high titer production | Pre-existing immunity |

## *6.4 Recombinant Vector Vaccines*

Recombinant vector vaccines adopt a similar strategy as DNA vaccines. The only difference is that the gene sequence is carried by a non-pathogenic or an attenuated vector. Majority of the currently studied HIV vaccine candidates belong to this category. In comparison to other vaccines, recombinant vector vaccines considered as safe and will not cause infection because they have one or few of the many HIV genes, making them incompetent to grow and replicate. Instead of using a DNA vaccine alone, it is believed that the presence of a vector makes the vaccine better in creating an immune reaction. Various viral vectors have shown their capability to be used as recombinant viral vectors. Some of them include adeno and adeno-associated viruses, modified Vaccinia Ankara (MVA), and Venezuelan Equine Encephalitis (VEE) virus. In a study with replication defective flavivirus vector-based recombinant HIV-1 vaccine candidate, high neutralizing antibody levels, high levels of IgG titres against viral antigens, T cell response, and ADCC responses were seen in the study recipients (Giel-Moloney et al. 2019). In a human study involving engineered canarypox virus expressing viral antigens as a priming vaccine and HIV envelope as a booster vaccine, 31% protection was observed with the induction of high levels of antibody response (Acevedo et al. 2015).

## *6.5 Vaccine with mRNA Technology*

Over the past few decades, advances in the field of molecular biology have led to the use of mRNA as a possible vaccine candidate (Pardi et al. 2018). Sequence optimized mRNA delivered either naked or via lipid encapsulation have shown protective antibodies against various agents like influenza virus, Zika virus, rabies virus, and others.

The effectiveness of mRNA vaccines is constrained by a number of fundamental disadvantages (Fan et al. 2018): (1) Nucleases can quickly break down free mRNA; (2) passive internalization of free mRNA by cells is not particularly effective since RNA has a negative charge; and (3) on entering the cell, foreign naked mRNA is quickly broken down in the endo-lysosomal compartments. The core idea behind several approaches to better deliver mRNA vaccines is to prevent mRNA from degrading while enhancing APC uptake.

Due to its various desirable features like high potency, ease of rapid development, low-cost manufacturing, and safe delivery, RNA-based immunogens might be an encouraging alternative to conventional vaccines (Esteban et al. 2021). The field of mRNA vaccines needs improvements in several key areas, including immunogens, prototype, immunogenicity, its chemical alterations, and its delivery strategies. mRNA vaccines can be classified into the following categories: (1) Non-replicating (NRM) constructs have an open frame containing the gene of interest (GOI), 5′ and 3′ untranslated regions (UTRs), and a terminal polyadenylation (A) region (2). Self-amplifying mRNA (SAM) construct.

A small number of studies have reported that modification of nucleosides can effectively boost mRNA stability and translational capability while reducing its immunogenicity. The first clinical trial for an HIV mRNA vaccine was carried out in 2009, after the instability problem was eliminated (Routy et al. 2010). A synthetic lipid nanoparticle-based formulation containing self-amplifying RNA (LNP/RNA) was tested, and it did show improvements in mRNA expression. This study demonstrated increased antigen synthesis and increased immunogenicity, without the necessity for an appropriate delivery mechanism (Geall et al. 2012).

An HIV vaccine based on mRNA technology that produces broadly neutralizing antibodies (bNAbs) that act against HIV-1 has been developed by Moderna in collaboration with the International AIDS Vaccine Initiative (IAVI) and Scripps Research. Results from a phase 1 experiment reveal that 97% of subjects who got the vaccine experienced the intended response (Leggat et al. 2022).

## *6.6 Additional Candidate Vaccines—Alternative Vaccine Designs/Strategies*

Because HIV is primarily transmitted through the mucosa, and despite the complexities of immune response induction and its evaluation, majority of vaccine strategies are being refocused on mucosal vaccines (Demberg and Robert-Guroff 2009). Since HIV spreads more quickly than a recall CTL response is initiated, it would be great if both IgG and IgA could kill initial infections (Wilkinson and Cunningham 2006; Shattock et al. 2008; Girard et al. 2008). Thus, an in-depth understanding of the mucosal immune system is of utmost important, and this knowledge will ultimately benefit vaccine design for other mucosal transmitted pathogens. A good mucosal vaccine should stimulate virus-specific secretory IgA (S-IgA) and mucosal CD8(+) lymphocytes as a first line of defence to protect against early stage of HIV infection. This will also inhibit the binding of virus into the secondary lymphoid organs in mucosal and systemic tissues (Yuki et al. 2007).

Conventional vaccine strategies have failed to provide protection against difficult viruses that cause chronic infections, like HIV-1 (Pardi et al. 2018). Additional candidate vaccines must be tested concurrently in industrialized and poor nations to hasten the development of an HIV vaccine. Important ethical questions will need to be resolved, and coordination and international cooperation will be needed. We should start preparing now for how to employ future HIV vaccines so that they contribute to the broader HIV/AIDS prevention effort (Esparza 2001). According to the National Institute of Allergy and Infectious Diseases (NIAID), "The progress for a secure and efficient vaccine still a challenge in order to end the ongoing HIV pandemic" (HIV vaccine development. National Institute of Allergy and Infectious Diseases. Accessed May 6, 2021. https://www.niaid.nih.gov/diseases-conditions/hiv-vaccine-development).

The long-term objective is to create an HIV vaccine that can shield people all around the world from contracting the disease. But even if a vaccine only partially protects those who receive it, or even if it only partially protects by decreasing the risk of infection, it could still have a major dent on the disease spread and will help in restricting the pandemic, especially in high risk individuals. A partially successful vaccination might lower the count of HIV infections, hence lowering the number of people who can spread the virus to others. We can end the epidemic by significantly lowering the number of new infections. For a successful cure of HIV, possibly there will be a need of a combination of different interventions (Deeks et al. 2016; Bailon et al. 2020 as cited by Esteban et al. 2021).

One of the biggest obstacles to treating HIV is the latent virus reservoir. The immune system can hardly see the silent virus that has gone integrated into the host genes. The reservoir will most likely not be eliminated even in the fictitious scenario that the greatest optimized therapeutic vaccination is discovered. Potential tactics include the co-administration of a curative mRNA vaccine with medicines that potentially reactivate the HIV reservoir, immunological checkpoint inhibitors, and antibodies that broadly neutralize the virus.

## 7 Ongoing Clinical Trials

Over the last few decades, extensive studies have been undertaken to understand the basic virology, immunology, pathogenesis of etiological agent as well as the disease. Significant amount of research has also been done for the development of antiretroviral drugs and effective vaccine candidate development. The Conference on Advances in AIDS Vaccine Development in October 1993 reviewed ongoing vaccine research, progress made and challenges faced. It is believed that about 12 candidate vaccines are under various stages of phase I/II clinical trials (Johnston et al. 1994).

In 1987, the first HIV vaccine phase I experiment was conducted. Since then, more than 60 phase I/II trials including more than 30 candidate vaccines have been conducted, involving about thousands of healthy volunteers. The majority of these trials were carried out in developed countries like USA and Europe, but few were also carried out in underdeveloped nations like (Brazil, China, Cuba, Haiti, Kenya, Peru, Thailand, Trinidad, and Uganda) (https://www.who.int/hiv/topics/vaccines/Vaccines/en/).

## 7.1 Major Vaccine Trials in the Past

Till date many novel vaccine candidates have been evaluated globally under several clinical trials. These vaccines are targeted to induce protective antibody responses. In early days, it was believed that neutralizing antibodies would provide immunity against HIV infection (Plotkin 2010). As discussed earlier, majority of these HIV vaccines targeted the envelope glycoproteins, gp120 or gp160 (Esparza 2013). In 1986, Zagury and colleagues conducted the first experimental immunization trial in human subjects against HIV in Congo (Previously Zaire) (Zagury et al. 1987).

A small number of persons received the vaccine via an engineered vaccinia vector expressing the viral gp160. It evaluated the vaccine candidate's capacity to stimulate cytotoxic T lymphocyte (CTL) and neutralizing antibody responses in HIV-negative people. This engineered vaccinia virus, was delivered via scarification, generated low immune responses. Further four additional booster doses helped to achieve desirable anamnestic immune responses and persisted for more than 1 year (Zagury et al. 1988). After this pioneering vaccine trial, over 250 clinical trials have been conducted worldwide, most of them being early-phase trials, i.e., phase 1 or 2 (Wang et al. 2015). These trials were carried out in United States, African countries, and Thailand. About 140 of such trials were carried out in the United States (Esparza 2013). In the 1990s, vaccine candidates were designed to target envelope proteins, and the pre-clinical tests were conducted on non-human primates followed by safety and immunogenicity trials in human subjects (Berman et al. 1990; Keefer et al. 1994; El-Amad et al. 1995; Gorse et al. 1995; Berman et al. 1996).

VaxSyn was the name of the first human trial that was carried out in the US. In this study, 72 HIV-negative, healthy adults were tested to determine the immunogenicity and safety of a rgpl60 produced in a baculovirus-insect system. The trial participants reported that the new vaccine was both safe and well tolerated. However, it showed a lackluster response in serum antibodies against HIV envelope proteins, failing to raise enough quantities of protective neutralizing antibodies (Dolin et al. 1991). After the VaxSyn trial, during the period between 1988 and 2003, 35 phase I trials were conducted to evaluate numerous envelope proteins as vaccine candidates. These vaccine candidates were found to induce long lasting binding and neutralizing antibodies but no visible CTL responses (Esparza 2013). In 1988, another engineered vaccinia virus expressing HIV gp160 (HIVAC-1e) was tested in phase 1 trials. It resulted to short-lived T-cell responses in the subjects, and no HIV-specific antibodies were detectable. In 1991, a combined approach known as prime-boost approach was studied. This phase 1 vaccine trial primed the subjects with HIVAC-1e and later boosted with VaxSyn. This approach proved to be beneficial as it was able to enhance overall immune responses and also induced neutralizing antibodies (Gallo and Montagnier 1987). Although the results of this trial were promising, but the utility of vaccinia as a suitable carrier has raised many red flags. In people who had previously had a smallpox vaccination, there was a noticeable decline in the vector's immunogenicity. Additionally, there were worries that giving the replicating vaccinia to immunocompromised people would cause serious illness (Redfield et al. 1987; Picard et al. 1991). Another engineered canarypox vectored vaccine candidate (ALVAC) that expressed gp160 was tested in 1993. It was examined using two different methods: both alone and in a prime-boost combination with an adjuvanted gp160 subunit. According to the findings, the novel vaccine candidate greatly primed the protein boost's neutralizing antibody response and caused a CTL response (Pialoux et al. 1995). Two vaccine contenders, having bivalent engineered gp120 and alum, first appeared in 1994. After being exposed to HIV in pre-clinical studies, these innovative vaccine candidates protected chimpanzees, and phase 1/2 clinical trials in people revealed that they were safe and immunogenic. These were moved forward to efficacy trials due to such good results (Berman et al. 1996). Sadly, data analysis in 2003 showed that the two vaccinations had little effect on preventing HIV infection (Pitisuttithum et al. 2006).

Antibody-based vaccines yielded no positive outcomes and pushed the research community to begin hunt for T cell-based vaccines. In the meantime, evidence showed that CD8+ T cell responses played pivotal role in curtailing HIV. The STEP trial, also known as the HVTN 502 trial (ClinicalTrials.gov Identifier: NCT00095576), was the first significant T cell-based vaccine trial. It was a phase 2b multicentric, double-blind, randomized, placebo-controlled "test-of-concept" study that examined the effectiveness of the MRKAd5 HIV-1gag/pol/nef vaccine. A total of 3000 individuals from the US, Australia, Canada, Brazil, the Dominican Republic, Jamaica, Haiti, Peru, and Puerto Rico were enrolled in the experiment as of December 2004. In this trial, participants were enrolled and randomly randomized to acquire three shots of the MRKAd5 HIV-1 gag/pol/nef vaccine or a placebo. This candidate vaccine has latter undergone phase 2b trials in 801 adult South African individuals (the Phambili/HVTN 503 trial; ClinicalTrials.gov Identifier: NCT00413725).

The purpose of this trial was to determine how well the vaccination prevented HIV in Southern Africa, where subtype C is more common. The STEP and Phambili trials were stopped in September 2007 due to preliminary evaluation reports which indicated the vaccine candidate had little efficacy (Cohen 2007). In 2009, another T cell-based vaccine HVTN 505 (ClinicalTrials.gov Identifier: NCT00865566) was tested in phase 2b randomized, placebo-controlled, efficacy trial using a prime-boost regimen. The trial included a total of 2504 males or transsexual women who have intercourse with men. A provisional analysis revealed that it was unable to stop the disease or lower participants HIV virus loads. Additionally, vaccine recipients allegedly experienced higher breakthrough infections than placebo control receivers. Due to this reason, it was terminated prematurely after 47 months (Esparza 2013).

After numerous setbacks seen in the field of HIV vaccine came the surprised success of RV144 trial (ClinicalTrials.govIdentifier: NCT00223080). There was moderate yet significant protection generated by the vaccine against HIV infection. This was a randomized, double-blind phase 3 trial that used the ALVAC-HIV (vCP1521) engineered canarypox vector vaccine, that expresses Env (clade E), group-specific antigen (Gag) (clade B), and protease (Pro) (clade B), as well as the adjuvanted AIDSVAX B/E and a bivalent HIV gp120 subunit vaccine (Watanaveeradej et al. 2006). So far only RV144 trial has shown 31.2% effectiveness even after 42 months of final shot. But fast lowering of protective antibody levels was reported in many of the trial participants. Hence, it was thought that late boosts of the vaccine would induce durable and protective immunity against HIV infection. Therefore, late boost studies, specifically RV305 and RV306, were created to evaluate the immunological responses produced in recently boosted vaccination recipients in comparison to RV144 recipients. The RV305 (ClinicalTrials.gov Identifier: NCT01435135) was a randomized, placebo-controlled, double-blind trial that investigated effect of late boosts with ALVAC-HIV (vCP1521) or AIDSVAX B/E, administered either alone or in combination in enhancing the immune parameters of defense.

In order to receive two booster shots administered 6–8 years after receiving the RV144 vaccine, the trial re-enrolled 162 HIV-negative, healthy Thai participants. Participants in the study were randomly assigned to three groups and given either the vaccine or a placebo. At weeks 0 and 24, group 1 was given ALVAC-HIV and AIDSVAX B/E; group 2 was given AIDSVAX B/E; and group 3 was given ALVAC-HIV or a placebo. The findings demonstrated that in groups 1 and 2, immunological correlates, such as plasma IgG, IgA, and neutralizing antibody titers at week 2, were all considerably increased in comparison to responses seen 2 weeks after the last RV144 immunization. Additionally, the boost led to greater antibody titers against several envelope proteins in plasma and mucosal secretions, surpassing those seen at the peak RV144 vaccination time point. However, CD4+ lymphocyte responses in RV305 recipients lasted for shorter time and successive boosts did not lead to any rise (Rerks-Ngarm et al. 2017). A randomized, placebo-controlled, double-blind investigation called the RV306 trial (ClinicalTrials.gov Identifier: NCT01435135) was carried out at three clinical locations in Thailand. Its major goal was to find out whether giving the RV144 regimen a boost at months 12, 15, or 18 after the initial immunization would enhance the immunological response. The findings showed that groups which received late boosts showed higher functionality and polyfunctionality scores compared to vaccine recipients without late boost. The outcome of these trials implied that boosting of the RV144 vaccine regimen with longer intervals between the initial vaccination and late boost improved the efficacy of the vaccine (Pitisuttithum et al. 2020). The necessity to evaluate RV144's effectiveness against other clades was brought on by the positive outcomes of its clinical trials. RV144 was altered to specifically target HIV clade C after being originally developed to guard against HIV subtype CRF01 AE infection. As a result, HVTN was created and carried out a number of clinical trials, including HVTN 097, HVTN 100, and HVTN 702 (Zhao et al. 2020).

Only a few HIV vaccine have undergone phase 2b trials over the past three decades. Clinical trials that can adapt are urgently needed. The development of an HIV vaccine will be accelerated by the rapid screening of novel candidates in human trials and the rapid advancement of promising candidates to efficacy studies. Some of the vaccine trials that have been finished and their timelines are displayed in Table 2.

## *7.2 Ongoing Clinical Trials in the World*

Rigorous and multi-faceted vaccine approach has resulted in development and clinical testing of more than 100 potential HIV vaccine candidates. There are few phase 2b efficacy trials that are currently ongoing, which includes HVTN 705/HPX2008, HVTN 706/HPX3002, and PrepVacc.

### 7.2.1 HVTN 705/HPX2008 (Imbokodo Study) Phase 2b Efficacy Trial

The HVTN 705/ HPX2008 (Imbokodo study) (ClinicalTrials.gov Identifier: NCT03060629) trial is a multi-centric randomized, controlled, double-blind phase 2b/3 efficacy trial. This trial commenced in November 2017, currently being carried out at 24 places in 5 African countries and is expected to be completed by May 2022. The main aim of this trial is to assess the efficacy, safety, and tolerability of a prime-boost regime for mitigating the disease. A total of 2600 healthy HIV-negative, sexually active women belonging to age-group of 18–35 years are enrolled in this trial. Participants in the study were given a placebo or the vaccine at random. An adjuvanted clade C gp140 and Mosaic gp140 HIV protein vaccine, as well as a tetravalent adenovirus vector vaccine, Ad26.Mos4.HIV (composed of Ad26.Mos1. Gag-Pol, Ad26.Mos2. Gag-Pol, Ad26.Mos1.Env, and Ad26.Mos2S.Env clade C) Ad26.Mos4.HIV was administered intramuscularly (IM) to vaccine recipients at months 0 and 3, followed by IM administrations of Ad26.Mos4.HIV and adjuvanted clade C gp140 at 6 and 12 months, respectively. Participants allocated to the placebo group will be administered a placebo intramuscularly. The main outcomes will be evaluations of the vaccine's effectiveness and adverse occurrences (AEs). The immunogenicity, immune response biomarkers as predictors of eventual infection, and detection of viral sequences from vaccine and placebo recipients will be the secondary objectives (Bekker et al. 2020).

**Table 2** Major vaccine trials that have been completed and their outcome

| Vaccine trial | Year | Vaccine candidate | Site | Study group | Result |
|---|---|---|---|---|---|
| VaxSyn | 1987 | Recombinant envelope glycoprotein subunit (rgp160) of HIV | Canada | 72 adults | No vaccine efficacy |
| HIVAC-1e | 1988 | Recombinant vaccinia virus designed to express HIV gp160 | USA | 35 male adults | No vaccine efficacy |
| Vax004 | 1998–2002 | AIDSVAX B/B gp120 with alum | North America | 5417 MSM and 300 women | No vaccine efficacy |
| Vax003 | 1999–2003 | AIDSVAX B/E gp120 with alum | Thailand | 2545 men and women | No vaccine efficacy |
| HVTN 505 | 2009–2013 | Three vaccinations with DNA encoding HIV clade B gag, pol and nef as well as env from HIV clades A, B and C followed by an Ad5 vector-based vaccine encoding clade B gag and pol as well as env from clades A, B and C | USA | 2504 men or transgender women who have sex with men | No vaccine efficacy |
| STEP/HVTN 502 trial MRKAd5 | 2004–2007 | HIV-1 gag/pol/nef trivalent vaccine | North America the Caribbean South America, and Australia | 3000 MSM and heterosexual men and women | No vaccine efficacy |
| Phambili/HVTN 503 trial | 2003–2007 | rAd5 (gag/pol/nef) | South Africa | 801 adults | No vaccine efficacy |
| RV144 | 2003–2009 | ALVAC-HIV (vCP1521) and AIDSVAX B/E vaccines | Thailand | 16,402 community-risk men and women | Vaccine efficacy–31.2% at 42 months |

### 7.2.2 HVTN 706/HPX3002/Mosaico Phase 3 Efficacy Trial

The HVTN 706/HPX3002 (ClinicalTrials.gov Identifier: NCT03964415) (https://clinicaltrials.gov/study/NCT03964415), or Mosaico trial is a multicentric, randomized, controlled, double-blind phase 3 efficacy trial, currently being carried out in North America, South America, and Europe. By September 2023, the study should be finished, having begun in October 2019. The aim of this study is to evaluate the safety and effectiveness of the Ad26.Mos4.HIV and adjuvanted clade C gp140 and Mosaic gp140 protein vaccine prime-boost vaccination regimen in apparently healthy, HIV-uninfected MSM and transgender individuals. A total of 3800 people, aged 18–60, have been enlisted in this trial and are being randomly assigned to receive the vaccine or a placebo. Evaluation of the vaccine's efficacy is the major objective. The number of subjects experiencing both requested and unrequested local or systemic adverse events (AEs), medically addressed adverse events (MAAEs), and serious adverse events (SAEs), as well as the frequency and intensity of both arms of immune responses (cellular and humoral) specific to the HIV Env, antibody titers against Ad26 antigen, precarious sexual behavior, and pre-exposure prophylaxis (PrEP) use are secondary endpoints. Initial findings presented at International AIDS Conference (2019) demonstrated proof of vaccine-induced immune responses to various HIV types currently prevalent globally (ClinicalTrials.gov 2020).

### 7.2.3 PrepVacc Phase 2b Trial

PrepVacc (ClinicalTrials.gov Identifier: NCT04066881) (https://clinicaltrials.gov/study/NCT04066881) is a multicentric randomized, controlled, double-blind phase 2b vaccine study, which commenced in January 2020 and is supposed to be over in March 2023. South Africa, Tanzania, Mozambique, and Uganda are the current locations of this study, where a combination of two HIV vaccination regimens (DNA/ AIDSVAX and DNA/CN54gp140+ MVA/CN54gp140) with PrEP (PrEPVacc) will be evaluated to determine its efficacy. The study has enrolled 1668 healthy HIV-uninfected adults between 18 and 40 years of age, who were equally randomized into six groups that will receive either the vaccine or placebo. Examination of infection in recipients and evaluation of adverse events related to receiving either the vaccine or a PrEP regimen that may result in the regimes being discontinued. The final objective is to ascertain whether the vaccine causes a significant reduction in HIV prevalence with sufficient community health importance to support the use of the combination regime (Prep-ClinicalTrials.gov 2020). Table 3: Ongoing clinical trials.

**Table 3** Ongoing HIV vaccine clinical trials

| Vaccine trial registered no. | Phase | Duration | Vaccine | Site | Target group |
|---|---|---|---|---|---|
| Imbokodo Study, ClinicalTrials.gov Identifier: NCT03060629 | Phase 2b Efficacy Trial | 2017–2022 | HVTN 705/HPX2008 | Sub-Saharan African countries | 2600 healthy HIV-uninfected sexually active women, aged 18–35 years |
| ClinicalTrials.gov Identifier: NCT03964415 | Phase 3 Efficacy Trial | 2019–2023 | HVTN 706/HPX3002/ Mosaico | Europe, North America, and South America | 3800 healthy HIV-uninfected adults, aged 18–60 years |
| PrepVacc ClinicalTrials.gov Identifier: NCT04066881 | Phase 2b Trial | 2020–2023 | PrepVacc | Mozambique, South Africa, Tanzania, and Uganda | 1668 healthy HIV-uninfected adults 18–40 years |

## 7.3 Ongoing Clinical Trials in India

### 7.3.1 P001 Trial

In 2009, the National Institute for Research in Tuberculosis in Chennai, India, and the National AIDS Research Institute in Pune, India, collaborated on an IAVI-sponsored Phase I HIV-1 subtype C preventive vaccination trial. The P001 trial (Clinical Trial registry CTRI/2009/091/000051) is the name of this study. Sixteen healthy adult males and females without HIV were included in this randomized, placebo-controlled, double-blind, phase I experiment at two locations. The study compared the safety and immunogenicity of a heterologous prime-boost regimen using ADVAX, a DNA-based vaccine made up of Chinese HIV-1 subtype C env gp160, gag, pol, and nef/tat genes cloned into a recombinant vector as the prime, and TBC-M4, a recombinant vector containing Indian HIV-1 subtype C env gp160, gag, RT, rev, tat, and nef genes as the boost. Post-trial studies carried out after 3 months showed that all study participants in both groups developed virus antibody responses against the viral proteins after the final booster dose. The study described the antibodies' capacity for neutralization as well. However, it was shown that 3 months following the final dose of vaccination, the titers of immunological markers steadily decreased. It was therefore determined that the vaccine did not result in the formation of strong, broadly neutralizing antibodies (Mehendale et al. 2013).

### 7.3.2 DNA Prime (ADVAX) and Modified Vaccinia Virus Ankara (TBC-M4) Boost

In order to evaluate a subtype C preventive vaccination in a prime-boost regimen, the IAVI conducted another clinical trial (Phase I) in India in 2009. This included an MVA (TBC-M4) boost and a DNA prime (ADVAX). It proved that the regimen was safe, well tolerated and also improved HIV-specific immune responses in study participants. The only early findings on vaccine-induced immune responses included neutralizing antibody analysis and the IFN-ELISPOT response. It was found that a sizable fraction of trial participants produced variable region-specific antibodies after immunization. These antibody titers were directly associated with the number of regulatory T cells (Treg) and T follicular helper cells (Tfh). This trial offered reassuring proof of the vaccine candidate strong immunogenicity (Munusamy Ponnan et al. 2018).

In spite of these vaccine trails, HIV vaccine development still remains a major scientific obstacle. The following sections go over the main obstacles to creating an HIV vaccine that is both safe and effective (Items 8.1–8.4).

## 8 Social Risks and Ethical Concerns in HIV Vaccine

Any vaccine trial involves ethical practice to ensure the safety and well-being of the trial subjects and protection of their rights. Over the years, there has been a rise in financing for prevention and treatment of HIV/AIDS, but the AIDS vaccine efforts still remain inadequate (Klausner et al. 2003). To commence a vaccine trial, a lot of effort is to be invested in careful reviewing the study by policymakers as well as researchers. Preparedness of the host country is also of utmost importance in commencing a trial. Before commencement of any HIV vaccine trial, international and national ethical committees must review the study to ensure the safety and protection of enrolled volunteers, appropriateness of the methods, risks, and benefits involved in the study and whether informed consent will be obtained from participants (Center for Immunization Research 2002). Potential volunteers or participants in a clinical trial should be made fully aware that they are being invited to take part in a research study to test an HIV vaccine. Volunteers have the right to refuse or withdraw from the trial at any given point of time. It is always committed and informed to the participants the number of visits to be made, procedures involved along with specimen collection. The participants should be made aware that the test vaccine is being tested on experimental basis in humans. Many trials involve use of placebo and in such instances, participants should be informed that they may or may not receive a placebo. Every trial involves some side-effects caused by the vaccine under study and it may clearly be informed to the study participants. In certain cases, HIV vaccine may be administered to an HIV infected individual, in this case, the most important aspect is participant confidentiality and has to be maintained throughout (Guenter et al. 2000).

The availability of an HIV vaccine does not guarantee its uptake. We are going to encounter many challenges in distributing HIV vaccines which may arise due to low vaccination rates in developing nations, poor uptake of highly effective and readily available vaccines in developed countries, and the expectation that the first HIV vaccines will only be partially effective.

Studies on HIV vaccine readiness have identified a number of obstacles to taking part in an HIV vaccination trial (Dhalla and Poole 2011). The recruitment of participants or the hunt for qualified volunteers is the main challenge for any HIV experiment. In addition to this, a host of social stigmas related to HIV/AIDS, such as the worry that a vaccine could cause HIV infection, the negative social consequences of testing positive for the virus, and mistrust of authorities and scientific findings, pose additional barriers to the widespread use of HIV vaccines.

The usefulness of partially effective vaccines in lowering HIV incidence can be compromised by increased high-risk behaviors brought on by the availability of HIV vaccines. HIV vaccine effectiveness trials also encounter substantial difficulties in finding enough participants for large phase 3 efficacy trials (Newman et al. 2004).

If the vaccine is found to be effective, a major challenge will be to ensure its sustained supply after a successful trial. There is a general agreement that once trial is over and vaccine proves to be effective, it is made available in nations where the trial was carried out at an affordable cost (Moodley 2002). The fairness concept calls for an equitable distribution of the advantages of research across people and communities.

## 9 Major Challenges Involved in Developing an HIV-1 Vaccine

HIV has established itself as a global epidemic, with currently over 40 million active infections (Ng'uni et al. 2020). Due to this reason, development of a safe and effective vaccine against HIV-1 vaccine remains a top global health priority. An ideal HIV-1 vaccine would completely prevent infection and with a reduced viral load should provide sterilizing immunity. Still we are no way near the development of an effective vaccine for HIV-1. The main reasons for this are both scientific challenges and challenges at the social levels. Some major challenges hampering that the development of an HIV-1 vaccine includes production of an immune response which can control virus replication, enormous genetic diversity of the virus and its influence on vaccine-induced immune responses, limitation of animal models to study HIV infection, destruction of various immune effector cells at an early phase of infection, vaccine testing, and its distribution.

### *9.1 Viral Diversity*

HIV is divided into two types: HIV-1 and HIV-2 with HIV-1 showing a tremendous amount of global sequence diversity While HIV-2 contains groups A-I, only groups A and D are now circulating globally. HIV-1 has four groups: M (Major), N (non-M, non-O), O (Outlier), and P. Data on HIV-2 subtypes is limited, and only few recombinants have been described (Ibe et al. 2010). The most common kind of HIV-1 infection globally is Group M, which is further split into nine subtypes (A-D, F-H, J, and K) and five sub-subtypes (A1-A3, F1, and F2) (Adhiambo et al. 2021). HIV-1 exhibits genetic diversity that ranges from 8 to 17% within a subtype and can reach 30%, whereas subtype diversity typically ranges from 17 to 35% and can reach 42%. Numerous recombinants, which includes 96 circulating recombinant forms (CRFs) and many distinct recombinant forms, have been produced as a result of frequent recombination events between HIV-1 subtypes (URFs) (Reis et al. 2019).

High frequency of recombination and mutation during reverse transcription leads to the rise of genetic diversity within an infected host (Liu et al. 2020; Fuhrmann et al. 2021). The virus is under tremendous amount of selection pressure from the host immune response and antiretroviral treatment. Along with these, lack of proofreading mechanism and a high rate of viral replication is also responsible for mutations. The considerable genetic variety identified in HIV-1 has a significant impact on its pathogenesis, clinical therapy, and transmission. The ability of HIV-1 to develop high levels of genetic diversity, which fuels the emergence of escape mutants, presents a significant obstacle to the development of vaccines (Boutwell et al. 2010). Due to this reason, it is important to study the quasispecies dynamics in an individual during infection.

After numerous vaccine failures mainly the AIDVAX trial in 2003, the STEP trial in 2007 and the HVTN505 trail in 2013 (Rerks-Ngarm et al. 2009; Altfeld and Goulder 2011; Hammer et al. 2013), the concept of mosaic-based regimen has reignited new hopes. In this approach, pieces of different HIV strains are taken and combined to elicit a broader immune response. In order to test the efficacy of a heterologous vaccination regimen using a tetravalent mosaic adenovirus serotype 26 (Ad26)-based HIV vaccine in combination with adjuvanted clade C gp140 protein (NCT03060629), the Imbokodo (HVTN 705/HPX2008) phase 2b proof of concept study was started in women in sub-Saharan Africa in 2017 (Barouch et al. 2018). The vaccination regimens included clade C Env gp160 protein-alum adjuvanted Ad26 or MVA vector boosting and Ad26 or MVA vector priming with mosaic Env/Gag/Pol antigens, respectively. All vaccination schedules showed good safety and acceptability. The highest immunogenic vaccination in humans was found to be the mosaic Ad26 prime/Ad26+gp140 boost vaccine, which produced T-cell responses in 100%, 80%, and 83% of recipients as well as Env-specific antibodies and antibody-dependent phagocytosis responses. In rhesus monkeys, this vaccination regimen elicited immune responses of a similar size, persistence, and profile and provided 66% protection against contracting infection after a series of six SHIV-SF162P3 challenges. The Imbokodo regimen involved 2 shots over 12 months, this would be difficult to implement. Due to this reason, a shorter regiment was optimized involving three vaccines over 6-month period (Stephenson et al. 2020). Both regimens were well-tolerated and safe. As we wait for the pending efficacy data from the ongoing trials, additional research is required to streamline the delivery of this vaccine. The mosaic vaccination trials have shown encouraging results, and they may be a feasible solution to cope with HIV-1 genetic variety.

## 9.2 *Animal Model Scarcity to Study HIV Infection*

In order to eliminate a disease, it is important to understand the etiology and disease progression using animal models. Animal models continue to be crucial for cutting down on the time, costs, and complications associated with disease prevention and drug discovery. In HIV vaccine discovery, animals' models play a very pivotal role. Since many HIV treatments, such as vaccines, cannot be directly tested on humans without any risks, it is essential to use animal models that imitate humans infected with HIV. Animal models, not only enable us to deconvolute many elements that are highly intricate in humans, it also offers us an in vivo system that aids in a deep investigation of the various anatomic reservoirs via invasive sampling. They also allow us to evaluate various interventions strategies aimed at modulating the immune systems or specific immune cell populations which may pave the way for HIV-1 cure. Non-human primates (NHPs), humanized mice, and to a lesser extent feline are the most often utilized animal models for HIV-1 research (Weichseldorfer et al. 2020; Akkina et al. 2020). Though none of these models exactly simulate HIV-1 infection in humans, they can be adapted to answer critical questions. Various strains of simian immunodeficiency viruses (SIV) are used to study different disease stated in NHPs. Studies in NHPs have helped us in identifying key questions on HIV infection pathogenesis and allowed us to test various therapeutic strategies which are essential for the development of HIV-1 vaccine.

The three macaques Macaca mulatta, Macaca nemestrina, and Macaca fascicularis are the main subjects of HIV-1 research (Thippeshappa et al. 2020). All of these macaque species are vulnerable to disease by different SIV types, and the course often varies from a controlled to a quick disease development, with mortality within a few months after infection (Folks et al. 1997). This variability in outcomes can be due to the pathogen type used for challenge, monkey species involved, and its genetic lineages (Evans and Silvestri 2013; Del Prete and Lifson 2013). Aside from these macaque species, research on HIV-1 has also been conducted on African NHPs, such as mandrills, sooty mangabeys, and African green monkeys (AGMs), all of which belong to the Cercocebus genus (Mandrillus sphinx) (Broussard et al. 2001). Despite being the SIV's natural hosts, it has been noted that SIV infection in these animals typically does not lead to AIDS. These African green monkeys are employed in comparison studies to determine the correlates of protection and the absence of disease progression. (Pandrea and Apetrei 2010; Sodora et al. 2009; Chahroudi et al. 2012).

HIV-1-infected humanized mice provide a less expensive, genetically similar model that simulates the human immune system (Karpel et al. 2015). However, they are subject to constraints in terms of production, breeding, tissue colonization, and tissue requirement in relation to size (Nischang et al. 2012). The other potential animal models are domestic felines infected with feline immunodeficiency virus (FIV) (Burkhard and Dean 2003). This model also shares its similarity in pathogenesis with HIV-1 in humans (Kanzaki and Looney 2004). The major limitations with FIV infected cats are its key anatomical differences with humans. SIV-infected monkeys are bigger than both mouse and feline models, but their physiology and disease course are similar to those of HIV-infected people. Researchers have tried to use small animals like rabbits, guinea pigs, and woodchuck for HIV-1 vaccines, but as their immune system is different from humans, they are of little help (Morrow et al. 1987; Menne and Cote 2007). As HIV-1 cannot infect or replicate efficiently in NHPs, researchers have developed a chimeric SIV (also called simian human immunodeficiency viruses (SHIVs)) which contains the envelope of HIV and rest all components from SIV (Kumudhini et al. 2018). These SHIVs have to be adapted in lab to bind to rhesus CD4 cells, then only they can infect them. Later Shaw Li and their colleagues found out that by changing a single amino acid (residue 375) in HIV envelope protein, a chimeric SIV can efficiently infect monkeys (Li et al. 2016). These findings have sparked the creation of effective models for examining

AIDS vaccine candidates. Due to their high maintenance expenses, difficulties in locating species that are sensitive, and usage of SIV for infection, NHPs pose a challenge for HIV-1 research. The research involving NHPs is also impeded by various animal welfare groups which stresses for social welfare, psychological well-being, and specific housing requirements for these animals.

## 9.3 Lack of Immune Correlates for Protective Immunity

HIV-1 binds to various immunological cells like CD4+ T and monocytes, leading to a decrease in number of CD4+ T cells. This progressive depletion of CD4+ T cells populations is the hall mark of acquiring AIDS, which leads to the decline in cell-mediated immunity, making the host more susceptible to opportunistic infections (Pegu et al. 2015). Although progression of disease in untreated HIV cases takes many years, massive C4+ T cell destruction occurs quite early. It has been noted that in majority of people, CD4+ memory T-cell regeneration stops this early destruction and maintains CD4+ T-cell numbers and functions. This process of regeneration of C4+ T cell is unable to restore all functionally important CD4+ T-cell populations and is also not stable for a long period. Finally, CD4+ memory T-cell homeostasis fails and the host becomes susceptible to opportunistic infections. CD8+ T-cells along with CD4+ T-cells constitute majority of T lymphocytes (Okoye and Picker 2013). Once activated, CD4+ T lymphocytes differentiate into distinct effector subtypes with regulates the immune system by secreting cytokines. Along with this, CD4+ T cells executes numerous other tasks, such as activating B-lymphocytes, cytotoxic T cells, non-immune cells, and innate immune system cells. They do play a critical part in the suppression of immunological response. Due to this depletion during the early phase of infection, developing an effective vaccine which will simulate and activate innate immune system for a better protection becomes difficult.

In a typical HIV infection progessor, within weeks of infection, viremia falls with a simultaneous induction of anti-HIV immune responses. This reduction is mainly due to the presence of anti-HIV MHC-1 restricted CD8+ cytotoxic T lymphocytes (Koup et al. 1994). CD8+ T lymphocytes are vital for the immune response against the virus during the latent phase of HIV infection by eliminating productively infected cells and controlling the viral load (Collins et al. 2020). Previous vaccine trials like ALVAC-HIV and Remune have failed to induce immune effector cells mainly CD8+ T lymphocytes and CD4+ T lymphocytes in either HIV-1 infected or non-infected subject and even if it is induced these responses were often transient (Nitayaphan et al. 2004). Another trial called the QUEST study, involved HIV-1 positive adults who have received ART for more than 2 years and have a HIV-RNA load of less than 50copies/ml. They received formulations containing either vaccines (ALVAC or Remune) or placebo. It was observed that subjects who received ALVAC with or without Remune, virus-specific CD4+ and CD8+ T lymphocyte responses were significantly increased. However, overall virus-specific T lymphocyte responses were low even after vaccination (Lampe et al. 2007).

So, we have learned many important lessons from these failed trials which will guide us in developing an effective therapeutic HIV-1 vaccine in future. Probably, the most important lesson is that our future vaccine should not only be immunogenic, but it must also induce a strong and effective immune response against HIV-1 replication (Betts et al. 2005). A lack of vaccine-induced immunity may restrict the vaccine's overall immunogenicity, which could give rise to weak virus-specific T lymphocyte responses that are unable to inhibit viral replication.

## 9.4 *HIV-1 Vaccine Testing and Its Distribution*

HIV vaccine research has been going on since last 30 years, still we have only five efficacy trials conducted so far (Esparza 2013). Of all these trials conducted only one trial involving RV144 showed protective immunity and that too at very modest level. Failures and disappointment brought on by the HVTN 505 and VaxGen gp120 trials ought to rekindle the process and prompt a review of the HIV vaccine approach in order to make it even better. These failures prompted the creation of the Global HIV Vaccine Enterprise, which refocused efforts on developing vaccines. Since there is a global need for an HIV vaccine, we want a well-coordinated international effort with open-minded people to reassess difficulties and uncover new opportunities. We have to innovate new methods to design and deliver the vaccine and learn from our failures which we have encountered in the past trials. One such example is the development of messenger RNA (mRNA) vaccines for COVID-19 (Jackson et al. 2020). In fact, the IAVI, which is researching cutting-edge HIV vaccine technologies including mRNA, replicating viral vector-based vaccines, and broadly neutralizing antibodies (bNAbs), has made this one of its primary research areas (Ura et al. 2014; Sok and Burton 2018). We need to involve new research collaboration and innovative minds to search for an HIV vaccine. So far, only major pharmaceutical industry in developed countries is involved in vaccine manufacturing, we should encourage companies in developing countries to get involve by announcing incentives or awards. People residing in developing countries must participate whole heartedly not only as a vaccine trial center, but as a responsible partner, as they are the once who will be benefited the most from an HIV vaccine.

Once a vaccine is developed, the real resistance will be its distribution and use. We have seen many such examples in history where a highly effective vaccine was developed and was made available in developed countries, but it came to developed country only recently. One such example is vaccine for hepatitis B (Wait et al. 2016). Public health professionals unanimously agree that delays in licensing and accessibility for an HIV vaccine in impoverished nations are unacceptable. One of the most difficult problems for government, NGOs, and policy makers to tackle is ensuring that HIV vaccine is available simultaneously in rich and poor countries. The cost of the vaccination will be another constraining factor (Leelahavarong et al. 2011). As these HIV vaccines are developed employing advanced technology, they will be expensive in comparison to other conventional vaccines. In addition to these limitations, poor health care infrastructure in many developing countries will make it difficult to distribute the vaccine, particularly in risk groups, which need it the most. International organizations like World Bank and IAVI are establishing purchase funds and intellectual property agreements which will facilitate the rapid access of HIV vaccine in poor countries (Clinton and Sridhar 2017). As about 70% of the entire world population lives in developing countries and if we want to stop this global epidemic of AIDS, we have to make sure that these people have full access to a HIV vaccine.

## 10 Future Trends/Prospects in HIV Vaccine

HIV is a global problem and must be handled as a global public health concern. The history of HIV vaccine trials is full to failures and disappointment. Still we should reinvigorate and further studies must be performed to improve on the findings of the previous trails. An illustration of this is the P5 (Pox Protein Public Private Partnership), which has proposed successive future investigations of RV144. The speed at which the SARS-CoV2 vaccines were created and produced has shown the power of international scientific innovation and what can be accomplished through public–private partnerships. Mosaic vaccines and cell-based vaccinations should be promoted since they have bright futures. Finally, to solve this global problem, regulatory organizations, pharmaceutical companies, governments, philanthropists, vaccine trialists, basic research scientists, and others must work together.

## Glossary

**AIDS** Acquired immunodeficiency syndrome (AIDS) is the most advanced stage of the disease due to human immunodeficiency virus (HIV) infection.

**Antiretroviral therapy** Antiretroviral therapy or ART is the pharmacological treatment for HIV.

**Candidate vaccines** The vaccines before the government's approval. The candidate vaccine is tested in a small number of healthy individuals to determine whether it is safe and can generate an immune response/memory.

**Deconvolute** To simplify a complex signal; a process that separates a heterogeneous mixture signal into its constituent components. Cellular deconvolution refers to computational techniques aiming at estimating the proportions of different cell types in samples collected from a tissue.

**Immunodeficiency** A failure or absence of elements of the immune system, including lymphocytes, phagocytes, and the complement system; these immunodeficiencies can be either primary, such as genetic, or secondary, as the one caused by HIV infection.

**Viral diversity** Variety of viruses.

## References

Acevedo R, Almagro JC, Alsaadi M, Appeldoorn CCM, Sassi AB, Bhambhani A, Bollard CM, Braz-Gomes K, Brezski RJ, Buchanan A, Casimiro D, Chandra D, Chiriva-Internati M, Coller B-A, Cruz CRY, de Boer M, De Rose R, D'Sa S, D'Souza B, D'Souza M, D'Souza MJ, D'Souza N, Ferrari M, Ferro VA, Gaillard PJ, Gala RP, Garside DA, Gebril A, Gladdines W, Gonzalez-Angulo AM, Hamblin P, He Q-Y, Heinrichs J, Jackson E, Karande P, Kent SJ, Kodangattil S, Lamprou D, Lundstrom K, Movassaghian S, Mulla NS, Mullen AB, Nagarkar R, Nair L, Nimmo N, Parenky AC, Oliver B, Oviedo-Orta E, Ramirez-Montagut T, Ranasinghe C, Reijerkerk A, Revell JD, Rip J, Shen H, Soliman H, Sun B, Suri K, ter Meulen J, Torchilin VP, Trasatti JP, Ubale RV, Visser CC, Vo TP, Wolfram J, Zughaier S (2015) In: Singh M, Salnikova M (eds) Novel approaches and strategies for biologics, vaccines and cancer therapies. Academic Press, Cambridge, MA, pp vii–viii. https://doi.org/10.1016/B978-0-12-416603-5.00024-

Adhiambo M, Makwaga O, Adungo F, Kimani H, Mulama DH, Korir JCK, Mwau M (2021) Human immunodeficiency virus (HIV) type 1 genetic diversity in HIV positive individuals on antiretroviral therapy in a cross-sectional study conducted in Teso, Western Kenya. Pan Afr Med J 38:335. https://doi.org/10.11604/pamj.2021.38.335.26357

Akkina R, Barber DL, Bility MT, Bissig KD, Burwitz BJ, Eichelberg K, Endsley JJ, Garcia JV, Hafner R, Karakousis PC, Korba BE, Koshy R, Lambros C, Menne S, Nuermberger EL, Ploss A, Podell BK, Poluektova LY, Sanders-Beer BE, Subbian S, Wahl A (2020) Small animal models for HIV, hepatitis B, and tuberculosis. Proceedings of the NIAID Workshop. Curr HIV Res 18(1):19–28. https://doi.org/10.2174/1570162X18666191223114019

Altfeld M, Goulder PJ (2011) The STEP study provides a hint that vaccine induction of the right CD8+ T cell responses can facilitate immune control of HIV. J Infect Dis 203(6):753–755. https://doi.org/10.1093/infdis/jiq119

Bailon L, Mothe B, Berman L, Brander C (2020) Novel approaches towards a functional cure of HIV/AIDS. Drugs 80(9):859–868. https://doi.org/10.1007/s40265-020-01322-y

Barouch DH (2008) Challenges in the development of an HIV-1 vaccine. Nature 455(7213):613–619. https://doi.org/10.1038/nature07352

Barouch DH, Tomaka FL, Wegmann F, Stieh DJ, Alter G, Robb ML, Michael NL, Peter L, Nkolola JP, Borducchi EN, Chandrashekar A, Jetton D, Stephenson KE, Li W, Korber B, Tomaras GD, Montefiori DC, Gray G, Frahm N, McElrath MJ, Baden L, Johnson J, Hutter J, Swann E, Karita E, Kibuuka H, Mpendo J, Garrett N, Mngadi K, Chinyenza K, Priddy F, Lazarus E, Laher F, Nitayapan S, Pitisuttithum P, Bart S, Campbell T, Feldman R, Lucksinger G, Borremans C, Callewaert K, Roten R, Sadoff J, Scheppler L, Weijtens M, Boer KF, van Manen D, Vreugdenhil J, Zahn R, Lavreys L, Nijs S, Hendriks J, Euler Z, Pau MGP, Schuitemaker H (2018) Evaluation of a mosaic HIV-1 vaccine in a multicenter, randomized, double-blind, placebo-controlled, phase 1/2a clinical trial (APPROACH) and in rhesus monkeys (NHP 13–19). Lancet 392(10143):232–243. https://doi.org/10.1016/S0140-6736(18)31364-3

Bekker LG, Tatoud R, Dabis F, Feinberg M, Kaleebu P, Marovich M et al (2020) The complex challenges of HIV vaccine development require renewed and expanded global commitment. Lancet 395(10221):384–388., 384–388. https://doi.org/10.1016/S0140

Berman PW, Gregory TJ, Riddle L, Nakamura GR, Champe MA, Porter JP, Wurm FM, Hershberg RD, Cobb EK, Eichberg JW (1990) Protection of chimpanzees from infection by HIV-1 after vaccination with recombinant glycoprotein gp120 but not gp160. Nature 345(6276):622–625. https://doi.org/10.1038/345622a0

Berman PW, Murthy KK, Wrin T, Vennari JC, Cobb EK, Eastman DJ, Champe M, Nakamura GR, Davison D, Powell MF, Bussiere J, Francis DP, Matthews T, Gregory TJ, Obijeski JF (1996) Protection of MN-rgp120-immunized chimpanzees from heterologous infection with a primary isolate of human immunodeficiency virus type 1. J Infect Dis 173(1):52–59. https://doi.org/10.1093/infdis/173.1.52

Betts MR, Exley B, Price DA, Bansal A, Camacho ZT, Teaberry V, West SM, Ambrozak DR, Tomaras G, Roederer M, Kilby JM, Tartaglia J, Belshe R, Gao F, Douek DC, Weinhold KJ, Koup RA, Goepfert P, Ferrari G (2005) Characterization of functional and phenotypic changes in anti-gag vaccine-induced T cell responses and their role in protection after HIV-1 infection. Proc Natl Acad Sci U S A 102(12):4512–4517. https://doi.org/10.1073/pnas.0408773102

Boutwell CL, Rolland MM, Herbeck JT, Mullins JI, Allen TM (2010) Viral evolution and escape during acute HIV-1 infection. J Infect Dis 202(Suppl. 2):S309–S314. https://doi.org/10.1086/655653

Broussard SR, Staprans SI, White R, Whitehead EM, Feinberg MB, Allan JS (2001 March) Simian immunodeficiency virus replicates to high levels in naturally infected African green monkeys without inducing immunologic or neurologic disease. J Virol. 75(5):2262–2275. https://doi.org/10.1128/JVI.75.5.2262-2275.2001

Buffalo CZ, Iwamoto Y, Hurley JH, Ren X (2019) How HIV Nef proteins hijack membrane traffic to promote infection. J Virol. 93(24):e01322–e01319. https://doi.org/10.1128/JVI.01322-19

Burkhard MJ, Dean GA (2003) Transmission and immunopathogenesis of FIV in cats as a model for HIV. Curr HIV Res 1(1):15–29. https://doi.org/10.2174/1570162033352101

Cabrera-Rodríguez R, Pérez-Yanes S, Estévez-Herrera J, Márquez-Arce D, Cabrera C, Espert L, Blanco J, Valenzuela-Fernández A (2021) The Interplay of HIV and Autophagy in Early Infection. Frontiers in microbiology 12:661446. https://doi.org/10.3389/fmicb.2021.661446

Center for Immunization Research, & Johns Hopkins University (1995–2002) (2002) Note for guidance for good clinical practice—Selected ICH Guidelines and Declaration of Helsinky, pp 1–49

Chahroudi A, Bosinger SE, Vanderford TH, Paiardini M, Silvestri G (2012) Natural SIV hosts: showing AIDS the door. Science 335(6073):1188–1193. https://doi.org/10.1126/science.1217550

Chang MO, Suzuki T, Suzuki H, Takaku H (2012) HIV-1 Gag-virus-like particles induce natural killer cell immune responses via activation and maturation of dendritic cells. J Innate Immun 4(2):187–200. https://doi.org/10.1159/000329226. Epub July 22 2011

Chen X, Zhang T, Zhang Y (2020) Endoplasmic reticulum stress and autophagy in HIV-1-associated neurocognitive disorders. J Neurovirol 26(6):824–833. https://doi.org/10.1007/s13365-020-00906-4

Cleghorn FR, Reitz MS, Popovic M, Gallo RC (2005) Human immunodeficiency viruses. In: Mandell GL, Bennett JE, Dolin R (eds) Principles and practice of infectious diseases, 6th edn. Churchill Livingstone, London, pp 2119–2133

Clinicaltrials.Gov (n.d.-a) A combination efficacy study in Africa of two DNAMVA- env protein or DNA-env protein HIV-1 vaccine regimens with PrEP (PrEPVacc). https://clinicaltrials.gov/ct2/show/NCT04066881?term=NCT04066881&draw=2&rank=1. Accessed May 22, 2020

Clinicaltrials.Gov (n.d.-b) A Study of Heterologous Vaccine Regimen of adenovirus Serotype 26 Mosaic4 human immunodeficiency virus (Ad26. Mos4.*HIV*), Adjuvanted Clade C gp140 and Mosaic gp140 to Prevent HIV-1 Infection Among Cis-gender Men and Transgender Individuals Who Have Sex with Cisgender Men and/or Transgender Individuals (MOSAICO). Available at: https://Clinicaltrials.gov/ct2/show/NCT03964415?term=NCT03964415&draw=2&rank=1. Accessed May 22, 2020

Clinton C, Sridhar D (2017) Who pays for cooperation in global health? A comparative analysis of WHO, the World Bank, the Global Fund to Fight HIV/AIDS, tuberculosis and Malaria, and Gavi, the Vaccine Alliance. Lancet 390(10091):324–332. https://doi.org/10.1016/S0140-6736(16)32402-3

Cohen J (2007) AIDS research. Promising AIDS vaccine's failure leaves field reeling. Science 318(5847):28–29. https://doi.org/10.1126/science.318.5847.28

Cohen YZ, Dolin R (2013) Novel HIV vaccine strategies: overview and perspective. Ther Adv Vaccines 1(3):99–112. https://doi.org/10.1177/2051013613494535

Collins DR, Gaiha GD, Walker BD (2020) CD8+ T cells in HIV control, cure and prevention. Nat Rev Immunology 20(8):471–482. https://doi.org/10.1038/s41577-020-0274-9

Cynthia N, Cornelissen M, Hobbs M, Muralidha S, Chawala R (2019) Lippincott, illustrated reviews microbiology. South Asian edition. Wolters Kluwer, Riverwoods, IL, pp 567–569

Deeks SG, Lewin SR, Ross AL, Ananworanich J, Benkirane M, Cannon P, Chomont N, Douek D, Lifson JD, Lo YR, Kuritzkes D, Margolis D, Mellors J, Persaud D, Tucker JD, Barre-Sinoussi F, International AIDS Society Towards a Cure Working Group, Alter G, Auerbach J, Autran B, Barouch DH, Behrens G, Cavazzana M, Chen Z, Cohen ÉA, Corbelli GM, Eholié S, Eyal N, Fidler S, Garcia L, Grossman C, Henderson G, Henrich TJ, Jefferys R, Kiem HP, McCune J, Moodley K, Newman PA, Nijhuis M, Nsubuga MS, Ott M, Palmer S, Richman D, Saez-Cirion A, Sharp M, Siliciano J, Silvestri G, Singh J, Spire B, Taylor J, Tolstrup M, Valente S, van Lunzen J, Walensky R, Wilson I, Zack J (2016) International AIDS Society global scientific strategy: towards an HIV cure 2016. Nat Med 22(8):839–850. https://doi.org/10.1038/nm.4108

Del Prete GQ, Lifson JD (2013) Considerations in the development of nonhuman primate models of combination antiretroviral therapy for studies of AIDS virus suppression, residual virus, and curative strategies. Curr Opin HIV AIDS 8(4):262–272. https://doi.org/10.1097/COH.0b013e328361cf40

Demberg T, Robert-Guroff M (2009) Mucosal immunity and protection against HIV/SIV infection: strategies and challenges for vaccine design. Int Rev Immunol 28(1):20–48. https://doi.org/10.1080/08830180802684331

Dhalla S, Poole G (2011) Barriers of enrolment in HIV vaccine trials: a review of HIV vaccine preparedness studies. Vaccine 29(35):5850–5859. https://doi.org/10.1016/j.vaccine.2011.06.055. Epub July 6 2011

Dolin R, Graham BS, Greenberg SB, Tacket CO, Belshe RB, Midthun K, Clements ML, Gorse GJ, Horgan BW, Atmar RL (1991) The safety and immunogenicity of a human immunodeficiency virus type 1 (HIV-1) recombinant gp160 candidate vaccine in humans. NIAID AIDS Vaccine Clinical Trials Network. Ann Intern Med 114(2):119–127. https://doi.org/10.7326/0003-4819-114-2-119

El-Amad Z, Murthy KK, Higgins K, Cobb EK, Haigwood NL, Levy JA, Steimer KS (1995) Resistance of chimpanzees immunized with recombinant gp120SF2 to challenge by HIV-1SF2. AIDS 9(12):1313–1322. https://doi.org/10.1097/00002030-199512000-00003

Esparza J (2001) An HIV vaccine: how and when? Bull World Health Organ 79(12):1133–1137

Esparza J (2013) A brief history of the global effort to develop a preventive HIV vaccine. Vaccine 31(35):3502–3518. https://doi.org/10.1016/j.vaccine.2013.05.018

Espert L, Varbanov M, Robert-Hebmann V, Sagnier S, Robbins I, Sanchez F, Lafont V, Biard-Piechaczyk M (2009) Differential role of autophagy in CD4 T cells and macrophages during x4 and R5 HIV-1 infection. PLoS One 4(6):e5787. https://doi.org/10.1371/journal.pone.0005787

Esteban I, Pastor-Quiñones CP, Usero L, Plana M, García F, Leal L (2021) In the Era of mRNA vaccines, Is there any hope for HIV functional cure? Viruses 13(3). https://doi.org/10.3390/v13030501

Evans DT, Silvestri G (2013) Non-human primate models in AIDS research. Curr Opin HIV AIDS 8(4):255–261. https://doi.org/10.1097/COH.0b013e328361cee8

Fan YN, Li M, Luo YL, Chen Q, Wang L, Zhang HB, Shen S, Gu Z, Wang J (2018) Cationic lipid-assisted nanoparticles for delivery of mRNA cancer vaccine. Biomater Sci 6(11):3009–3018. https://doi.org/10.1039/c8bm00908b

Folks T, Rowe T, Villinger F, Parekh B, Mayne A, Anderson D, McClure H, Ansari AA (1997) Immune stimulation may contribute to enhanced progression of SIV induced disease in rhesus macaques. J Med Primatol 26(4):181–189. https://doi.org/10.1111/j.1600-0684.1997.tb00050.x

Fuhrmann L, Jablonski KP, Beerenwinkel N (2021) Quantitative measures of within-host viral genetic diversity. Curr Opin Virol 49:157–163. https://doi.org/10.1016/j.coviro.2021.06.002

Gallo RC, Montagnier L (1987) The chronology of AIDS research. Nature 326(6112):435–436. https://doi.org/10.1038/326435a0

Geall AJ, Verma A, Otten GR, Shaw CA, Hekele A, Banerjee K, Cu Y, Beard CW, Brito LA, Krucker T, O'Hagan DT, Singh M, Mason PW, Valiante NM, Dormitzer PR, Barnett SW,

Rappuoli R, Ulmer JB, Mandl CW (2012) Non-viral delivery of self-amplifying RNA vaccines. Proc Natl Acad Sci U S A 109(36):14604–14609. https://doi.org/10.1073/pnas.1209367109

Genescà M, McChesney MB, Miller CJ (2009) Antiviral CD8+ T cells in the genital tract control viral replication and delay progression to AIDS after vaginal SIV challenge in rhesus macaques immunized with virulence attenuated SHIV 89.6. J Intern Med 265(1):67–77. https://doi.org/10.1111/j.1365-2796.2008.02051.x

Genescà M, Ma ZM, Wang Y, Assaf B, Qureshi H, Fritts L, Huang Y, McChesney MB, Miller CJ (2012) Live-attenuated lentivirus immunization modulates innate immunity and inflammation while protecting rhesus macaques from vaginal simian immunodeficiency virus challenge. J Virol 86(17):9188–9200. https://doi.org/10.1128/JVI.00532-12. Epub June 13 2012

German Advisory Committee Blood (Arbeitskreis Blut), Subgroup 'Assessment of Pathogens Transmissible by Blood' (2016) HIV. Transfus Med Hemother 43(3):203–222. https://doi.org/10.1159/000445852

Giel-Moloney M, Esteban M, Oakes BH, Vaine M, Asbach B, Wagner R, Mize GJ, Spies AG, McElrath J, Perreau M, Roger T, Ives A, Calandra T, Weiss D, Perdiguero B, Kibler KV, Jacobs B, Ding S, Tomaras GD, Montefiori DC, Ferrari G, Yates NL, Roederer M, Kao SF, Foulds KE, Mayer BT, Bennett C, Gottardo R, Parrington M, Tartaglia J, Phogat S, Pantaleo G, Kleanthous H, Pugachev KV (2019) Recombinant HIV-1 vaccine candidates based on replication-defective Flavivirus vector. Sci Rep 9(1):20005. https://doi.org/10.1038/s41598-019-56550-4

Girard MP, Osmanov SK, Kieny MP (2006) A review of vaccine research and development: the HIV. Vaccine 24(19):4062–4081. https://doi.org/10.1016/j.vaccine.2006.02.031

Girard MP, Bansal GP, Pedroza-Martins L, Dodet B, Mehra V, Schito M, Mathieson B, Delfraissy JF, Bradac J (2008) Mucosal immunity and HIV/AIDS vaccines. Report of an International Workshop, 28-30 October 2007. Vaccine 26(32):3969–3977. https://doi.org/10.1016/j.vaccine.2008.04.051. Report of an International Workshop

Girard MP, Osmanov S, Assossou OM, Kieny MP (2011) HIV immunopathogenesis and vaccine development: a review. Vaccine 29(37):6191–6218. https://doi.org/10.1016/j.vaccine.2011.06.085

Gorse GJ, Rogers JH, Perry JE, Newman FK, Frey SE, Patel GB, Belshe RB (1995) HIV-1 recombinant gp160 vaccine induced antibodies in serum and saliva. The NIAID AIDS Vaccine Clinical Trials Network. Vaccine 13(2):209–214. https://doi.org/10.1016/0264-410x(95)93138-y

Guenter D, Esparza J, Macklin R (2000) Ethical considerations in international HIV vaccine trials; summary of a consultative process conducted by the Joint United Nations Programme on HIV/AIDS [UNAIDS]. J Med Ethics 26(1):37–43. https://doi.org/10.1136/jme.26.1.37

Hammer SM, Sobieszczyk ME, Janes H, Karuna ST, Mulligan MJ, Grove D, Koblin BA, Buchbinder SP, Keefer MC, Tomaras GD, Frahm N, Hural J, Anude C, Graham BS, Enama ME, Adams E, DeJesus E, Novak RM, Frank I (2013) Efficacy trial of a DNA/rAd5 HIV-1 preventive vaccine. N Engl J Med 369(22):2083–2092. https://doi.org/10.1056/NEJMoa1310566

Hartana CA, Yu XG (2021) Immunological effector mechanisms in HIV-1 elite controllers. Curr Opin HIV AIDS 16(5):243–248. https://doi.org/10.1097/COH.0000000000000693

Hokello J, Sharma AL, Tyagi M (2021) An update on the HIV DNA vaccine strategy. Vaccine 9(6):605. https://doi.org/10.3390/vaccines9060605

https://www.who.int/hiv/topics/vaccines/Vaccines/en/

Ibe S, Yokomaku Y, Shiino T, Tanaka R, Hattori J, Fujisaki S, Iwatani Y, Mamiya N, Utsumi M, Kato S, Hamaguchi M, Sugiura W (2010) HIV-2 CRF01_AB: first circulating recombinant form of HIV-2. J Acquir Immune Defic Syndr 54(3):241–247. https://doi.org/10.1097/QAI.0b013e3181dc98c1

Ireton RC, Gale M Jr (2011) RIG-I like receptors in antiviral immunity and therapeutic applications. Viruses 3(6):906–919. https://doi.org/10.3390/v3060906. Epub June 23 2011

Jackson LA, Anderson EJ, Rouphael NG, Roberts PC, Makhene M, Coler RN, McCullough MP, Chappell JD, Denison MR, Stevens LJ, Pruijssers AJ, McDermott A, Flach B, Doria-Rose NA, Corbett KS, Morabito KM, O'Dell S, Schmidt SD, Swanson PA 2nd. (2020) mRNA-1273 study group. An mRNA vaccine against SARS-CoV-2—Preliminary [Report]. N Engl J Med. 383(20):1920–1931. https://doi.org/10.1056/NEJMoa2022483. Epub July 14 2020

Johnson RP (2002) Mechanisms of protection against simian immunodeficiency virus infection. Vaccine 20(15):1985–1987. https://doi.org/10.1016/s0264-410x(02)00083-x

Johnston MI, Noe JS, Killen JY (1994) Recent advances in AIDS vaccine research and development. AIDS Res Hum Retrovir 10(Suppl. 2):S317–S323

Kanzaki LI, Looney DJ (2004) Feline immunodeficiency virus: a concise review. Front Biosci 9(January):370–377. https://doi.org/10.2741/1235

Karpel ME, Boutwell CL, Allen TM (2015) BLT humanized mice as a small animal model of HIV infection. Curr Opin Virol 13(August):75–80. https://doi.org/10.1016/j.coviro.2015.05.002

Keefer MC, Graham BS, Belshe RB, Schwartz D, Corey L, Bolognesi DP, Stablein DM, Montefiori DC, McElrath MJ, Clements ML (1994) Studies of high doses of a human immunodeficiency virus type 1 recombinant glycoprotein 160 candidate vaccine in HIV type 1-seronegative humans. The AIDS Vaccine Clinical Trials Network. AIDS Res Hum Retrovir 10(12):1713–1723. https://doi.org/10.1089/aid.1994.10.1713

Klausner RD, Fauci AS, Corey L, Nabel GJ, Gayle H, Berkley S, Haynes BF, Baltimore D, Collins C, Douglas RG, Esparza J, Francis DP, Ganguly NK, Gerberding JL, Johnston MI, Kazatchkine MD, McMichael AJ, Makgoba MW, Pantaleo G, Wasserheit JN (2003) The need for a global vaccine enterprise. Science 300(5628):2036–2039. https://doi.org/10.1126/science.1086916

Koup RA, Safrit JT, Cao Y, Andrews CA, McLeod G, Borkowsky W, Farthing C, Ho DD (1994) Temporal association of cellular immune responses with the initial control of viremia in primary human immunodeficiency virus type 1 syndrome. J Virol 68(7):4650–4655. https://doi.org/10.1128/JVI.68.7.4650-4655.1994

Kumarasamy N, Solomon S, Flanigan TP, Hemalatha R, Thyagarajan SP, Mayer KH (2003) Natural history of human immunodeficiency virus disease in Southern India. Clin Infect Dis 36(1):79–85. https://doi.org/10.1086/344756

Kumudhini PH, Hajduczki A, Pampusch MS, Haran KP, Hajduczki A, Pampusch MS, Mwakalundwa G, Vargas-Inchaustegui DA, Rakasz EG, Connick E, Berger EA, Skinner PJ (2018) Simian immunodeficiency virus (SIV)—Specific chimeric antigen receptor-T cells engineered to target B cell follicles and suppress SIV replication. Front Immunol 9:492. https://doi.org/10.3389/fimmu.2018.00492

Lampe FC, Porter K, Kaldor J, Law M, Kinloch-de Loes SK, Phillips AN, CASCADE Collaboration (2007) Effect of transient antiretroviral treatment during acute HIV infection: comparison of the Quest trial results with Cascade natural history study. Antivir Ther 12(2):189–193

Leelahavarong P, Teerawattananon Y, Werayingyong P, Akaleephan C, Premsri N, Namwat C, Peerapatanapokin W, Tangcharoensathien V (2011) Is a HIV vaccine a viable option and at what price? An economic evaluation of adding HIV vaccination into existing prevention programs in Thailand. BMC Public Health 11:534. https://doi.org/10.1186/1471-2458-11-534

Leggat, D. J., Cohen, K. W., Willis, J. R., Fulp, W. J., deCamp, A. C., Kalyuzhniy, O., Cottrell, C. A., Menis, S., Finak, G., Ballweber-Fleming, L., Srikanth, A., Plyler, J. R., Schiffner, T., Liguori, A., Rahaman, F., Lombardo, A., Philiponis, V., Whaley, R. E., Seese, A., Brand, J., … Schief, W. R. (2022). Vaccination induces HIV broadly neutralizing antibody precursors in humans. Science (New York, N.Y.), 378(6623):eadd6502. https://doi.org/10.1126/science.add6502

Lemiale F, Asefa B, Ye D, Chen C, Korokhov N, Humeau L (2010) An HIV-based lentiviral vector as HIV vaccine candidate: immunogenic characterization. Vaccine 28(8):1952–1961. https://doi.org/10.1016/j.vaccine.2009.10.089

Li H, Wang S, Kong R, Ding W, Lee FH, Parker Z, Kim E, Learn GH, Hahn P, Policicchio B, Cofano EB, Deleage C, Hao X, Chuang GY, Gorman J, Gardner M, Lewis MG, Santra TSH, Apetrei C, Alam SM, Liao HX, Shen X, Tomaras GD, Farzan M, Chertova E, Keele BF, Estes JD, Doms RW, Montefiori DC, Haynes BF, Sodroski JG, Kwong PD, Hahn BH, Shaw GM (2016) Envelope residue 375 substitutions in simian–human immunodeficiency viruses enhance CD4 binding and replication in rhesus macaques. Proc Natl Acad Sci U S A 113(4):3413–3422

Liu Y, Jia L, Su B, Li H, Li Z, Han J, Zhang Y, Zhang T, Li T, Wu H, Li J, Li L (2020) The genetic diversity of HIV-1 quasispecies within primary infected individuals. AIDS Res Hum Retrovir 36(5):440–449. https://doi.org/10.1089/AID.2019.0242

Mehendale S, Thakar M, Sahay S, Kumar M, Shete A, Sathyamurthi P et al (2013) Safety and immunogenicity of DNA and MVA HIV-1 subtype C vaccine prime-boost regimens: a phase I randomised trial in HIV uninfected Indian volunteers. PLoS One. 8(2):e55831. https://doi.org/10.1371/journal.pone.0055831

Menne S, Cote PJ (2007) The woodchuck as an animal model for pathogenesis and therapy of chronic hepatitis B virus infection. World J Gastroenterol 13(1):104–124. https://doi.org/10.3748/wjg.v13.i1.104

Miller CJ, Abel K (2005) Immune mechanisms associated with protection from vaginal SIV challenge in rhesus monkeys infected with virulence-attenuated SHIV 89.6. J Med Primatol 34(5–6):271–281. https://doi.org/10.1111/j.1600-0684.2005.00125.x

Moodley K (2002) HIV Vaccine Trial participation in South Africa—An ethical assessment. J Med Philos 27(2):197–215. https://doi.org/10.1076/jmep.27.2.197.2986

Morrow WJW, Wharton M, Lau D, Levy JA (1987) Small animals are not susceptible to human immunodeficiency virus infection. J Gen Virol 68(8):2253–2257. https://doi.org/10.1099/0022-1317-68-8-2253

Munusamy Ponnan S, Swaminathan S, Tiruvengadam K, K K V, Cheedarla N, Nesakumar M, Kathirvel S, Goyal R, Singla N, Mukherjee J, Bergin P, Kopycinski TJ, Gilmour J, Prasad Tripathy S, Luke HE (2018) Induction of circulating T follicular helper cells and regulatory T cells correlating with HIV-1 gp120 variable loop antibodies by a subtype C prophylactic vaccine tested in a Phase I trial in India. PLoS One 13(8):e0203037. https://doi.org/10.1371/journal.pone.0203037

Nath A (2010) HIV/AIDS vaccine: an update. Indian J Community Med 35(2):222–225. https://doi.org/10.4103/0970-0218.66862

Newman PA, Duan N, Rudy ET, Anton PA (2004) Challenges for HIV vaccine dissemination and clinical trial recruitment: If we build it, will they come? AIDS Patient Care STDs 18(12):691–701., 691–701. https://doi.org/10.1089/apc.2004.18.691

Ng'uni T, Chasara C, Ndhlovu ZM (2020) Major scientific hurdles in HIV vaccine development: historical perspective and future directions. Front Immunol 11:590780. https://doi.org/10.3389/fimmu.2020.590780

Nischang M, Gers-Huber G, Audigé A, Akkina R, Speck RF (2012) Modeling HIV infection and therapies in humanized mice. Swiss Med Wkly 142:w13618. https://doi.org/10.4414/smw.2012.13618

Nitayaphan S, Pitisuttithum P, Karnasuta C, Eamsila C, de Souza M, Morgan P, Polonis V, Benenson M, VanCott T, Ratto-Kim S, Kim J, Thapinta D, Garner R, Bussaratid V, Singharaj P, el-Habib R, Gurunathan S, Heyward W, Birx D, McNeil J, Brown AE, Thai AIDS Vaccine Evaluation Group (2004) Safety and immunogenicity of an HIV subtype B and E prime-boost vaccine combination in HIV-negative Thai adults. J Infect Dis 190(4):702–706. https://doi.org/10.1086/422258

Okoye AA, Picker LJ (2013) CD4+ T cell depletion in HIV infection: Mechanisms of immunological failure. Immunol Rev 254(1):54–64. https://doi.org/10.1111/imr.12066

Onomoto K, Onoguchi K, Yoneyama M (2021) Regulation of RIG-I-like receptor-mediated signaling: Interaction between host and viral factors. Cell Mol Immunol 18(3):539–555. https://doi.org/10.1038/s41423-020-00602-7

Pandrea I, Apetrei C (2010) Where the wild things are: pathogenesis of SIV infection in African nonhuman primate hosts. Curr HIV/AIDS Rep 7(1):28–36. https://doi.org/10.1007/s11904-009-0034-8

Pardi N, Hogan MJ, Porter FW, Weissman D (2018) mRNA vaccines—A new era in vaccinology. Nat Rev Drug Discov 17(4):261–279. https://doi.org/10.1038/nrd.2017.243

Pegu A, Asokan M, Wu L, Wang K, Hataye J, Casazza JP, Guo X, Shi W, Georgiev I, Zhou T, Chen X, O'Dell S, Todd JP, Kwong PD, Rao SS, Yang ZY, Koup RA, Mascola JR, Nabel GJ (2015)

Activation and lysis of human CD4 cells latently infected with HIV-1. Nat Commun 6:8447. https://doi.org/10.1038/ncomms9447
Pialoux G, Excler JL, Rivière Y, Gonzalez-Canali G, Feuillie V, Coulaud P, Gluckman JC, Matthews TJ, Meignier B, Kieny MP (1995) A prime-boost approach to HIV preventive vaccine using a recombinant canarypox virus expressing glycoprotein 160 (MN) followed by a recombinant glycoprotein 160 (MN/LAI). The AGIS Group, and l'Agence Nationale de Recherche sur le Sida. AIDS Res Hum Retrovir 11(3):373–381. https://doi.org/10.1089/aid.1995.11.373
Picard O, Lebas J, Imbert JC, Bigel P, Zagury D (1991) Complication of intramuscular/subcutaneous immune therapy in severely immunocompromised individuals. J Acquir Immune Defic Syndr 4:641–643
Pitisuttithum P, Gilbert P, Gurwith M, Heyward W, Martin M, van Griensven F, Hu D, Tappero JW, Choopanya K, Bangkok Vaccine Evaluation Group (2006) Randomized, double-blind, placebo-controlled efficacy trial of a bivalent recombinant glycoprotein 120 HIV-1 vaccine among injection drug users in Bangkok, Thailand. J Infect Dis 194(12):1661–1671. https://doi.org/10.1086/508748
Pitisuttithum P, Nitayaphan S, Chariyalertsak S, Kaewkungwal J, Dawson P, Dhitavat J, Phonrat B, Akapirat S, Karasavvas N, Wieczorek L, Polonis V, Eller MA, Pegu P, Kim D, Schuetz A, Jongrakthaitae S, Zhou Y, Sinangil F, Phogat S, Diazgranados CA, Tartaglia J, Heger E, Smith K, Michael NL, Excler JL, Robb ML, Kim JH, O'Connell RJ, Vasan S, RV306 study group (2020) Late boosting of the RV144 regimen with AIDSVAX B/E and ALVAC-HIV in HIV-uninfected Thai volunteers: A double-blind, randomised controlled trial. Lancet HIV 7(4):e238–e248. https://doi.org/10.1016/S2352-3018(19)30406-0
Plotkin SA (2010) Correlates of protection induced by vaccination. Clin Vaccine Immunol 17(7):1055–1065. https://doi.org/10.1128/CVI.00131-10
Plotkin SA (2015) Increasing complexity of vaccine development. J Infect Dis. 212(Suppl 1):S12–S16. https://doi.org/10.1093/infdis/jiu568
Ratner L, Fisher A, Jagodzinski LL, Mitsuya H, Liou RS, Gallo RC, Wong-Staal F (1987) Complete nucleotide sequences of functional clones of the AIDS virus. AIDS Res Hum Retrovir 3(1):57–69. https://doi.org/10.1089/aid.1987.3.57
Redfield RR, Wright DC, James WD, Jones TS, Brown C, Burke DS (1987) Disseminated vaccinia in a military recruit with HIV disease. N Engl J Med 316(11):673–676. https://doi.org/10.1056/NEJM198703123161106
Reis MNG, Guimarães ML, Bello G, Stefani MMA (2019) Identification of new HIV-1 circulating recombinant forms CRF81_cpx and CRF99_BF1 in central Western Brazil and of unique BF1 recombinant forms. Front Microbiol. 10:97. https://doi.org/10.3389/fmicb.2019.00097
Rerks-Ngarm S, Pitisuttithum P, Nitayaphan S, Kaewkungwal J, Chiu J, Paris R, Premsri N, Namwat C, de Souza M, Adams E, Benenson M, Gurunathan S, Tartaglia J, McNeil JG, Francis DP, Stablein D, Birx DL, Chunsuttiwat S, Khamboonruang C, Thongcharoen P et al (2009) Vaccination with ALVAC and AIDSVAX to prevent HIV-1 infection in Thailand. The New England journal of medicine 361(23):2209–2220. https://doi.org/10.1056/NEJMoa0908492
Rerks-Ngarm S, Pitisuttithum P, Excler JL, Nitayaphan S, Kaewkungwal J, Premsri N, Kunasol P, Karasavvas N, Schuetz A, Ngauy V, Sinangil F, Dawson P, de Camp AC, Phogat S, Garunathan S, Tartaglia J, DiazGranados C, Ratto-Kim S, Pegu P, Eller M, Karnasuta C, Montefiori DC, Sawant S, Vandergrift N, Wills S, Tomaras GD, Robb ML, Michael NL, Kim JH, Vasan S, O'Connell RJ, RV305 Study Team (2017) Randomized, double-blind evaluation of late boost strategies for HIV-uninfected vaccine recipients in the RV144 HIV vaccine efficacy trial. J Infect Dis. 215(8):1255–1263. https://doi.org/10.1093/infdis/jix099
Rosenberg ZF, Fauci AS (1989) Immunology of AIDS: approaches to understanding the immunopathogenesis of HIV infection. Ric Clin Lab 19(3):189–209. https://doi.org/10.1007/BF02871809
Routy JP, Boulassel MR, Yassine-Diab B, Nicolette C, Healey D, Jain R, Landry C, Yegorov O, Tcherepanova I, Monesmith T, Finke L, Sékaly RP (2010) Immunologic activity and safety of

autologous HIV RNA-electroporated dendritic cells in HIV-1 infected patients receiving antiretroviral therapy. Clin Immunol 134(2):140–147. https://doi.org/10.1016/j.clim.2009.09.009

Sandstrom TS, Ranganath N, Angel JB (2017) Impairment of the type I interferon response by HIV-1: Potential targets for HIV eradication. Cytokine Growth Factor Rev 37:1–16. https://doi.org/10.1016/j.cytogfr.2017.04.004

Shattock RJ, Haynes BF, Pulendran B, Flores J, Esparza J, Working Group convened by the Global HIV Vaccine Enterprise (2008) Improving defences at the portal of HIV entry: mucosal and innate immunity. PLoS Med 5(4):e81. https://doi.org/10.1371/journal.pmed.0050081

Sodora DL, Allan JS, Apetrei C, Brenchley JM, Douek DC, Else JG, Estes JD, Hahn BH, Hirsch VM, Kaur A, Kirchhoff F, Muller-Trutwin M, Pandrea I, Schmitz JE, Silvestri G (2009) Toward an AIDS vaccine: lessons from natural simian immunodeficiency virus infections of African nonhuman primate hosts. Nat Med 15(8):861–865. https://doi.org/10.1038/nm.2013

Sok D, Burton DR (2018) Recent progress in broadly neutralizing antibodies to HIV. Nat Immunol 19(11):1179–1188. https://doi.org/10.1038/s41590-018-0235-7

Solis M, Nakhaei P, Jalalirad M, Lacoste J, Douville R, Arguello M, Zhao T, Laughrea M, Wainberg MA, Hiscott J (2011) RIG-I-mediated antiviral signalling is inhibited in HIV-1 infection by a protease-mediated sequestration of RIG-I. J Virol 85(3):1224–1236. https://doi.org/10.1128/JVI.01635-10

Stephenson KE, Wegmann F, Tomaka F, Walsh SR, Tan CS, Lavreys L, Ansel JL, Kanjilal DG, Jaegle K, Nkolola JP, Peter L, Fogel R, Bradshaw C, Tyler A, Makoni T, Howe L, Quijada D, Chandrashekar A, Bondzie EA, Borducchi EN, Yanosick KE, Hendriks J, Nijs S, Truyers C, Tolboom J, Zahn RC, Seaman MS, Alter G, Stieh DJ, Pau MG, Schuitemaker H, Barouch DH (2020) Comparison of shortened mosaic HIV-1 vaccine schedules: a randomised, double blind, placebo-controlled phase-1 trail (IPCAVD010/HPX1002) and a preclinical study in rhesus monkeys (NHP 17–22). Lancet HIV 7(6):e410–e421. https://doi.org/10.1016/S2352-3018(20)30001-1

Thippeshappa R, Kimata JT, Kaushal D (2020) Toward a macaque model of HIV-1 infection: roadblocks, progress, and future strategies. Front Microbiol 11:882. https://doi.org/10.3389/fmicb.2020.00882

Ura T, Okuda K, Shimada M (2014) Developments in viral vector-based vaccines. Vaccine 2(3):624–641. https://doi.org/10.3390/vaccines2030624

Vahlne A (2009) A historical reflection on the discovery of human retroviruses. Retrovirology 6:40. https://doi.org/10.1186/1742-4690-6-40

Vanangamudi M, Nair PC, Engels SEM, Palaniappan S, Namasivayam V (2021) Structural insights to human immunodeficiency virus (HIV-1) targets and their inhibition. Adv Exp Med Biol 1322:63–95. https://doi.org/10.1007/978-981-16-0267-2_3

Virot E, Duclos A, Adelaide L, Miailhes P, Hot A, Ferry T, Seve P (2017) Autoimmune diseases and HIV infection: A cross-sectional study. Medicine 96(4):e5769. https://doi.org/10.1097/MD.0000000000005769

Wait S, Kell E, Hamid S, Muljono DH, Sollano J, Mohamed R, Shah S, Mamun-Al-Mahtab MA, Abbas Z, Johnston J, Tanwandee T, Wallace J (2016) Hepatitis B and hepatitis C in southeast and southern Asia: challenges for governments. Lancet Gastroenterol Hepatol 1(3):248–255. https://doi.org/10.1016/S2468-1253(16)30031-0

Wang HB, Mo QH, Yang Z (2015) HIV vaccine research: The challenge and the way forward. J Immunol Res 2015:503978. https://doi.org/10.1155/2015/503978

Watanaveeradej V, Benenson MW, Souza MD, Sirisopana N, Nitayaphan S, Tontichaivanich C, Amphaipit R, Renzullo PO, Brown AE, McNeil JG, Robb ML, Birx DL, Tovanabutra S, Carr JK, McCutchan FE (2006) Molecular epidemiology of HIV Type 1 in preparation for a Phase III prime-boost vaccine trial in Thailand and a new approach to HIV Type 1 genotyping. AIDS Res Hum Retroviruses 22(8):801–807. https://doi.org/10.1089/aid.2006.22.801

Web link (n.d.-a). https://www.niaid.nih.gov/diseases-conditions/hiv-vaccine-development. Retrieved May 6, 2021

Web link (n.d.-b). https://www.sciencedirect.com/topics/nursing-and-health-professions/hiv-vaccine

Weichseldorfer M, Heredia A, Reitz M, Bryant JL, Latinovic OS (2020) Use of humanized mouse models for studying HIV-1 infection, pathogenesis and persistence. J AIDS HIV Treat 2(1):23–29

Whitney JB, Ruprecht RM (2004) Live attenuated HIV vaccines: pitfalls and prospects. Curr Opin Infect Dis 17(1):17–26. https://doi.org/10.1097/00001432-200402000-00004

Wilkinson J, Cunningham AL (2006) Mucosal transmission of HIV-1: first stop dendritic cells. Curr Drug Targets 7(12):1563–1569. https://doi.org/10.2174/138945006779025482

Yuki Y, Nochi T, Kiyono H (2007) Progress towards an AIDS mucosal vaccine: an overview. Tuberculosis 87(Suppl. 1):S35–S44. https://doi.org/10.1016/j.tube.2007.05.005

Zagury D, Léonard R, Fouchard M, Réveil B, Bernard J, Ittelé D, Cattan A, Zirimwabagabo L, Kalumbu M, Justin W (1987) Immunization against AIDS in humans. Nature 326(6110):249–250. https://doi.org/10.1038/326249a0

Zagury D, Bernard J, Cheynier R, Desportes I, Leonard R, Fouchard M, Reveil B, Ittele D, Lurhuma Z, Mbayo K (1988) A group specific anamnestic immune reaction against HIV-1 induced by a candidate vaccine against AIDS. Nature 332(6166):728–731. https://doi.org/10.1038/332728a0

Zandman-Goddard G, Shoenfeld Y (2002) HIV and autoimmunity. Autoimmun Rev 1(6):329–337. https://doi.org/10.1016/s1568-9972(02)00086-1

Zhao LP, Fiore-Gartland A, Carpp LN, Cohen KW, Rouphael N, Fleurs L, Dintwe O, Zhao M, Moodie Z, Fong Y, Garrett N, Huang Y, Innes C, Janes HE, Lazarus E, Michael NL, Nitayaphan S, Pitisuttithum P, Rerks-Ngarm S, Robb ML, De Rosa SC, Corey L, Gray GE, Seaton KE, Yates NL, MJ ME, Frahm N, Tomaras GD, Gilbert PB (2020) Landscapes of binding antibody and T-cell responses to pox-protein HIV vaccines in Thais and South Africans. PLoS One 15(1):e0226803. https://doi.org/10.1371/journal.pone.0226803

# Neurobiological Mechanisms of Maternal Immune Activation-Induced Neurodevelopmental Disorders: A Serotonin Model

**Janak Raj Awasthi, Toru Takumi, and Priyadarshi S. Sahu**

**Abstract** Serotonin (5-HT) is a significant regulator of neurodevelopment and behavior. Any perturbation in 5-HT system during pre- or early postnatal brain development has been established as a causative factor for the emergence of neurodevelopmental disorders, such as autism spectrum disorders (ASDs) and schizophrenia. Maternal immune activation (MIA) during pregnancy can alter the 5-HT system, leading to emergence of these disorders. The exact molecular mechanism by which MIA induces changes in the 5-HT system to cause neurodevelopmental disorders is not well understood. The SERT gene which codes for the serotonin transporter has been implicated in ASDs. Additionally, transiently SERT-expressing neurons have been identified in various brain areas during development, and their disruption has been linked to derailing of cortical synaptic architectural trajectory, leading to persistent alterations in cortical circuits. The disruption of the 5-HT system and along with transiently SERT-expressing neuronal system, particularly during critical periods of brain development, is a likely cause of diverse autistic features. Understanding the precise mechanisms by which alterations in serotonin

---

Janak Raj Awasthi and Toru Takumi are contributed equally to this work.

---

J. R. Awasthi (✉)
National Institute of Drug Abuse, National Institute of Health, Baltimore, MD, USA

Department of Neuroscience, University of Medicine and Health Sciences, Basseterre, St. Kitts & Nevis

T. Takumi
Department of Physiology and Cell Biology, Kobe University School of Medicine, Kobe, Japan
e-mail: takumit@med.kobe-u.ac.jp

P. S. Sahu
Department of Microbiology & Immunology, School of Medicine, Medical University of the Americas, Nevis, St. Kitts and Nevis
e-mail: p.sahu@mua.edu



A. Das Mohapatra, P. S. Sahu (eds.), *Advances in Immunology and Immuno-techniques*, Interdisciplinary Biotechnological Advances,
https://doi.org/10.1007/978-981-97-5508-0_6

levels lead to ASDs could help in the development of targeted therapies to mitigate the impact of these disorders.

**Keywords** Maternal immune activation · Serotonin · Serotonin transporter · Neurodevelopmental disorders · Autism spectrum disorders

## 1 Introduction

Fetal developmental events that occur in utero have been linked to postnatal health outcomes of the offspring. Neurodevelopment, which encompasses a series of intricately orchestrated processes, including neuron genesis, differentiation, distribution, synapse formation, and apoptosis, renders the developing brain susceptible to external stressors (Bale et al. 2010). Prenatal insults due to maternal infection, deficiency of nutrition, chemical toxicity, and other factors can have prolonged effects on the neurodevelopment and overall health of the offspring. This chapter aims to examine the effect of prenatal maternal immune activation (MIA) on the development of fetal brain and its association with brain disorders that are rooted in the serotonergic system.

## 2 Epidemiological Evidence Linking MIA and Neurodevelopmental Disorders

A large number of population studies have revealed a strong correlation between heightened MIA and disorders in offspring, such as schizophrenia (Canetta et al. 2014; Brown and Derkits 2010; Debost et al. 2017), autism spectrum disorders (ASD) (Brown et al. 2014; Patterson 2011; Jiang et al. 2016; Abdallah et al. 2012; Walder et al. 2014; Beversdorf et al. 2005), bipolar disorder (Parboosing et al. 2013; Oliveira et al. 2017), and cerebral palsy (Antonelli 2015).

The likelihood that newborn may develop these disorders varies depending on the time of exposure, the type of infecting organism, and the strength of mother's immunological response. A Danish study found heightened risk of ASD after a hospitalized case of viral and bacterial illness in the first and second trimester, respectively (Atladóttir et al. 2010). Untreated fever during pregnancy has also been associated with increased ASD risk (Zerbo et al. 2013). A further study discovered that severe maternal illness identified in a hospital was associated with a heightened ASD risk (Zerbo et al. 2015). These data suggest that for a subgroup of pregnant women, infection during pregnancy may raise the risk of ASD or other neurodevelopmental disorders in their newborns (Brown 2012; Brown and Derkits 2010).

Additionally, a California cohort study showed that early pregnancy maternal influenza exposure is linked with a threefold rise in the risk of developing schizophrenia (Brown et al. 2004a). Elevated T. gondii IgG and maternal genital/reproductive infections have also been linked to an increased schizophrenia risk (Brown et al. 2005; Babulas et al. 2006). Maternal exposure to factors such as HSV-2, influenza infection, and neonatal parasitic (T. gondii) and viral infections (cytomegalovirus) have been linked to affective and non-affective psychosis in adult stage including bipolar disorder (Blomström et al. 2016; Canetta et al. 2014; Parboosing et al. 2013; Xiao et al. 2009).

## 3 MIA as Disease Primer

Healthy pregnancies require a well-regulated, non-inflammatory environment free of pathogens to prevent rejection of the maternal-fetal unit (Yang et al. 2019). Maternal inflammatory conditions during pregnancy associated with infection, mental stress, psychopathology, and obesity have been linked to the disruption of the in-utero environment, leading to impaired neurodevelopment and neurological disease (Knuesel et al. 2014).

An intriguing question is why neurodevelopmental disorders only occur in a subset of cases rather than in all pregnant women exposed to infection. Studies show that infection in some subgroup of pregnant women may act as a "disease primer" when paired with genetic and epigenetic variables which may alter the course of embryonic neurodevelopment and increase the chance of developing brain disorders (Meyer 2014; Knuesel et al. 2014). Several other factors like genetic susceptibility, timing and intensity of infection, and exposure to postnatal stressors may also influence the neurodevelopment and disease phenotype that are expressed such as ASD, schizophrenia, or any other disorders (Careaga et al. 2017).

Genome-wide association studies show that genes involved in both immune and neurodevelopmental functions, including those within the major histocompatibility complex locus (Grunze et al. 2009) and complement C4 (Sekar et al. 2016) have a small impact on psychiatric outcomes such as schizophrenia (Schizophrenia Working Group of the Psychiatric Genomics Consortium 2014) and ASD (Weiss et al. 2009), except for rare copy number variants (Takumi and Tamada 2018). However, prenatal immune activation, combined with specific risk genes, can result in unique neurobehavioral characteristics not observed in animals harboring the genetic variation alone (Ayhan et al. 2016).

Therefore, the prenatal immunological mileu can affect early brain development at various levels, ranging from direct disruption of neurogenesis to epigenetic changes that increase vulnerability to future stressors, known as the "second hit" and potentially lead to the emergence of disorders (Boulanger 2009; Bilbo and Schwarz 2012; Giovanoli et al. 2013; Estes and McAllister 2016; Coulthard et al. 2018). These basic-science findings have been translated in a Danish epidemiological study that demonstrated that boys who had both prenatal infection and subsequent psychological traumatic experience in pubertal age had a noticeably increased chance of developing schizophrenia compared to those who experienced either injury alone (Debost et al. 2017).

## 4 Animal Models of MIA

The epidemiological data have been replicated in rodents and primates, such as rhesus monkeys. Animal models provide a valuable platform for understanding the cellular and molecular mechanisms of various neurological disorders, establishing causality, and tracking biological traits across time from fetal development to adulthood that an epidemiological approach cannot achieve.

MIA animal models entail inducing the maternal immunological reaction during fetal development using artificial immunogens such as lipopolysaccharide and polyinosinic: polycytidylic acid that mimic bacterial and viral infection, with a goal of identifying alterations in the offspring's neural and behavioral development (Meyer and Feldon 2012). Prenatal exposure to these immunogens can induce autistic behavioral abnormalities, such as deficits in social communication, restrictions and repetitive behaviors (Choi et al. 2016; Malkova et al. 2012; Schwartzer et al. 2017), and depression-like behaviors (Khan et al. 2014; Ronovsky et al. 2016) in the offspring of the MIA rodent model. These manipulations can also bring about ASD-associated morphological and cellular abnormalities in the brain, such as aberrant development of the cerebellum, decreased reelin expression, and alteration in synaptic density and neural connection (Estes and McAllister 2016; Shi et al. 2009). Notably, some of these findings in rodents have been supported by similar observations in rhesus monkeys regarding both behavior and brain structure (Bauman et al. 2013; Bauman et al. 2014; Willette et al. 2011; Machado et al. 2015; Careaga et al. 2017).

The MIA animal model has also been shown to have neuronal and behavioral anomalies that are similar to those seen in schizophrenia and related psychotic disorders. This includes behavioral, neurochemical, morphological, and gene expression changes associated with schizophrenia following influenza virus exposure and other manifestations such as deficits in sensorimotor gating, attention, working memory, hyper-responsiveness to psychotomimetic drugs, and maturational delay in various other rodent models of MIA (Estes and McAllister 2016; Meyer 2014; Harvey and Boksa 2012; Meyer and Feldon 2010).

## 5 Pathophysiology of MIA-Induced Neurodevelopmental Disorders

Epidemiological studies revealed a correlation between maternal infections with pathogens (such as influenza, herpes, T. gondii, rubella, and bacteria) and an elevated level of inflammatory biomarkers (such as cytokines and C-reactive protein) in the development of schizophrenia, ASDs, and bipolar disorder (Canetta et al. 2014; Brown and Derkits 2010; Brown et al. 2004b; Brown and Meyer 2018). Cytokines are critical for normal fetal brain developmental processes such as neuron proliferation and synaptogenesis (Burns et al. 1993), but their high levels can disrupt these processes and lead to abnormal neurodevelopment (Gayle et al. 2004).

An inflammatory mediator, interleukin-6 (IL-6), can penetrate the placental barrier and trigger the additional immunological reactions in the embryonic compartment (Zaretsky et al. 2004; Hauguel-de Mouzon and Guerre-Millo 2006) leading to abnormal fetal brain development, including changes in white matter microstructure and abnormality in sensory processing and impulse control (Rasmussen et al. 2019; Graham et al. 2018). In addition, disproportionate expression of other inflammatory factors, such as IL-1α, IL-10, tumor necrosis factor α (TNFα), and C-reactive protein, may also trigger offspring's psychopathologies (Smith et al. 2007; Girard et al. 2010; Canetta et al. 2014).

## 6 MIA and ASDs

Recent research has focused on the potential link between pregnancy-related maternal illness or inflammation and the development of ASDs (Estes and McAllister 2015; Patterson 2011). This is because signaling molecules in immune system play a crucial role in neurodevelopment, and its disruptions through infection in the feto-maternal unit, may impair neural development and increase the risk of brain disorders (Deverman and Patterson 2009; Garay 2010; Estes and McAllister 2016). One key unanswered question is: what could be the most plausible pathway that accounts for the diverse clinical manifestations of ASDs?

## 7 The Disruption of Synapse Excitation-Inhibition (E/I) Balance in ASDs

A high ratio of excitation to inhibition, causing excessive excitement in cortical circuits, has been proposed as a primary contributor to the hallmark symptoms of ASD (Rubenstein and Merzenich 2003). The E/I imbalance at the synaptic or circuit level has been replicated in the animal model of ASD (Nakatani et al. 2009; Tamada et al. 2021; Nelson and Valakh 2015; Lee et al. 2017). The high ASD-epilepsy comorbidity also supports it due to synchronized neuronal overactivity (Canitano 2007; Tuchman et al. 2010). Experiments in mouse have shown that optogenetic stimulation of glutamatergic neurons in the medial prefrontal cortical area impaired its social behavior (Yizhar et al. 2011). Altered social interaction linked to increased NMDA activity in the hippocampus was observed in mice with a deficit in the excitatory signaling-related IRSp53 gene (Chung et al. 2015a; b). The next big question is: how does MIA result in an E/I imbalance in the brain?

# 8 Microglia and ASDs

Evidence shows that microglia are activated in various brain areas involved in autistic behavior, such as the prefrontal cortex, anterior cingulate cortex, and cerebellum (Suzuki et al. 2013). Their density is reported to increase in the dorsolateral prefrontal cortex, fronto-insular cortex, and visual cortex of individuals with autism (Morgan et al. 2010; Tetreault et al. 2012). Heightened microglia–neuron interaction in the dorsolateral prefrontal cortex has been noted to occur from very young age in people with autism (Morgan et al. 2012). Elevated levels of proinflammatory chemokines released by activated microglia, such as MCP-1 (monocyte chemoattractant protein-1), have been found in the cerebrospinal fluid of young autism patients (Vargas et al. 2005). These findings suggest the implication of microglia in ASD, but the exact mechanism by which they contribute to the pathology of synaptic E/I imbalance remains largely unknown.

Some studies have shed initial light on the potential role of microglia in disorder. Through their phagocytic activity, microglia play an important role in synaptic pruning during brain development, shaping both the synaptic and circuitry of the brain. Microglia also release inflammatory molecules such as interleukins and TNF-α, which regulate synaptic maturation and plasticity (Paolicelli et al. 2011; Schafer et al. 2012). The formation, pruning, and maturation of neuronal synapses and circuitry are crucial for the development of a balanced E/I ratio in the brain (Rubenstein and Merzenich 2003). Thus, any factors causing microglial activation or abnormalities can impair synaptic pruning, maturation, and plasticity, leading to neurodevelopmental disorders.

Prenatal exposure to maternal inflammatory reactions results in the activation of excitatory neurons and/or loss of inhibitory neurons, causing abnormal behavior in offspring (Shin Yim et al. 2017). Moreover, an MIA-activated state of microglia in offspring can cause heightened inflammation during postnatal periods, potentially contributing to the development of ASD (Knuesel et al. 2014). Microglia have been linked to neuroinflammation in patients with both ASD and schizophrenia (Takahashi and Sakurai 2013; Chung et al. 2015a; b). A harmful variant of CX3CR1, a chemokine receptor expressed in brain microglia that regulates microglial recruitment to sites of neuroinflammation, is linked with a heightened risk for both disorders (Ishizuka et al. 2017). Transcriptomics has revealed molecular abnormalities related to the immune system and glia in ASD (Ansel et al. 2016; Reilly et al. 2017).

Finally, if an increased synapse E/I ratio in the ASD brain is caused by impaired microglia-mediated synaptic pruning, maturation, and plasticity, what mechanisms are involved in linking the MIA, microglia, and the brain neurotransmitter system? Here, we illustrate this using the serotonin model.

## 9 MIA, Serotonin, and ASDs

Serotonin (5-hydroxytryptamine; 5-HT) plays a vital role in regulating cortical plasticity in adults and in promoting various developmental processes such as neurogenesis, migration, neuronal differentiation, and apoptosis during early brain development (Jenkins et al. 2016). 5-HT is released by volume transmission, and its function is concluded by reuptake into the presynaptic neuron by the serotonin transporter (SERT). SERT is also found on cells in the peripheral tissue, such as in the enteric nervous system, platelets, mast cells, and the placenta (Rudnick 2007). Any transient alterations in serotonin levels have the potential to permanently impact the intricate brain connections and adult behavior (Gaspar et al. 2003).

MIA causes a decrease in levels of cerebral 5-HT, tph2 (5-HT producing enzyme), and slc6a4 (5-HT transporter) genes and 5-HT neurons in the brainstem raphe nuclei, causing anxious behaviors in offspring during adulthood (Hsueh et al. 2017). However, it leads to an increase in serotonin 5-HT(2A) and metabotropic glutamate 2 (mGlu2) receptors in the fetal frontal cortex, a key area for perception, cognition, and mood, causing behavior changes consistent with schizophrenia (Holloway et al. 2013). It can epigenetically alter the expression of the SERT gene in the hippocampus and contribute to the development of major depressive disorder (Reisinger et al. 2016). MIA-induced upregulation of SERT activity in the frontal cortex has been noted to elicit short-lasting depression-like behavior even in adulthood (Schwamborn et al. 2016).

The exact molecular mechanism by which MIA induces changes in the 5-HT system is largely unexplored. Virtually all immune cells exhibit the presence of at least one component related to serotonin (Herr et al. 2017). MIA induces specific long-term alterations in 5-HT activity through the alternative activation of the oxidative kynurenine pathway, resulting in the formation of neurotoxic compounds and compromising 5-HT synthesis (MacDowell et al. 2021). Imbalance in the 5-HT-SERT neural pathway has been noted to occur in cases of exposure to stressors such as disrupted maternal diet and air pollution during pregnancy, resulting in the postnatal appearance of neurodevelopmental disorders (Ye et al. 2021).

Prenatal stress is a risk factor for ASD, but it does not necessarily cause ASD to manifest itself, implying that factors such as the maternal genotype of the promoter region of the serotonin transporter gene (5-HTTLPR) affect stress resistance and the risk of ASD when exposed to prenatal stress (Hecht et al. 2016; Lister 2021). Studies have demonstrated that the combination of the 5-HTTLPR polymorphism of the SERT gene and prenatal maternal stress related to the disaster has been involved in the development of ASD (Laplante et al. 2019). The interaction between the SERT gene and the quality of maternal care during development can also influence social behavior in juveniles (Madrid et al. 2018). This shows the susceptibility of SERT activity to prenatal stress. If so, what could be the most plausible pathway for the diverse clinical manifestations of ASD?

## 10 The Serotonin Model of MIA-Induced ASDs

The disruption in the serotonin system in conjunction with a transiently SERT-expressing neuronal system during brain development is the most likely cause of diverse autistic features.

The serotonin neurons are widely projected in the whole brain (Awasthi et al. 2020a; b), and their development continues even postnatally (Awasthi et al. 2020a, b). Transiently SERT-expressing neurons that use glutamate as a neurotransmitter are found in brain areas such as sensory relay nuclei in the thalamus, lateral and medial geniculate nuclei, some hypothalamic nuclei, the pituitary gland, the hippocampus, the septum, the amygdala, olfactory areas, cortical areas such as somatosensory, visual, auditory, prefrontal, cingulate cortices and in the retina, sensory cranial nerve nuclei, superior olivary nucleus, and cochlear nucleus between embryonic (E) day 15 and postnatal (P) third week with variable expression timing among different brain areas (D'Amato et al. 1987; Lebrand et al. 1996, 1998; Brüning and Liangos 1997; Hansson et al. 1998; Narboux-Nême et al. 2008). In our unpublished data, we noted the presence of transiently SERT-expressing neurons in brain areas that could be responsible for all autistic behavior, including auditory and visual hypersensitivity in case of their altered circuitry.

The internalized serotonin by the transient-SERT-glutamatergic neurons may exert a trophic influence on the development of these neurons or may facilitate neurodevelopment by uptaking excessive extrasynaptic serotonin, preventing excessive 5-HT signals at extrasynaptic targets (Gaspar et al. 2003; Jafari et al. 2011). Moreover, the transient SERT expression reflects critically important periods in brain development. Perturbation of these transient SERT expressions in developing thalamocortical neurons has shown derailment of cortical synaptic architectural trajectory that persists into adulthood (Chen et al. 2015; De Gregorio et al. 2020), suggesting that these neurons could be responsible for the sensory and cognitive abnormalities linked to neurodevelopmental disorders.

Inflammatory cytokines such as IL-1β can enhance SERT activity via a p38α MAPK-dependent pathway (Stilley and Blakely 2021) which suggests that MIA can cause the functional abnormality associated with transiently SERT-expressing glutamatergic neurons. This may disrupt neurodevelopment in many ways. 1) Increased SERT activity may result in less availability of extrasynaptic serotonin, which can lessen the trophic effect of 5-HT in the development of neuronal circuitry. (2) Enhanced SERT activity in transiently SERT-expressing glutamatergic neurons may make them more resilient to microglial-mediated synaptic pruning due to the trophic effect of absorbed serotonin. (3) The MIA-induced abnormalities in microglia may add to impaired synaptic pruning in the brain. This may explain the pathophysiology behind increased SERT expression in children with autism (Abdulamir et al. 2018), an increased ratio between excitation and inhibition leading to autistic behavior (Rubenstein and Merzenich 2003), and behavioral differences in individuals with ASD. It is summarized in Fig. 1. Further understanding of

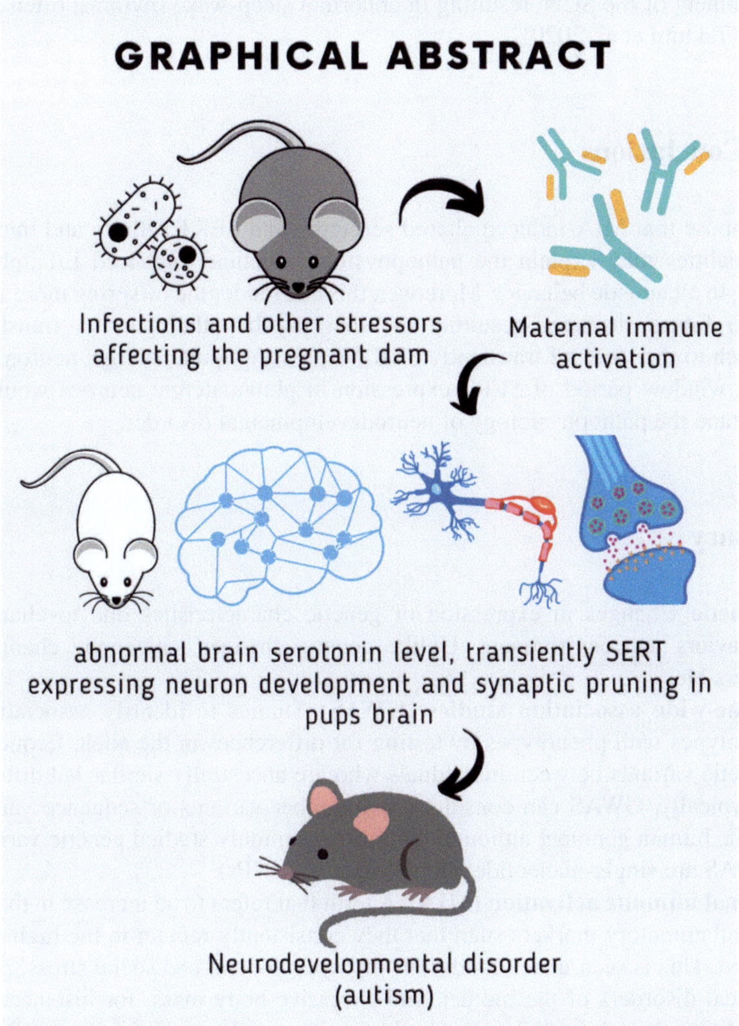


**Fig. 1** Hypothesized model for maternal immune activation mediated autism pathophysiology based on serotonin and serotonin transporter (SERT) system

this critical window of transiently SERT-expressing-glutamatergic neurons using the MIA model is likely promising in unraveling the etiology of ASD.

The serotonergic neurons also strongly project the suprachiasmatic nucleus (SCN), the mammalian circadian center (Awasthi et al. 2020a, b). Our data show that adequate 5-HT innervation in the SCN occurred during the postnatal second week in mice, and that should have a significant role in the development of SCN physiology. MIA may affect this 5-HTergic innervation and cause improper development of the SCN, resulting in abnormal sleep-wake rhythms, often seen in ASDs (Takumi et al. 2020).

## 11 Conclusions

We propose that MIA-induced altered serotonin and SERT activity and microglial abnormalities may explain the pathophysiology behind enhanced E/I imbalance leading to all autistic behavior. Moreover, this can render the offspring more susceptible to future stressors, resulting in other psychopathologies. A translational approach to the study of transiently SERT-expressing glutamatergic neurons and a critical window period of SERT expression in glutamatergic neurons would help understand the pathophysiology of neurodevelopmental disorder.

## Glossary

**Epigenetic** Changes in expression of genetic characteristics due to changes in behaviors and environment. Unlike genetic changes, epigenetic changes are reversible.

**Genome-wide association studies (GWAS)** Studies to identify associations of genotypes with phenotypes by testing for differences in the allele frequency of genetic variants between individuals who are ancestrally similar but differ phenotypically. GWAS can consider copy-number variants or sequence variations in the human genome, although the most commonly studied genetic variants in GWAS are single-nucleotide polymorphisms (SNPs).

**Maternal immune activation (MIA)** A term that refers to an increase in the levels of inflammatory markers such that they consistently remain in the high normal range. This is seen under conditions of psychological and social stress, psychological disorders of the mother, and excessive body mass, for instance. These conditions have been linked to a higher incidence of impaired neurodevelopment and mental illness in the offspring over time.

**Neurodevelopmental disorders** A group of conditions in which the growth and development of the brain are affected.

**Prenatal** Occurring or existing before birth. Prenatal care is the regular health care women should receive from an obstetrician or midwife during pregnancy.

**Serotonin** A neurotransmitter that carries messages between nerve cells in the brain (central nervous system) and throughout the human body (peripheral nervous system).

**Synapse** The points of contact between neurons (nerve cells) where information is passed from one neuron to the next. Synapses most often form between axons and dendrites and consist of a presynaptic neuron, synaptic cleft, and a postsynaptic neuron.

## References


Abdallah MW, Larsen N, Grove J, Nørgaard-Pedersen B, Thorsen P, Mortensen EL, Hougaard DM (2012) Amniotic fluid chemokines and autism spectrum disorders: an exploratory study utilizing a Danish Historic Birth Cohort. Brain Behav Immun 26:170–176. https://doi.org/10.1016/j.bbi.2011.09.003

Abdulamir HA, Abdul-Rasheed OF, Abdulghani EA (2018) Serotonin and serotonin transporter levels in autistic children. Saudi Med J 39:487–494. https://doi.org/10.15537/smj.2018.5.21751

Ansel A, Rosenzweig JP, Zisman PD, Melamed M, Gesundheit B (2016) Variation in gene expression in autism spectrum disorders: an extensive review of transcriptomic studies. Front Neurosci 10:601. https://doi.org/10.3389/fnins.2016.00601

Antonelli MC (ed) (2015) Perinatal programming of neurodevelopment, advances in neurobiology. Springer, New York. https://doi.org/10.1007/978-1-4939-1372-5

Atladóttir HÓ, Thorsen P, Østergaard L, Schendel DE, Lemcke S, Abdallah M, Parner ET (2010) Maternal infection requiring hospitalization during pregnancy and autism spectrum disorders. J Autism Dev Disord 40:1423–1430. https://doi.org/10.1007/s10803-010-1006-y

Awasthi JR, Tamada K, Overton ETN, Takumi T (2020a) Comprehensive topographical map of the serotonergic fibers in the male mouse brain. J Comp Neurol. https://doi.org/10.1002/cne.25027

Awasthi JR, Tamada K, Overton ETN, Takumi T (2020b) Development of serotonergic projections to the suprachiasmatic nucleus in the mouse brain. Neurosci Lett 739:135438. https://doi.org/10.1016/j.neulet.2020.135438

Ayhan Y, McFarland R, Pletnikov MV (2016) Animal models of gene-environment interaction in schizophrenia: a dimensional perspective. Prog Neurobiol 136:1–27. https://doi.org/10.1016/j.pneurobio.2015.10.002

Babulas V, Factor-Litvak P, Goetz R, Schaefer CA, Brown AS (2006) Prenatal exposure to maternal genital and reproductive infections and adult schizophrenia. Am J Psychiatry 163:927–929. https://doi.org/10.1176/ajp.2006.163.5.927

Bale TL, Baram TZ, Brown AS, Goldstein JM, Insel TR, McCarthy MM, Nemeroff CB, Reyes TM, Simerly RB, Susser ES, Nestler EJ (2010) Early life programming and neurodevelopmental disorders. Biol Psychiatry 68:314–319. https://doi.org/10.1016/j.biopsych.2010.05.028

Bauman MD, Iosif A-M, Ashwood P, Braunschweig D, Lee A, Schumann CM, Van de Water J, Amaral DG (2013) Maternal antibodies from mothers of children with autism alter brain growth and social behavior development in the rhesus monkey. Transl Psychiatry 3:e278–e278. https://doi.org/10.1038/tp.2013.47

Bauman MD, Iosif A-M, Smith SEP, Bregere C, Amaral DG, Patterson PH (2014) Activation of the maternal immune system during pregnancy alters behavioral development of rhesus monkey offspring. Biol Psychiatry 75:332–341. https://doi.org/10.1016/j.biopsych.2013.06.025

Beversdorf DQ, Manning SE, Hillier A, Anderson SL, Nordgren RE, Walters SE, Nagaraja HN, Cooley WC, Gaelic SE, Bauman ML (2005) Timing of prenatal stressors and autism. J Autism Dev Disord 35:471–478. https://doi.org/10.1007/s10803-005-5037-8

Bilbo SD, Schwarz JM (2012) The immune system and developmental programming of brain and behavior. Front Neuroendocrinol 33:267–286. https://doi.org/10.1016/j.yfrne.2012.08.006

Blomström Å, Karlsson H, Gardner R, Jörgensen L, Magnusson C, Dalman C (2016) Associations between maternal infection during pregnancy, childhood infections, and the risk of subsequent psychotic disorder - a Swedish cohort study of nearly 2 million individuals. Schizophr Bull 42:125–133. https://doi.org/10.1093/schbul/sbv112

Boulanger LM (2009) Immune proteins in brain development and synaptic plasticity. Neuron 64:93–109. https://doi.org/10.1016/j.neuron.2009.09.001

Brown AS (2012) Epidemiologic studies of exposure to prenatal infection and risk of schizophrenia and autism. Dev Neurobiol 72:1272–1276. https://doi.org/10.1002/dneu.22024

Brown AS, Derkits EJ (2010) Prenatal infection and schizophrenia: a review of epidemiologic and translational studies. Am J Psychiatry 167:261–280. https://doi.org/10.1176/appi.ajp.2009.09030361

Brown AS, Meyer U (2018) Maternal immune activation and neuropsychiatric illness: a translational research perspective. Am J Psychiatry 175:1073–1083. https://doi.org/10.1176/appi.ajp.2018.17121311

Brown AS, Begg MD, Gravenstein S, Schaefer CA, Wyatt RJ, Bresnahan M, Babulas VP, Susser ES (2004a) Serologic evidence of prenatal influenza in the etiology of schizophrenia. Arch Gen Psychiatry 61:774. https://doi.org/10.1001/archpsyc.61.8.774

Brown AS, Hooton J, Schaefer CA, Zhang H, Petkova E, Babulas V, Perrin M, Gorman JM, Susser ES (2004b) Elevated maternal interleukin-8 levels and risk of schizophrenia in adult offspring. Am J Psychiatry 161:889–895. https://doi.org/10.1176/appi.ajp.161.5.889

Brown AS, Schaefer CA, Quesenberry CP, Liu L, Babulas VP, Susser ES (2005) Maternal exposure to toxoplasmosis and risk of schizophrenia in adult offspring. Am J Psychiatry 162:767–773. https://doi.org/10.1176/appi.ajp.162.4.767

Brown AS, Sourander A, Hinkka-Yli-Salomäki S, McKeague IW, Sundvall J, Surcel H-M (2014) Elevated maternal C-reactive protein and autism in a national birth cohort. Mol Psychiatry 19:259–264. https://doi.org/10.1038/mp.2012.197

Brüning G, Liangos O (1997) Transient expression of the serotonin transporter in the developing mouse thalamocortical system. Acta Histochem 99:117–121. https://doi.org/10.1016/S0065-1281(97)80016-5

Burns TM, Clough JA, Klein RM, Wood GW, Berman NE (1993) Developmental regulation of cytokine expression in the mouse brain. Growth Factors 9:253–258. https://doi.org/10.3109/08977199308991585

Canetta S, Sourander A, Surcel H-M, Hinkka-Yli-Salomäki S, Leiviskä J, Kellendonk C, McKeague IW, Brown AS (2014) Elevated maternal C-reactive protein and increased risk of schizophrenia in a national birth cohort. Am J Psychiatry 171:960–968. https://doi.org/10.1176/appi.ajp.2014.13121579

Canitano R (2007) Epilepsy in autism spectrum disorders. Eur Child Adolesc Psychiatry 16:61–66. https://doi.org/10.1007/s00787-006-0563-2

Careaga M, Murai T, Bauman MD (2017) Maternal immune activation and autism spectrum disorder: from rodents to nonhuman and human primates. Biol Psychiatry 81:391–401. https://doi.org/10.1016/j.biopsych.2016.10.020

Chen X, Ye R, Gargus JJ, Blakely RD, Dobrenis K, Sze JY (2015) Disruption of transient serotonin accumulation by non-serotonin-producing neurons impairs cortical map development. Cell Rep 10:346–358. https://doi.org/10.1016/j.celrep.2014.12.033

Choi GB, Yim YS, Wong H, Kim S, Kim H, Kim SV, Hoeffer CA, Littman DR, Huh JR (2016) The maternal interleukin-17a pathway in mice promotes autism-like phenotypes in offspring. Science 351:933–939. https://doi.org/10.1126/science.aad0314

Chung W, Choi SY, Lee E, Park H, Kang J, Park H, Choi Y, Lee D, Park S-G, Kim R, Cho YS, Choi J, Kim M-H, Lee JW, Lee S, Rhim I, Jung MW, Kim D, Bae YC, Kim E (2015a) Social deficits in IRSp53 mutant mice improved by NMDAR and mGluR5 suppression. Nat Neurosci 18:435–443. https://doi.org/10.1038/nn.3927

Chung W-S, Welsh CA, Barres BA, Stevens B (2015b) Do glia drive synaptic and cognitive impairment in disease? Nat Neurosci 18:1539–1545. https://doi.org/10.1038/nn.4142

Coulthard LG, Hawksworth OA, Woodruff TM (2018) Complement: the emerging architect of the developing brain. Trends Neurosci 41:373–384. https://doi.org/10.1016/j.tins.2018.03.009

D'Amato RJ, Blue ME, Largent BL, Lynch DR, Ledbetter DJ, Molliver ME, Snyder SH (1987) Ontogeny of the serotonergic projection to rat neocortex: transient expression of a dense innervation to primary sensory areas. Proc Natl Acad Sci 84:4322–4326. https://doi.org/10.1073/pnas.84.12.4322

De Gregorio R, Chen X, Petit EI, Dobrenis K, Sze JY (2020) Disruption of transient SERT expression in thalamic glutamatergic neurons alters trajectory of postnatal interneuron development in the mouse cortex. Cereb Cortex 30:1623–1636. https://doi.org/10.1093/cercor/bhz191

Debost J-CPG, Larsen JT, Munk-Olsen T, Mortensen PB, Meyer U, Petersen L (2017) Joint effects of exposure to prenatal infection and peripubertal psychological trauma in schizophrenia. Schizophr Bull 43:171–179. https://doi.org/10.1093/schbul/sbw083

Deverman BE, Patterson PH (2009) Cytokines and CNS development. Neuron 64:61–78. https://doi.org/10.1016/j.neuron.2009.09.002

Estes ML, McAllister AK (2015) Immune mediators in the brain and peripheral tissues in autism spectrum disorder. Nat Rev Neurosci 16:469–486. https://doi.org/10.1038/nrn3978

Estes ML, McAllister AK (2016) Maternal immune activation: implications for neuropsychiatric disorders. Science 353:772–777. https://doi.org/10.1126/science.aag3194

Garay P (2010) Novel roles for immune molecules in neural development: implications for neurodevelopmental disorders. Front Synaptic Neurosci 2. https://doi.org/10.3389/fnsyn.2010.00136

Gaspar P, Cases O, Maroteaux L (2003) The developmental role of serotonin: news from mouse molecular genetics. Nat Rev Neurosci 4:1002–1012. https://doi.org/10.1038/nrn1256

Gayle DA, Beloosesky R, Desai M, Amidi F, Nuñez SE, Ross MG (2004) Maternal LPS induces cytokines in the amniotic fluid and corticotropin releasing hormone in the fetal rat brain. Am J Physiol Regul Integr Comp Physiol 286:R1024–R1029. https://doi.org/10.1152/ajpregu.00664.2003

Giovanoli S, Engler H, Engler A, Richetto J, Voget M, Willi R, Winter C, Riva MA, Mortensen PB, Feldon J, Schedlowski M, Meyer U (2013) Stress in puberty unmasks latent neuropathological consequences of prenatal immune activation in mice. Science 339:1095–1099. https://doi.org/10.1126/science.1228261

Girard S, Tremblay L, Lepage M, Sébire G (2010) IL-1 receptor antagonist protects against placental and neurodevelopmental defects induced by maternal inflammation. J Immunol 184:3997–4005. https://doi.org/10.4049/jimmunol.0903349

Graham AM, Rasmussen JM, Rudolph MD, Heim CM, Gilmore JH, Styner M, Potkin SG, Entringer S, Wadhwa PD, Fair DA, Buss C (2018) Maternal systemic interleukin-6 during pregnancy is associated with newborn amygdala phenotypes and subsequent behavior at 2 years of age. Biol Psychiatry 83:109–119. https://doi.org/10.1016/j.biopsych.2017.05.027

Grunze H, Vieta E, Goodwin GM, Bowden C, Licht RW, Moller H-J, Kasper S (2009) The World Federation of Societies of Biological Psychiatry (WFSBP) guidelines for the biological treatment of bipolar disorders: update 2009 on the treatment of acute mania. World J Biol Psychiatry 10:85–116. https://doi.org/10.1080/15622970902823202

Hansson SR, Mezey E, Hoffman BJ (1998) Serotonin transporter messenger RNA in the developing rat brain: early expression in serotonergic neurons and transient expression in non-serotonergic neurons. Neuroscience 83:1185–1201. https://doi.org/10.1016/s0306-4522(97)00444-2

Harvey L, Boksa P (2012) Prenatal and postnatal animal models of immune activation: relevance to a range of neurodevelopmental disorders. Dev Neurobiol 72:1335–1348. https://doi.org/10.1002/dneu.22043

Hauguel-de Mouzon S, Guerre-Millo M (2006) The placenta cytokine network and inflammatory signals. Placenta 27:794–798. https://doi.org/10.1016/j.placenta.2005.08.009

Hecht PM, Hudson M, Connors SL, Tilley MR, Liu X, Beversdorf DQ (2016) Maternal serotonin transporter genotype affects risk for ASD with exposure to prenatal stress. Autism Res 9:1151–1160. https://doi.org/10.1002/aur.1629

Herr N, Bode C, Duerschmied D (2017) The effects of serotonin in immune cells. Front Cardiovasc Med 4:48. https://doi.org/10.3389/fcvm.2017.00048

Holloway T, Moreno JL, Umali A, Rayannavar V, Hodes GE, Russo SJ, González-Maeso J (2013) Prenatal stress induces schizophrenia-like alterations of serotonin 2A and metabotropic glutamate 2 receptors in the adult offspring: role of maternal immune system. J Neurosci 33:1088–1098. https://doi.org/10.1523/JNEUROSCI.2331-12.2013

Hsueh P-T, Wang H-H, Liu C-L, Ni W-F, Chen Y-L, Liu J-K (2017) Expression of cerebral serotonin related to anxiety-like behaviors in C57BL/6 offspring induced by repeated subcutaneous prenatal exposure to low-dose lipopolysaccharide. PLoS One 12:e0179970. https://doi.org/10.1371/journal.pone.0179970

Ishizuka K, Fujita Y, Kawabata T, Kimura H, Iwayama Y, Inada T, Okahisa Y, Egawa J, Usami M, Kushima I, Uno Y, Okada T, Ikeda M, Aleksic B, Mori D, Someya T, Yoshikawa T, Iwata N, Nakamura H, Yamashita T, Ozaki N (2017) Rare genetic variants in CX3CR1 and their contribution to the increased risk of schizophrenia and autism spectrum disorders. Transl Psychiatry 7:e1184. https://doi.org/10.1038/tp.2017.173

Jafari G, Xie Y, Kullyev A, Liang B, Sze JY (2011) Regulation of extrasynaptic 5-HT by serotonin reuptake transporter function in 5-HT-absorbing neurons underscores adaptation behavior in Caenorhabditis elegans. J Neurosci 31:8948–8957. https://doi.org/10.1523/JNEUROSCI.1692-11.2011

Jenkins TA, Nguyen JCD, Polglaze KE, Bertrand PP (2016) Influence of tryptophan and serotonin on mood and cognition with a possible role of the gut-brain axis. Nutrients 8:56. https://doi.org/10.3390/nu8010056

Jiang H, Xu L, Shao L, Xia R, Yu Z, Ling Z, Yang F, Deng M, Ruan B (2016) Maternal infection during pregnancy and risk of autism spectrum disorders: a systematic review and meta-analysis. Brain Behav Immun 58:165–172. https://doi.org/10.1016/j.bbi.2016.06.005

Khan D, Fernando P, Cicvaric A, Berger A, Pollak A, Monje FJ, Pollak DD (2014) Long-term effects of maternal immune activation on depression-like behavior in the mouse. Transl Psychiatry 4:e363–e363. https://doi.org/10.1038/tp.2013.132

Knuesel I, Chicha L, Britschgi M, Schobel SA, Bodmer M, Hellings JA, Toovey S, Prinssen EP (2014) Maternal immune activation and abnormal brain development across CNS disorders. Nat Rev Neurol 10:643–660. https://doi.org/10.1038/nrneurol.2014.187

Laplante DP, Simcock G, Cao-Lei L, Mouallem M, Elgbeili G, Brunet A, Cobham V, Kildea S, King S (2019) The 5-HTTLPR polymorphism of the serotonin transporter gene and child's sex moderate the relationship between disaster-related prenatal maternal stress and autism spectrum disorder traits: The QF2011 Queensland flood study. Dev Psychopathol 31:1395–1409. https://doi.org/10.1017/S0954579418000871

Lebrand C, Cases O, Adelbrecht C, Doye A, Alvarez C, El Mestikawy S, Seif I, Gaspar P (1996) Transient uptake and storage of serotonin in developing thalamic neurons. Neuron 17:823–835. https://doi.org/10.1016/S0896-6273(00)80215-9

Lebrand C, Cases O, Wehrlé R, Blakely RD, Edwards RH, Gaspar P (1998) Transient developmental expression of monoamine transporters in the rodent forebrain. J Comp Neurol 401:506–524

Lee E, Lee J, Kim E (2017) Excitation/inhibition imbalance in animal models of autism spectrum disorders. Biol Psychiatry 81:838–847. https://doi.org/10.1016/j.biopsych.2016.05.011

Lister A (2021) Investigating the effect of reduced serotonin transporter expression in a maternal immune activation model of neurodevelopmental disorders. Open Access Victoria University of Wellington | Te Herenga Waka. https://doi.org/10.26686/wgtn.16613965.v1

MacDowell KS, Munarriz-Cuezva E, Meana JJ, Leza JC, Ortega JE (2021) Paliperidone reversion of maternal immune activation-induced changes on brain serotonin and kynurenine pathways. Front Pharmacol 12:682602. https://doi.org/10.3389/fphar.2021.682602

Machado CJ, Whitaker AM, Smith SEP, Patterson PH, Bauman MD (2015) Maternal immune activation in nonhuman primates alters social attention in juvenile offspring. Biol Psychiatry 77:823–832. https://doi.org/10.1016/j.biopsych.2014.07.035

Madrid JE, Mandalaywala TM, Coyne SP, Ahloy-Dallaire J, Garner JP, Barr CS, Maestripieri D, Parker KJ (2018) Adaptive developmental plasticity in rhesus macaques: the serotonin transporter gene interacts with maternal care to affect juvenile social behaviour. Proc Biol Sci 285:20180541. https://doi.org/10.1098/rspb.2018.0541

Malkova NV, Yu CZ, Hsiao EY, Moore MJ, Patterson PH (2012) Maternal immune activation yields offspring displaying mouse versions of the three core symptoms of autism. Brain Behav Immun 26:607–616. https://doi.org/10.1016/j.bbi.2012.01.011

Meyer U (2014) Prenatal poly(I:C) exposure and other developmental immune activation models in rodent systems. Biol Psychiatry 75:307–315. https://doi.org/10.1016/j.biopsych.2013.07.011

Meyer U, Feldon J (2010) Epidemiology-driven neurodevelopmental animal models of schizophrenia. Prog Neurobiol 90:285–326. https://doi.org/10.1016/j.pneurobio.2009.10.018

Meyer U, Feldon J (2012) To poly(I:C) or not to poly(I:C): advancing preclinical schizophrenia research through the use of prenatal immune activation models. Neuropharmacology 62:1308–1321. https://doi.org/10.1016/j.neuropharm.2011.01.009

Morgan JT, Chana G, Pardo CA, Achim C, Semendeferi K, Buckwalter J, Courchesne E, Everall IP (2010) Microglial activation and increased microglial density observed in the dorsolateral prefrontal cortex in autism. Biol Psychiatry 68:368–376. https://doi.org/10.1016/j.biopsych.2010.05.024

Morgan JT, Chana G, Abramson I, Semendeferi K, Courchesne E, Everall IP (2012) Abnormal microglial–neuronal spatial organization in the dorsolateral prefrontal cortex in autism. Brain Res 1456:72–81. https://doi.org/10.1016/j.brainres.2012.03.036

Nakatani J, Tamada K, Hatanaka F, Ise S, Ohta H, Inoue K, Tomonaga S, Watanabe Y, Chung YJ, Banerjee R, Iwamoto K, Kato T, Okazawa M, Yamauchi K, Tanda K, Takao K, Miyakawa T, Bradley A, Takumi T (2009) Abnormal behavior in a chromosome- engineered mouse model for human 15q11-13 duplication seen in autism. Cell 137:1235–1246. https://doi.org/10.1016/j.cell.2009.04.024

Narboux-Nême N, Pavone LM, Avallone L, Zhuang X, Gaspar P (2008) Serotonin transporter transgenic (SERTcre) mouse line reveals developmental targets of serotonin specific reuptake inhibitors (SSRIs). Neuropharmacology 55:994–1005. https://doi.org/10.1016/j.neuropharm.2008.08.020

Nelson SB, Valakh V (2015) Excitatory/inhibitory balance and circuit homeostasis in autism spectrum disorders. Neuron 87:684–698. https://doi.org/10.1016/j.neuron.2015.07.033

Oliveira J, Oliveira-Maia AJ, Tamouza R, Brown AS, Leboyer M (2017) Infectious and immunogenetic factors in bipolar disorder. Acta Psychiatr Scand 136:409–423. https://doi.org/10.1111/acps.12791

Paolicelli RC, Bolasco G, Pagani F, Maggi L, Scianni M, Panzanelli P, Giustetto M, Ferreira TA, Guiducci E, Dumas L, Ragozzino D, Gross CT (2011) Synaptic pruning by microglia is necessary for normal brain development. Science 333:1456–1458. https://doi.org/10.1126/science.1202529

Parboosing R, Bao Y, Shen L, Schaefer CA, Brown AS (2013) Gestational influenza and bipolar disorder in adult offspring. JAMA Psychiatry 70:677. https://doi.org/10.1001/jamapsychiatry.2013.896

Patterson PH (2011) Maternal infection and immune involvement in autism. Trends Mol Med 17:389–394. https://doi.org/10.1016/j.molmed.2011.03.001

Rasmussen JM, Graham AM, Entringer S, Gilmore JH, Styner M, Fair DA, Wadhwa PD, Buss C (2019) Maternal Interleukin-6 concentration during pregnancy is associated with variation in frontolimbic white matter and cognitive development in early life. NeuroImage 185:825–835. https://doi.org/10.1016/j.neuroimage.2018.04.020

Reilly J, Gallagher L, Chen JL, Leader G, Shen S (2017) Bio-collections in autism research. Mol Autism 8:34. https://doi.org/10.1186/s13229-017-0154-8

Reisinger SN, Kong E, Khan D, Schulz S, Ronovsky M, Berger S, Horvath O, Cabatic M, Berger A, Pollak DD (2016) Maternal immune activation epigenetically regulates hippocampal serotonin transporter levels. Neurobiol Stress 4:34–43. https://doi.org/10.1016/j.ynstr.2016.02.007

Ronovsky M, Berger S, Molz B, Berger A, Pollak DD (2016) Animal models of maternal immune activation in depression research. Curr Neuropharmacol 14:688–704. https://doi.org/10.2174/1570159x14666151215095359

Rubenstein JLR, Merzenich MM (2003) Model of autism: increased ratio of excitation/inhibition in key neural systems. Genes Brain Behav 2:255–267. https://doi.org/10.1034/j.1601-183x.2003.00037.x

Rudnick G (2007) SERT, serotonin transporter. In: Enna SJ, Bylund DB (eds) XPharm: The comprehensive pharmacology reference. Elsevier, New York, pp 1–6. https://doi.org/10.1016/B978-008055232-3.60442-8

Schafer DP, Lehrman EK, Kautzman AG, Koyama R, Mardinly AR, Yamasaki R, Ransohoff RM, Greenberg ME, Barres BA, Stevens B (2012) Microglia sculpt postnatal neural circuits in an activity and complement-dependent manner. Neuron 74:691–705. https://doi.org/10.1016/j.neuron.2012.03.026

Schizophrenia Working Group of the Psychiatric Genomics Consortium (2014) Biological insights from 108 schizophrenia-associated genetic loci. Nature 511:421–427. https://doi.org/10.1038/nature13595

Schwamborn R, Brown E, Haase J (2016) Elevation of cortical serotonin transporter activity upon peripheral immune challenge is regulated independently of p38 mitogen-activated protein kinase activation and transporter phosphorylation. J Neurochem 137:423–435. https://doi.org/10.1111/jnc.13596

Schwartzer JJ, Careaga M, Coburn MA, Rose DR, Hughes HK, Ashwood P (2017) Behavioral impact of maternal allergic-asthma in two genetically distinct mouse strains. Brain Behav Immun 63:99–107. https://doi.org/10.1016/j.bbi.2016.09.007

Sekar A, Bialas AR, de Rivera H, Davis A, Hammond TR, Kamitaki N, Tooley K, Presumey J, Baum M, Van Doren V, Genovese G, Rose SA, Handsaker RE, Schizophrenia Working Group of the Psychiatric Genomics Consortium, Daly MJ, Carroll MC, Stevens B, McCarroll SA (2016) Schizophrenia risk from complex variation of complement component 4. Nature 530:177–183. https://doi.org/10.1038/nature16549

Shi L, Smith SEP, Malkova N, Tse D, Su Y, Patterson PH (2009) Activation of the maternal immune system alters cerebellar development in the offspring. Brain Behav Immun 23:116–123. https://doi.org/10.1016/j.bbi.2008.07.012

Shin Yim Y, Park A, Berrios J, Lafourcade M, Pascual LM, Soares N, Yeon Kim J, Kim S, Kim H, Waisman A, Littman DR, Wickersham IR, Harnett MT, Huh JR, Choi GB (2017) Reversing behavioural abnormalities in mice exposed to maternal inflammation. Nature 549:482–487. https://doi.org/10.1038/nature23909

Smith SEP, Li J, Garbett K, Mirnics K, Patterson PH (2007) Maternal immune activation alters fetal brain development through interleukin-6. J Neurosci 27:10695–10702. https://doi.org/10.1523/JNEUROSCI.2178-07.2007

Stilley SE, Blakely RD (2021) Rare opportunities for insights into serotonergic contributions to brain and bowel disorders: studies of the SERT Ala56 mouse. Front Cell Neurosci 15:677563

Suzuki K, Sugihara G, Ouchi Y, Nakamura K, Futatsubashi M, Takebayashi K, Yoshihara Y, Omata K, Matsumoto K, Tsuchiya KJ, Iwata Y, Tsujii M, Sugiyama T, Mori N (2013) Microglial activation in young adults with autism spectrum disorder. JAMA Psychiatry 70:49–58. https://doi.org/10.1001/jamapsychiatry.2013.272

Takahashi N, Sakurai T (2013) Roles of glial cells in schizophrenia: possible targets for therapeutic approaches. Neurobiol Dis 53:49–60. https://doi.org/10.1016/j.nbd.2012.11.001

Takumi T, Tamada K (2018) CNV biology in neurodevelopmental disorders. Curr Opin Neurobiol 48:183–192. https://doi.org/10.1016/j.conb.2017.12.004

Takumi T, Tamada K, Hatanaka F, Nakai N, Bolton PF (2020) Behavioral neuroscience of autism. Neurosci Biobehav Rev 110. https://doi.org/10.1016/j.neubiorev.2019.04.012

Tamada K, Fukumoto K, Toya T, Nakai N, Awasthi JR, Tanaka S, Okabe S, Spitz F, Saitow F, Suzuki H, Takumi T (2021) Genetic dissection identifies Necdin as a driver gene in a mouse model of paternal 15q duplications. Nat Commun 12:4056. https://doi.org/10.1038/s41467-021-24359-3

Tetreault NA, Hakeem AY, Jiang S, Williams BA, Allman E, Wold BJ, Allman JM (2012) Microglia in the cerebral cortex in autism. J Autism Dev Disord 42:2569–2584. https://doi.org/10.1007/s10803-012-1513-0

Tuchman R, Cuccaro M, Alessandri M (2010) Autism and epilepsy: Historical perspective. Brain Dev 32:709–718. https://doi.org/10.1016/j.braindev.2010.04.008. SPECIAL SECTION: Epilepsy in Autism Spectrum Disorders and Related Conditions

Vargas DL, Nascimbene C, Krishnan C, Zimmerman AW, Pardo CA (2005) Neuroglial activation and neuroinflammation in the brain of patients with autism. Ann Neurol 57:67–81. https://doi.org/10.1002/ana.20315

Walder DJ, Laplante DP, Sousa-Pires A, Veru F, Brunet A, King S (2014) Prenatal maternal stress predicts autism traits in 6½ year-old children: project Ice Storm. Psychiatry Res 219:353–360. https://doi.org/10.1016/j.psychres.2014.04.034

Weiss LA, Arking DE, Gene Discovery Project of Johns Hopkins & the Autism Consortium, Daly MJ, Chakravarti A (2009) A genome-wide linkage and association scan reveals novel loci for autism. Nature 461:802–808. https://doi.org/10.1038/nature08490

Willette AA, Lubach GR, Knickmeyer RC, Short SJ, Styner M, Gilmore JH, Coe CL (2011) Brain enlargement and increased behavioral and cytokine reactivity in infant monkeys following acute prenatal endotoxemia. Behav Brain Res 219:108–115. https://doi.org/10.1016/j.bbr.2010.12.023

Xiao J, Buka SL, Cannon TD, Suzuki Y, Viscidi RP, Torrey EF, Yolken RH (2009) Serological pattern consistent with infection with type I Toxoplasma gondii in mothers and risk of psychosis among adult offspring. Microbes Infect 11:1011–1018. https://doi.org/10.1016/j.micinf.2009.07.007

Yang F, Zheng Q, Jin L (2019) Dynamic function and composition changes of immune cells during normal and pathological pregnancy at the maternal-fetal interface. Front Immunol 10:2317. https://doi.org/10.3389/fimmu.2019.02317

Ye X, Shin B-C, Baldauf C, Ganguly A, Ghosh S, Devaskar SU (2021) Developing brain glucose transporters, serotonin, serotonin transporter, and oxytocin receptor expression in response to early-life hypocaloric and hypercaloric dietary, and air pollutant exposures. Dev Neurosci 43:27–42. https://doi.org/10.1159/000514709

Yizhar O, Fenno LE, Prigge M, Schneider F, Davidson TJ, O'Shea DJ, Sohal VS, Goshen I, Finkelstein J, Paz JT, Stehfest K, Fudim R, Ramakrishnan C, Huguenard JR, Hegemann P, Deisseroth K (2011) Neocortical excitation/inhibition balance in information processing and social dysfunction. Nature 477:171–178. https://doi.org/10.1038/nature10360

Zaretsky MV, Alexander JM, Byrd W, Bawdon RE (2004) Transfer of inflammatory cytokines across the placenta. Obstet Gynecol 103:546–550. https://doi.org/10.1097/01.AOG.0000114980.40445.83

Zerbo O, Iosif A-M, Walker C, Ozonoff S, Hansen RL, Hertz-Picciotto I (2013) Is maternal influenza or fever during pregnancy associated with autism or developmental delays? Results from the CHARGE (CHildhood Autism Risks from Genetics and Environment) Study. J Autism Dev Disord 43:25–33. https://doi.org/10.1007/s10803-012-1540-x

Zerbo O, Qian Y, Yoshida C, Grether JK, Van de Water J, Croen LA (2015) Maternal infection during pregnancy and autism spectrum disorders. J Autism Dev Disord 45:4015–4025. https://doi.org/10.1007/s10803-013-2016-3

# Immunopharmacogenomics: Challenges and Future Prospects

**Saranya Vilvanathan and Swaminathan Ravichandran**

**Abstract** A detailed understanding of the pharmacokinetics and pharmacodynamics of drugs along with the enormous advancement and availability of genomic resources has led us to the era of pharmacogenomics and personalized medicine where the prediction of drug efficacy and safety based on genetic testing is made possible; this exciting new area of science is still at infancy in clinical practice due to various challenges. With the availability of sophisticated genomic tools including next-generation sequencing and microarray analysis, it has become possible to sequence DNA nucleotides and assess the expression of genes in the various pathological states elucidating the exact molecular pathogenesis of diseases including cancer and autoimmune disorders. The critical role of the immune system in various pathological conditions and the latest developments in immunotherapy have highlighted the importance of host immune cells and their influence on the pathogenesis and therapy of various disorders. The variations in expressions and mutations of immune-specific genes were associated with pathological immune-mediated diseases as well as immune-related adverse drug reactions. Not surprisingly, the role of immune cells and the effects of genetic mutations affecting the molecular pathogenesis and drug response in these disorders were specifically highlighted to be the area that required extensive research on our way to finding an efficient therapeutic cure for these disorders.

**Keywords** Pharmacogenomics · Pharmacodynamics · Adverse drug reactions · Tailored therapy · Immunotherapy

---

S. Vilvanathan (✉)
Department of Pharmacology, School of Medicine, Medical University of the Americas, Nevis, St. Kitts & Nevis
e-mail: s.vilvanathan@mua.edu

S. Ravichandran
Department of Pathology, School of Medicine, Medical University of the Americas, Nevis, St. Kitts & Nevis



A. Das Mohapatra, P. S. Sahu (eds.), *Advances in Immunology and Immuno-techniques*, Interdisciplinary Biotechnological Advances,
https://doi.org/10.1007/978-981-97-5508-0_7

# 1 Introduction

The extensive and consolidated understanding of pharmacokinetics and pharmacodynamics of drugs along with the enormous advancement and availability of genomic resources has led us to the era of pharmacogenomics and personalized medicine where in the prediction of drug efficacy and safety based on genetic testing is made possible, and this field is still trying to deepen its roots in clinical practice due to the various challenges that hinder the transition of these available scientific resources from bench to routine bedside (Cecchin and Stocco 2020). The major branch of pharmacogenomics that had seen a remarkable advancement in the past decade is the immunopharmacogenomics wherein the genomic tools are utilized to analyze the genetics of immune cells to help predict the drug response especially with respect to immune-related adverse reactions, cancer genetics, organ transplantation, autoimmune disorders, understanding of which paves us to the further advancement in the field of immunotherapy (Nakamura 2015).

With the availability of sophisticated genomic tools including next-generation sequencing and microarray analysis, it has become possible to sequence DNA nucleotides and assess the expression of genes in the various pathological states elucidating the exact molecular pathogenesis of diseases including cancer and autoimmune disorders (Hurd and Nelson 2009). The variations in expressions and mutations of immune-specific genes were associated with pathological immune-mediated diseases as well as immune-related adverse drug reactions. Not surprisingly, the role of immune cells and the effects of genetic mutations affecting the molecular pathogenesis and drug response in these disorders were specifically highlighted to be the area that required extensive research on our way to finding an efficient therapeutic cure for these disorders.

# 2 Progression and Advancements in Pharmacogenomic Research

The remarkable milestone in genetic research started and progressed rapidly with the elucidation of human genome sequence under the Human Genome Project in April 2003. Consequently, with the latest technological advances in bioinformatics, the field has shown remarkable advancement in the past decade including the identification and characterization of numerous genetic disorders, from the rarely identified monogenic disorders to the more commonly identified genetically complex disorders including diabetes, cancer, and cardiovascular disorders. The application of DNA sequencing to the study of inter- and intra-individual variations in drug responses has been the initial steppingstone to the field of pharmacogenomics. With exploration of the influence of RNA, proteins, and metabolites in disease pathogenesis, pharmacogenomics evaluates the molecular determinants of drug response at the genome, transcriptome, and proteome-wide levels. This field has become a promising approach as it has taken the future of medicine from empirical, trial-and-error, population-based approach to a more personalized and individually tailored therapy. Pharmacogenetic testing is applied in drug and clinical trials to facilitate the development of newer molecules and effectively strategize the dosage of medications based on the efficacy and risk of adverse events.

Though the advancements and availability of the scientific updates, in the field of pharmacogenomics, are certainly piling up, the translation of these evidences into clinical practice has faced enormous challenges and barriers. Consequently, several committees and organizations including the CPIC [the Clinical Pharmacogenetics Implementation Consortium], DPWG [The Dutch Pharmacogenetics Working group], CPNDS [the Canadian Pharmacogenomics Network for Drug Safety], and RNPGx [the French National Network (Reseau) of Pharmacogenetics] have developed guidelines with recommendations to help and guide the clinicians to understand and implement the valuable pharmacogenetic knowledge in everyday practice to optimize therapy of individual patients. Though the guidelines were developed through similar methodologies by these committees, their defined objectives were different at the initiation of the project which determines and defines the unique strength and profile of their guidelines. CPIC and DPWG guidelines are mainly focused on updating the genotype-phenotype information and the pharmacogenetic recommendations of a large number of drugs in a standard format. CPIC levels are assigned based on PharmGKB Clinical Annotation Levels of Evidence for FDA-approved drug labels. DPWG, CPNDS, and RNPGx recommend performing genetic testing in daily clinical practice, and RNPGx even describes specific clinical settings or medical conditions where genotyping is recommended (Abdullah-Koolmees et al. 2021).

Though the identification of different genetic biomarkers has raised our hopes of personalized medicine, the challenges faced and the practical limitations in implementations are quite disappointing so far. Most of the genetic biomarkers identified has not progressed beyond the stage of identification (Pirmohamed 2011), and there is not much research data available to associate these biomarkers with health and diseases. This lack of progress may partly be due to replication failure due to the inability to partially or completely replicate the genetic biomarkers which makes it hard to progress further to establish the clinical validity of these genetic biomarkers. The complex interplay of genetic, environmental, and gene–environment interactions in influencing the phenotype of drug response cannot be completely neglected. The estimate is that only 10–15% of the genetic biomarkers have been established to have a direct effect on drug response (Ventola 2011). Since not enough information is available regarding the interplay of genetic and environmental influences on pharmacological response, the practice of personalized medicine appears incomplete.

Conventionally, most of the pharmacogenetic traits were attributed to single nucleotide polymorphisms (SNPs), variations of alleles in a single genetic locus. With the expansion of pharmacogenetic studies, variations in multiple genetic loci were associated with drug response. Overlooking the contribution of other genetic variations can significantly reduce the predictive value and applicability of pharmacogenetic testing (Scott 2011; Vijverberg et al. 2010). This calls in for the comprehensive survey of human genome that identifies multiple genetic variations that could help in predicting the pharmacokinetics and pharmacodynamics of medications for better tailoring the dosage to improve the efficacy and decrease the incidence of adverse events.

Optimal integration of pharmacogenomics in clinical practice also basically needs adequate training of health care professionals who practically manage the clinical prescription of drug regimens based on the available clinical and genetic test results (Guy et al. 2020).

A summary table listing over 400 drugs and associated biomarkers was developed by FDA and is available in DRUGS@FDA. This includes the essential pharmacogenetic information that are found in the drug labeling information (Khan and Phillips 2020) which can clinically help practitioners predict the responders and also help determine the drug dosage, reduce side effects, and personally tailor the medications for every patient based on their genetic profile to promote safe therapeutic care (Guy et al. 2020). Implementation of reactive pharmacogenetic testing on gene-by-gene basis, while there is a need to prescribe the specific high-risk medication, has a slow turn-around time and at times may also sound irrelevant when rapid prescription is necessary. As technology advances, it becomes increasingly recognized that broad screening of multiple pharmacogenes is required, as is the availability of recommendations for dosing in electronic health records and drug prescription systems (Krebs and Milani 2019).

In the year 2013, the National Institute of Health (NIH) and the National Human Genome Research Institute (NHGRI) in collaboration initiated the Implementation of GeNomics In pracTicE (IGNITE) network which is dedicated to support the genomics implementation in healthcare. The network is tasked to find ways for incorporating the genomic research information into the electronic health records and developing clinical decision support for providers across diverse health care settings (Levy et al. 2020). The status of this implementation program shows there is a lag in the adoption of pharmacogenetic tests in clinical practice even at sites where the leading experts are conducting research. Clinical research supporting the CPIC level A CPIC evidence is important for provider and consumer in order for these tests to be extensively implemented.

## 3 Implications of Immunopharmacogenomics in Genomic Research

The critical role of immune system in various pathological conditions and the latest developments in immunotherapy has highlighted the importance of host immune cells and their influence in the pathogenesis and therapy of various disorders.

The complexity of the immune system has started to unravel with the utilization of genomic tools, including next-generation sequencing to elucidate the structure of

T-cell receptors (TCR), B-cell receptors (BCR), and HLA molecules (Buermans and den Dunnen 2014). TCRs are the surface molecules of T-cells with a heterodimeric structure that plays an important role in recognizing an antigen. Genetic rearrangements in the sequence encoding the TCR have an important role in characterizing the inter-individual variations in T-cell responses, which in turn can be implicated in the pathogenesis of autoimmune disorders and also affect the response to various drug treatments. Millions of TCRs have been characterized by next-generation sequencing technology with genomic DNA or complementary cDNA-based TCR-sequencing. TCR analysis in cancer patients has led to the labeling of T-cell repertoire in tumor tissues and blood in patients with certain cancer such as advanced colorectal cancer, ovarian cancer, follicular lymphoma, and muscle-invasive bladder cancer (Zewde et al. 2018).

Immunopharmacogenomic analysis and prediction of TCR repertoire have been extended for analysis of immunogenomic background in patients with different autoimmune disorders including systemic lupus erythematosus, rheumatoid arthritis, type 1 diabetes mellitus, and so on (Song et al. 2021). Changes in the TCR repertoire have also been associated with the breakdown of peripheral immune tolerance, providing a load of biomarkers for the disease and expanding the strategies of immune-targeted therapy in these disorders.

In the field of transplantation medicine, quantitative characterization of TCR repertoire in the allogenic hematogenic stem cell recipients have increased the scope of therapeutic treatment options in patients with hematological malignancies. Graft-versus-host disease (GVHD) caused by the donor's T-cells has been the most challenging toxicity which affects the overall treatment outcome in patients undergoing allogenic stem cell transplantation. Next-generation sequencing has helped in defining the functional kinetics of different T-cell clones including, TCRA and TCRB clones, changes in T-cell diversity with the identification of CDR3 sequences and also the prediction of the extent of clinical advancement of certain T-cells. The study has concluded that higher diversity of T-cell repertoire early after transplant can probably indicate a lower risk of both GVHD and relapse subsequent to transplantation (Yew et al. 2015).

The extension of pharmacogenomic research into the field of immunological practice has been uplifted with the identification and characterization of various HLA genetic biomarkers associated with drug responses. The immune-specific genetic variations in HLA genes, which are involved in the presentation of an antigen to the T-cells, affecting the variation in drug disposition and response has led to the expansion of the field of immunogenomics which has majorly contributed to the development of next-generation immunopharmacology. The field of applied immunopharmacology is mainly focused on the development of immunomodulatory agents that are therapeutically useful in clinical medicine as well as helpful as a basic research tool in bench medicine.

Immunopharmacogenomics involves the blending of the knowledge base of pharmacogenomics and immunogenomics and has the potential to facilitate the selection and monitoring of immunopharmacological therapies predominantly in cancer, autoimmune and transplantation medicine. The elucidation of detailed molecular pathogenesis of cancer and the understanding of immune microenvironment in cancer has led to the identification of neoantigens and development of cancer vaccines, increased therapeutic efficacy of existing chemo- and radiotherapy, and development of newer innovative therapies for cancer. The advancements in transplantation medicine for patients with end-organ damage has taken a newer turn with the availability of novel immunosuppressive therapies and the development of tolerogens which produces a state of immune-specific non-responsiveness on exposure to specific antigens or allergens. Immunopharmacogenomics can help understand the role of genomic biomarkers and their role in allergy and immunology practice, revolutionizing the role of immunosuppressive medications used in asthma and other allergic immunological disorders to predict various drug interactions associated with the use of certain drugs in individuals with specific immunopharmacogenetic variations. Various pharmacogenetic biomarkers in drug labeling information listed out by USFDA show that the biomarkers that are related to immunological practice could cover various health care areas including psychiatry, neurology, hematology, infectious disease, oncology, rheumatology, pulmonary, dermatology, endocrinology, and so on (Khan and Phillips 2020). The interpretation of the molecular mechanisms of immunological drug hypersensitivity reactions is one of the most important applications of immunopharmacogenomics, which has led to updates in preclinical drug screening techniques to help in prediction and prevention of these reactions.

The implications of advances made in the field of immunopharmacogenomics have been so vast in the field of drug discovery and therapeutics. Advances in the genome sequencing technology together with the computational and informatics support has helped the pharmaceutical companies to make better molecular models of diseases to identify potential drug targets. For the past several decades, the major setback in the field of drug discovery has been the longer time taken for each drug in trial to reach the market. It has been estimated that the average time taken by a drug molecule to finish the clinical trials and reach market would be around 12 years. With the availability of advances in the field of genetic sequencing and computational technology, this process has been fastened. Genome sequencing has not only allowed researchers to develop drug therapies at a faster rate than ever before but has made it easier for the clinicians to make a diagnosis and identify/predict rare genetic diseases at a quicker pace which in turn allows time for designing appropriate personalized therapeutic management. There has also been a drastic advancement in the prediction of drug-responsive patients in clinical trials which could increase the reliability and success rate of a research study. Application of immunopharmacogenomic tools in clinical trials can increase the rate of prediction of risk factors as well as reduce the cost of trial. With clear guidelines from FDA regarding the submission of pharmacogenetic data to be considered as a potential valid genetic biomarker, the clinical trials to accumulate the evidence that support the physiology, pharmacology, toxicology, and clinical efficacy of the drug molecules has been revised and reorganized.

The implications of immunopharmacogenetic biomarkers in different fields of clinical practice have been discussed here.

### 3.1 *Immunopharmacogenetic Biomarkers in Cancer Therapeutics*

The growth and research findings in the field of cancer genetics, are considerable and the scope of future research is extensive. The comprehensive genomic characterization of various types of human cancer by the International Cancer Genome Consortium (ICGC) and The Cancer Genome Atlas (TCGA) has successfully characterized the genetic drivers of the disease (Shyr and Liu 2013; Cancer Genome Atlas Research Network et al. 2013; International Cancer Genome Consortium et al. 2010). This along with the utilization of advances in technology such as next-generation sequencing immensely increases the scope of immunopharmacogenomics. The understanding of tumor immune microenvironment has led to better implementation of chemotherapeutic regimens and identification of novel cancer vaccines and other immunotherapies/tolerogens for cancer prevention and treatment and has led us to the era of personalized immunotherapy wherein based on the specific immunogenomic biomarkers of an individual the host immune cells can be utilized to attack the cancer cells with minimum adverse reactions and maximum therapeutic response. Moreover, using the host immune cells to attack the cancer cells has been the most investigated subject in cancer research (Kiyotani et al. 2018, 2021).

In 2006, Galon et al. studied the role of tumor-infiltrating immune cells in colorectal cancer and affirmed the prediction that adaptive immune response plays a significant role in controlling the growth and recurrence of tumor and are the better predictors of tumor outcome than the histopathological methods currently used to stage colorectal cancer. In 2012, Erdag et al. have studied the infiltration of tumor microenvironment, melanoma metastases, by immune cells and assessed the composition and organization of immune cells based on their histology cells pattern of infiltration. It was reported that the majority of immune cell infiltration was seen in the regions proximal to the intratumoral blood vessels and that the predominant type of immune cells were T-cells (53%), B-cells (33%), and macrophages (13%). Increased densities of CD8 (+) T-cells were associated best with the survival and higher densities of CD45 (+) leukocytes; T-cells and B-cells were correlated with increased survival. Microenvironmental variations in the immune homing receptors and ligands affect the immune cell recruitment to the tumor microenvironment.

The mechanism by which the immune cells, especially the T-cells, recognize and attack the tumor antigens and not the host cells was not well understood.

De Plaen et al. (1997) initially stated that the T-cells in tumor microenvironment can specifically recognize the abnormal peptides derived from tumor-specific mutations in a methylchlanthrene (MCA)-induced mouse model. They further discovered the first T-cell recognized neoantigens, using cDNA library screen. Later, a series of neoantigens were identified to be derived from somatic mutations in different human cancers, including melanoma and renal cell carcinoma. The genomic instability of cancer cells leads to increased incidence of non-synonymous mutations that can induce the synthesis of tumor-specific antigens called neoantigens

(Peng et al. 2019). These neoantigens are specifically expressed only in the cancer cells, not in the benign host cells (and are less likely to trigger autoimmunity), and were found to be highly immunogenic, activating T-cells and B-cells generating profound immune response. The concept of neoantigens was later explored with the whole-exome sequencing technology and was used in immunotherapeutic researches to identify vaccines and define therapeutic approaches in cancer management. A growing number of researches later highlighted the relationship between neoantigens and T-cell recognition of cancer cells (Zhang et al. 2021). A breakthrough in the understanding of neoantigens was when Huang et al. reported the near-complete regression of melanoma following administration of a cell product with high proportion of neoantigen-reactive T-cells. Kenner's et al. reported that the T-cells from melanoma patients were reactive to five different mutated peptides synthesized from somatic point mutations. These reactive T-cells against mutated proteins were clearly more prevalent over the response to tumor-associated antigens.

Immunopharmacogenomic studies, especially the relevance of next-generation sequencing (NGS) technology, have given a more and more thoughtful understanding of neoantigens. The technology was utilized to identify the immunogenic neoantigens and to study the effect of immunotherapeutic neoantigen vaccines in different tumor models. Studies, later on, showed that neoantigen T-cell responses were observed in different human cancers, including melanoma, colorectal cancer, squamous cell carcinoma of head and neck, and cholangiocarcinoma.

The major challenge to the success of immunotherapy would be the process of immune editing of cancer cells, where the tumor cells that activate the immune cells are specifically screened out. Also the release of various substances in the tumor microenvironment, including the inflammatory mediators that are predominantly immunosuppressive, can selectively suppress the immune cells around the tumor tissue. This also explains the selective inhibition of immune response around the tumor tissue than the systemic immune response by various immunotherapeutic measures (Shemesh et al. 2021). Despite the challenges, various researches are underway in the field of cancer immunotherapy to try and find out ways to overcome these challenges and the concept of neoantigens has rapidly expanded the field of promising personalized immune-based treatments for solid tumors. The major role of neoantigens in immune escape, immune editing, and sensitivity to immune checkpoint inhibitors was reported (Jiang et al. 2019).

In order to overcome the challenge of immune escape, where the tumor cells sensitive to immune attack are selectively screened and eliminated during the process of tumor progression was to a certain extent overcome by the concept of immune checkpoint blockade, which has become one of the emerging and promising therapeutic choices for patients in the history of cancer treatment (Hargadon et al. 2018). The initially identified negative regulator of T-cell activation in tumor microenvironment was CTLA-4, a co-inhibitory receptor that is not regulated in the early stages of T-cell activation. Activation of these receptors by ligands impedes the acquisition of T-cell effector function by various mechanisms including endocytosis and degradation of ligands needed for co-stimulation; inhibition of IL-2 synthesis and secretion blocking the T-cell proliferation and differentiation; induction of anergy and T-cell tolerance. Monoclonal antibodies targeting the CTLA-4 antigens were found to effectively inhibit the regulatory T-cell-associated immunosuppression and promote the T-cell effector function in both CD4+ and CD8+ population. Ipilimumab, a fully human IgG1 anti-CTLA-4 monoclonal antibody, was the first immune checkpoint inhibitor approved by the USFDA, following the outcome of a clinical trial demonstrating improved survival in patients with stage III/IV melanoma. Although currently approved only for the management of melanoma, clinical trials have shown modest improvement in survival outcomes, when this medication was included in the regimens used for the management of other different types of cancer including renal cell carcinoma, non-small cell lung carcinoma, prostate cancer, and other solid tumors that are not able to repair errors in its DNA that occur when the DNA is copied during the proliferation of cancer cells. Tremelimumab, a fully human IgG2 anti-CTLA-4 monoclonal antibody, is currently being under investigation in clinical trials.

The other checkpoint protein that was widely researched and therapeutically utilized would be the PD-1 (programmed death-1) molecules present on the surface of T-cells that links PD-L1 on tumor cells to help suppress the immunological attack on tumor cells.

Pembrolizumab and nivolumab are the monoclonal antibodies that target PD-1 and have shown promising results in melanoma, non-small cell lung carcinoma, bladder cancer, relapsed/refractory Hodgkin's lymphoma, and triple negative breast cancer.

Since 2011, around seven checkpoint inhibitors have been approved by the USFDA and many more are being evaluated either as monotherapy or in addition to other drug combinations in the treatment of cancer. The major therapeutic challenge with the mainstream initiation of the immune check inhibitors in cancer treatment would be the low response rate and development of immune-related adverse drug reactions, most of which are reported to have a delayed onset, prolonged duration, and are known to affect any organ or system. Autoantibody production, complement-mediated cellular damage, or drastic decline in regulatory T-cell count were proposed to be the mechanisms behind the development of these immune-mediated adverse events, and most of these are managed with steroids as first-line therapy for severe reactions followed by therapy with non-steroidal synthetic immunosuppressive agents or intravenous immunoglobulins are being used. There are also studies that are ongoing, looking out for biomarkers that can predict the risk or incidence of these adverse reactions including assessment pretreatment levels of autoantibodies. Much more studies are required to characterize the risk and manifestations of adverse outcomes and to standardize the usage of immune check point inhibitors (Ramos-Casals et al. 2020).

### 3.2 Immunopharmacogenetic Biomarkers in Transplantation Medicine

Understanding the process of immune response and the process of immune reconstitution after transplantation with the help of TCR analysis will help strategize the therapeutic regimens to decrease incidence of relapse and GVHD.

It has been proposed that restoration and quicker recovery immunogenic T-cells after transplant can lower the risk of infections and rejection.

The process of immune recovery following transplantation is a dynamic process the kinetics of which correlates with the spectrum of infections reported in post-transplant recipients. The T-cell proliferation responses to specific and most common infections have been predicted to be absent in about 20–50% of patients (Mehta and Rezvani 2016). Prediction and characterization of TCR repertoires and close follow-up of the diversity in the T-cell repertoire after allogenic bone marrow transplantation can possibly provide clinical biomarkers to monitor the immune recovery and restoration reducing the incidence of associated infections risks in this population (Tickotsky-Moskovitz et al. 2021).

TCR diversity is one of the most investigated research topics with the next-generation sequencing technology. Discrimination of self and non-self antigens by the T-cells, after the interaction of TCR and various ligands presented by the MHC complex, forms the foundation of adaptive immune response. Initially, it was predicted that the TCR diversity is generated by the rearrangements of V and J segment in TCR alpha and V, D, and J segments in TCR beta in the thymus usually is declined in the elderly due to thymic involution. Qi et al. (2014) has shown that highly diverse repertoires were maintained despite thymic involution and that peripheral fitness selection. TCR diversity plays a very important role in immune recovery following transplantation. It has been postulated that the higher the TCR diversity, the lower the incidence of GVHD and relapse. TCR diversity was later investigated in different post-transplant recipients. One such study by Alachkar and Nakamura (2015) characterized TCR repertoires in patients who received the haplo-cord and matched-donor transplant.

In patients who cannot find a matched-donor for hematopoietic stem cell transplantation, haplo-cord transplant is considered as the most reliable approach. It was found that a higher TCR diversity was found in patients with a higher percentage of cord cells post-transplant, and this suggests a lower risk of GVHD and relapse.

### 3.3 Immunopharmacogenetic Biomarkers in Autoimmune Disorder

Autoimmune disorders are an entity of disorders that clearly have established the pathological role of immune cells when they recognize and react destructively to self-antigens. The molecular and genetic characterization of immune cells helps with a better understanding of the pathogenesis of autoimmune disorders as well has an enormous potential to lead us to achieve a better cure for these disorders. A few of these autoimmune disorders such as Crohn's disease and rheumatoid arthritis are discussed below.

Crohn's disease is a chronic intestinal inflammatory disorder, resorted to having an infectious etiology and has now been redefined to have pathogenesis associated with abnormalities of T-cell function. Genetic and environmental factors along with the host immune response and gut flora play a crucial role in the pathogenesis of this disorder (Segal 2019). It is postulated that a specific clone of T-cells was expanded in immune cells amid a decrease in T-cell diversity which plays an important role in the incidence and pathogenesis of this disorder in susceptible individuals. Chapman et al. (2016) performed RNA extraction followed by cDNA synthesis which was used to characterize TCR of the mucosa and the peripheral blood mononuclear cells of Crohn's disease patients, using next-generation sequencing. They concluded that the TCR diversity was low in mucosa compared to peripheral mononuclear cells, which indicates the expansion of the monoclonal T-cell population in the inflamed intestinal mucosa. The abundance of the specific clone of T-cells was also associated with the severity of recurrence of the disease. The authors have also concluded that for rapidly analyzing the in-depth, real-time T-cell repertoire assessment, the deep TCR sequencing is a powerful tool. Steroids, TNF-alpha inhibitors, immunomodulators, and 5-aminosalicylic acid derivatives are extensively utilized medications to maintain remission in Crohn's disease. Pharmacogenetic association studies have so far described potential genetic biomarkers in relation to inter-individual variability in drug response, toxicity, and disease severity. Various drug-related markers are being utilized to decide if the patient will be a responder to a particular therapy and also to predict and prevent the incidence of adverse events in the high-risk population (Yamamoto-Furusho 2017). There are a lot more queries and important in-depth explanations that demand to be addressed in this area to effectively strategize the usage of immunosuppressive medications in the management of this disorder.

With the availability of newer inputs to the pathogenesis using the molecular characterization of TCR, further revolutionizing immunotherapies are yet to be formulated to achieve complete remission or cure for this disorder.

## *3.4 Immunopharmacogenetic Biomarkers in Rheumatology*

Rheumatoid arthritis is a progressive disabling autoimmune disorder that causes inflammation, pain, and swelling in and around joints usually symmetrical in small joints initially. The lack of specific biomarkers for a personalized therapeutic approach has led to the extensive trial-and-error approach being practiced to optimize therapy for this disorder. With advances in the field of immunogenomics, emerging evidence shows some promise in the availability of genetic biomarkers and prediction of treatment response to various biological response modifiers such as TNF-alpha inhibitors used in the management of rheumatoid arthritis (Szekanecz et al. 2013). Utilization of pharmacogenomic biomarkers can help predict treatment response and failures, drug toxicities, and hence can provide greatest benefit to individual patients. Biomarker-driven studies could help in identifying and predicting responders to a specific therapy and hence could lead to more successful therapies coming to market in shorter time frame. REASON (RhEumatoid Arthritis Synovial tissue Network) consortium has an established framework for patient recruitment, curation of clinical data, ultrasound-guided synovial biopsies, cell sorting and RNA sequencing, and computational analyses. It was shown that macrophages isolated from the synovial tissue biopsies in patients with RA had some transcriptional differences in the RNA sequences across patients. Another consortium, AMP (Accelerating Medicines Partnership) has identified four populations of monocytes/macrophages using single cell RNA-sequence. It is recommended that examination of patient's synovial tissue prior to and following therapy using transcriptional approach can help predict the genetic expression profiles that can predict the therapeutic efficacy of medications. Humbly et al. described the distinct synovial subtypes (lymphoma-myeloid, diffuse-myeloid, and pauci-immune fibroid) that are associated with disease activity/severity and treatment response. Identification of biomarkers in the serum to correlate with the synovial phenotypes is the critical research question in future.

## 3.5 *Immunopharmacogenetic Biomarkers in Allergy and Immunology Practice*

In allergy and immunology practice, various pharmacogenetic biomarkers have been summarized along with their potential implications by the FDA in drug labeling information. The influence of pharmacogenetics in determining the efficacy of inhalational corticosteroids in asthmatics is a special area of interest among AI (Allergy & Immunology) specialists. Polymorphisms in genes including TBX21, FCER2, and CRHR1 were associated with response to inhalational corticosteroids (Khan and Phillips 2020). Immunosuppressive and anti-inflammatory medications utilized in the management of disorders such as urticaria, and common variable immunodeficiency were shown to have pharmacogenetic biomarkers that can predict the risk of adverse effects with the use of these agents. Azathioprine is an immunosuppressive medication that predominantly acts by decreasing the T-cell population. Genetic polymorphisms in TPMT (thiopurine S-methyl transferase) and NUDT15 (nucleoside diphosphate-linked moiety X motif 15) were linked with an increased risk of serious bone marrow toxicities in patients treated with azathioprine. Mutations in the TPMT gene can considerably decrease the enzyme activity leading to increased accumulation of azathioprine metabolites precipitating toxicity whereas NUDT15 is involved in inactivation of thiopurine metabolites and hence patients with loss-of-activity variants are associated with an increased risk of toxicities. Numerous guidelines including the clinical pharmacogenetics implementation consortium, CPIC, can help with dose reduction guidelines in patients who are treated with azathioprine, once the pharmacogenetic data is available.

Mycophenolic acid is another potential immunosuppressive medication that is to be avoided in patients with a deficiency of HGPRT-ase [Hypoxanthine-guanine phosphoribosyl-transferase] enzyme implicated in causing Lesch-Nyhan syndrome, as it may lead to overproduction and accumulation of uric acid. Patients presenting with cutaneous reactions to the use of morphine due to non-IgE-related mast cell activation have also been associated with polymorphisms in *CYP2D6 gene*. The ultra-rapid metabolizers of *CYP2D6 gene polymorphism* can rapidly metabolize codeine and other opioid drugs to active morphine metabolites leading to an increased risk of cutaneous reactions, sedation, and respiratory depression.

The other major areas of interest with respect to identification of potential immunopharmacogenetic biomarkers in the future would be related to the prediction of drug-induced hypersensitivity reactions which frequently requires the determination of specific HLA-associated alleles to avoid the incidence of potential serious toxicities. Some of these drug-associated immunopharmacogenetic biomarkers linked with hypersensitivity reactions are discussed below.

## *3.6 Immunopharmacogenetic Biomarkers in Prediction and Prevention of Drug-Induced Hypersensitivity Reactions*

One of the leading causes of increased mortality that has a huge impact on health care costs would be the incidence of adverse drug reactions to the common medications used in clinical practice. There is a specific category of adverse drug reactions called idiosyncratic reactions where the mechanism of how these reactions occur is not clearly elucidated. Variations in the HLA genes are implicated in certain reactions to induce drug hypersensitivity reactions.

The HLA genes are located in chromosome 6 and encode the cell surface glycoproteins that present antigenic peptides to T-cell receptors. Variations and structural mutations in the HLA genes can cause inter-individual differences in drug efficacy and incidence of adverse drug reactions. Variations in HLA genes have been associated with the incidence of drug-induced hypersensitivity reactions, and these reactions and associations are foreseeable with the advancements in immunopharmacogenomics (Karnes et al. 2019).

With the strong association between drug-induced hypersensitivity and the HLA alleles and the understanding that most of these causative drugs stimulate the TCR at sites of autoimmune destruction, further interest in analyzing the changes in TCR sequences due to the presentation of viral antigens and drug molecules was done by immunopharmacogenetic analysis. The association between viruses and drug hypersensitivity reactions were established and the hypothesis that the cytotoxic

T-cells specific for viral antigens cross-react with self-peptides complexed to drug and HLA molecules was extensively researched. Viral sequences like the immunogenic self-peptide were identified. Later on, abacavir-specific T-cell clones were used to determine if the virus peptides presented in the context of abacavir stimulate T-cell responsiveness. The research by Yerly et al. (2017) suggests that the polyclonal response of T cell to abacavir includes multiple subsets, which include the T-cells that recognize self-peptides/HLA-B*57:01 complexes that cross-react with viral peptide HLA-B*57:01 complexes due to similarity in TCR contact residues. Recently, Miyagawa and Asada (2021) has proposed that the CD4 T-cells mainly initiate the response to drug antigens, and then the virus-specific CD8 T-cells that target virus-infected cells emerge, resulting in tissue damage. The study concludes that further investigation of HLA risk alleles for drug-induced hypersensitivity, which can precisely describe how drugs stimulate the T-cells and what triggers viral reactivation is to be done to facilitate the preclinical prediction and prevention of these adverse events.

Some of these biomarkers and their potential implications are discussed here, in order to get an in-depth understanding of the influence of genetic biomarkers associated with immune cells affecting drug response.

### 3.6.1 Abacavir-Induced Hypersensitivity Reaction

Abacavir is a guanosine analog used as a part of the HAART (Highly Active Anti-Retroviral Therapy) regimen, as a nucleoside reverse transcriptase inhibitor. The drug gets actively metabolized into abacavir triphosphate, an active intracellular agent that acts as a synthetic substrate for the enzyme reverse transcriptase and competitively inhibits it. These medications primarily prevent infection of susceptible cells (cells with CD4 receptors including T-cells, macrophages, dendritic cells) but do not eradicate the virus that has already integrated the pro-viral DNA with the host DNA. Abacavir monotherapy is shown to produce up to 99% decrease in the viral RNA levels along with a modest increase in CD4 lymphocyte levels. Though abacavir is generally well-tolerated, in around 5% of population, a unique and potentially fatal hypersensitivity reaction (HSR) has been documented. The presence of concurrent fever, rash, and abdominal pain within 6 weeks of initiating therapy with abacavir should raise the suspicion of HSR and immediate discontinuation of the medication is necessary as the condition may worsen with continued treatment, unlike other hypersensitivity syndromes. The mortality rate is even higher in sensitive individuals who are treated with abacavir, as there will a rapid recurrence of symptoms with hypotension and shock-like state and possibly death.

The cause for this HSR was found to be a genetically mediated immune response linked with HLA-B *5701 following which the pharmacogenetic screening for *HLA-B*57:01* was done in patients before initiation of abacavir therapy. With the advancement in the field of immunopharmacogenetic analysis, the unique mechanism by which abacavir can potentially trigger this unique HSR has been elucidated. In a lymphoblastic cell model, Norcross MA et al., proposed that the binding of abacavir to the peptide binding groove of *HLA-B*57:01* peptide is found to influence the binding of other peptides which when presented to the immune system triggers the hypersensitive response. Additionally, Pompeu et al. found that the usage of novel peptides with abacavir to stimulate T-cells of hypersensitive patients produced an immune response, measured in terms of interferon gamma production (Pompeu et al. 2012). The final outcome of the alteration of peptide binding profile after exposure to abacavir is the presentation of novel self or a novel conformation of constitutive self, which denotes that there is presentation of peptides for which there is no host tolerance (Illing et al. 2013).

Inter-ethnic variations in HLA-B region were found to significantly influence the incidence of HSS and researchers conducted various studies to assess the impact of HLA-B screening to predict abacavir-induced HSS. PREDICT-I was a prospective study which concluded that the incidence of HSS in the HLA-screening group was significantly lower (3.4%) than that with no-HLA screening group with immunologically confirmed disease. These observations strongly support the mandatory pharmacogenetic screening for *HLA-B*57:01* before abacavir treatment.

### 3.6.2 Carbamazepine-Induced SJS/TEN

Carbamazepine is an iminostilbene that was initially approved as an anti-seizure medication and is also increasingly being prescribed for bipolar disorder and neuropathic pain including trigeminal neuralgia. The serious adverse effects associated with the clinical use of this drug include bone marrow toxicity (aplastic anemia, agranulocytosis) and hypersensitivity reactions (severe fatal skin reactions, eosinophilia, lymphadenopathy, and splenomegaly). The term DRESS was initially proposed by Bocquet et al. (1996) to refer to a series of drug-induced skin reactions, eosinophilia, and systemic symptoms. It is a kind of severe hypersensitivity reaction that begins around 2–6 weeks after drug exposure. During the early phase, the reaction generally presents with fever, rash, hyper eosinophilia, generalized lymphadenopathy, and hypogammaglobulinemia. It is a diagnosis of exclusion that requires ruling out of other eosinophilia disorders, malignancies, or parasitic infestations. The cutaneous hypersensitive reactions can range in severity from benign mild urticaria to severe lethal disorders including Steven-Johnson syndrome (SJS) and toxic epidermal necrolysis (TEN) which carries a mortality risk of around 30% and requires prompt withdrawal of the medication. The strong association of mutation in HLA-B gene, *HLA-B*1502* and carbamazepine-induced SJS/TEN was first reported by Fan WL et al., based on a case-control retrospective study on patients reported with the drug reaction and comparing it with the genotyping data of 157 single-nucleotide polymorphisms, *CYP*-genes as well as alleles of HLA-A, -B, -C, and DRB1 genes. Following this various reports in Chinese population associating the incidence of SJS/Ten with HLA-B*1502 were surging up and the screening test for this HLA allele was made mandatory in Han Chinese population. Due to increased prevalence of HLA-B*1502 in Asians, though not enough published data to strongly support association, the FDA has made a recommendation for pharmacogenetic screening of Asians before initiating carbamazepine therapy in order to decrease the incidence of adverse outcomes. Studies in certain ethnic populations such as the Caucasians have not elucidated any strong correlations, mainly due to the fact that the incidence of this alleles and the adverse outcome of the medication, is relatively rare in this population (Ferrell Jr and McLeod 2008). The pharmacogenetic information in the FDA label for the medication has clearly stated that it should be avoided in patients positive for *HLA-B*1502* unless the benefits outweigh the risks. There were two other HLA alleles, associated with the incidence of carbamazepine-induced SJS/TEN; *HLA-A*31:01* is found to be more strongly associated with carbamazepine-induced DRESS and MPE, and this allele is found in most populations worldwide while *HLA-B*15:11* is the other allele associated with SJS/TEN. The recommendations by CPIC includes avoiding the usage of carbamazepine in patients with *HLA-B*15:02* positive results and any of those with *HLA-A*31:01* genotype (Dean 2012).

## 4 Conclusion

Though the identification of different genetic biomarkers has raised our hopes of personalized medicine, the challenges faced and the practical limitations in implementations are quite disappointing so far. Most of the genetic biomarkers identified have not progressed beyond the stage of identification, and there is not much research data available to associate these biomarkers with health and diseases. This lack of progress may partly be due to replication failure due to the inability to partially or completely replicate the genetic biomarkers which makes it hard to progress further to establish the clinical validity of these genetic biomarkers. The complex interplay of genetic, environmental, and gene–environment interactions influencing the phenotype of drug response cannot be completely neglected. Since not enough information is available regarding the interplay of genetic and environmental influences on pharmacological response, the practice of personalized medicine appears incomplete. Optimal integration of pharmacogenomics in clinical practice also basically needs adequate training of health care professionals who practically manage the clinical prescription of drug regimens based on the available clinical and genetic test results.

## Glossary

**Adverse drug reactions** An appreciably harmful or unpleasant reaction resulting from an intervention related to the use of a medicinal product.

**Checkpoint inhibitors** Immune checkpoint inhibitors are a type of immunotherapy that blocks immune checkpoint proteins from binding with partner proteins.

**Chemotherapeutic regimens** Treatment that uses drugs to stop the growth of cancer cells, either by killing the cells or by stopping them from dividing.

**Haplo-cord transplant** Haploidentical (haplo)-cord transplantation.

**Immunoglobulins** A protein that is secreted by activated B-cells or plasma cells and helps the body fight infection or provide protection.

**Immunotherapy** A type of therapy that uses substances to stimulate or suppress the immune system to help the body fight cancer, infection, and other diseases.

**Matched-donor transplant** Donors and patients are matched by their HLA type; a brother or sister is most likely to be a match.

**Microarray analysis** A technique that scientists use to determine whether genes are on or off.

**Monotherapy** A therapy that uses only one type of treatment modality to treat a certain disease or condition.

**Next-generation sequencing** A technology of massively parallel sequencing that offers ultra-high throughput, scalability, and speed.

**Personalized medicine** An approach to treatment based on each person's unique genetic makeup; also called as precision medicine.

**Pharmacodynamics** A drug's molecular, biochemical, and physiologic effects or actions.

**Pharmacogenomics** The study of how genes affect a person's response to drugs.

**Pharmacokinetics** The study of how the body interacts with administered medications for the entire duration of exposure.

**Tailored therapy** A therapy to cure a patient who suffers from a specific disease with an effective and safe drug, based on the complex interactions among patient's characteristics, disease physiopathology, and drug metabolism.

**Transplantation** A medical procedure in which an organ, tissue, or group of cells are removed from one person (the donor) and transplanted into another person (the recipient), or moved from one site to another in the same person.

**Trigeminal neuralgia** Sudden, severe facial pain; it is often described as a sharp shooting pain or like having an electric shock in the jaw, teeth, or gums.

## References


Abdullah-Koolmees H, van Keulen AM, Nijenhuis M, Deneer VHM (2021) Pharmacogenetics guidelines: overview and comparison of the DPWG, CPIC, CPNDS, and RNPGx guidelines. Front Pharmacol 11:595219. https://doi.org/10.3389/fphar.2020.595219

Alachkar H, Nakamura Y (2015) Deep-sequencing of the T-cell receptor repertoire in patients with haplo-cord and matched-donor transplants. Chimerism 6(3):47–49. https://doi.org/10.1080/19381956.2015.1128624

Bocquet H, Bagot M, Roujeau JC (1996) Drug-induced pseudolymphoma and drug hypersensitivity syndrome (drug rash with eosinophilia and systemic symptoms: DRESS). Semin Cutan Med Surg 15(4):250–257

Buermans HP, den Dunnen JT (2014) Next generation sequencing technology: advances and applications. Biochim Biophys Acta 1842(10):1932–1941. https://doi.org/10.1016/j.bbadis.2014.06.015

Cancer Genome Atlas Research Network, Weinstein JN, Collisson EA, Mills GB, Shaw KR, Ozenberger BA, Ellrott K, Shmulevich I, Sander C, Stuart JM (2013) The cancer genome atlas pan-cancer analysis project. Nat Genet 45(10):1113–1120

Cecchin E, Stocco G (2020) Pharmacogenomics and personalized medicine. Genes (Basel) 11(6):679. https://doi.org/10.3390/genes11060679. PMID: 32580376; PMCID: PMC7348959.

Chapman CG, Yamaguchi R, Tamura K, Weidner J, Imoto S, Kwon J, Fang H, Yew PY, Marino SR, Miyano S, Nakamura Y, Kiyotani K (2016) Characterization of T-cell receptor repertoire in inflamed tissues of patients with Crohn's disease through deep sequencing. Inflamm Bowel Dis 22(6):1275–1285. https://doi.org/10.1097/MIB.0000000000000752. PMID: 27135481

De Plaen E, Lurquin C, Lethé B, van der Bruggen P, Brichard V, Renauld JC, Coulie P, Van Pel A, Boon T (1997) Identification of genes coding for tumor antigens recognized by cytolytic T lymphocytes. Methods 12(2):125–142. https://doi.org/10.1006/meth.1997.0462. PMID: 9184377.

Dean L (2012) Carbamazepine therapy and HLAGenotype. 2015 Oct 14 [updated 2018 Aug 1]. In: Pratt VM, Scott SA, Pirmohamed M, Esquivel B, Kane MS, Kattman BL, Malheiro AJ, editors. Medical genetics summaries [Internet]. National Center for Biotechnology Information (US), Bethesda. PMID: 28520367.

Erdag G, Schaefer JT, Smolkin ME, Deacon DH, Shea SM, Dengel LT, Patterson JW, Slingluff CL Jr (2012) Immunotype and immunohistologic characteristics of tumor-infiltrating immune cells are associated with clinical outcome in metastatic melanoma. Cancer Res 72(5):1070–1080. https://doi.org/10.1158/0008-5472.CAN-11-3218. Epub 2012 Jan 19. PMID: 22266112; PMCID: PMC3306813

Ferrell PB Jr, McLeod HL (2008) Carbamazepine, HLA-B*1502 and risk of Stevens-Johnson syndrome and toxic epidermal necrolysis: US FDA recommendations. Pharmacogenomics 9(10):1543–1546. https://doi.org/10.2217/14622416.9.10.1543. PMID: 18855540; PMCID: PMC2586963

Galon J, Costes A, Sanchez-Cabo F, Kirilovsky A, Mlecnik B, Lagorce-Pagès C, Tosolini M, Camus M, Berger A, Wind P, Zinzindohoué F, Bruneval P, Cugnenc PH, Trajanoski Z, Fridman WH, Pagès F (2006) Type, density, and location of immune cells within human colorectal tumors predict clinical outcome. Science 313(5795):1960–1964. https://doi.org/10.1126/science.1129139. PMID: 17008531.

Guy JW, Patel I, Oestreich JH (2020) Clinical application and educational training for pharmacogenomics. Pharmacy (Basel) 8(3):163. https://doi.org/10.3390/pharmacy8030163

Hargadon KM, Johnson CE, Williams CJ (2018) Immune checkpoint blockade therapy for cancer: an overview of FDA-approved immune checkpoint inhibitors. Int Immunopharmacol 62:29–39. https://doi.org/10.1016/j.intimp.2018.06.001. Epub 2018 Jul 2. PMID: 29990692.

Hurd PJ, Nelson CJ (2009) Advantages of next-generation sequencing versus the microarray in epigenetic research. Brief Funct Genomic Proteomic 8(3):174–183. https://doi.org/10.1093/bfgp/elp013

Illing PT, Vivian JP, Purcell AW, Rossjohn J, McCluskey J (2013) Human leukocyte antigen-associated drug hypersensitivity. Curr Opin Immunol 25(1):81–89

International Cancer Genome Consortium, Hudson TJ, Anderson W, Artez A, Barker AD, Bell C, Bernabé RR et al (2010) International network of cancer genome projects. Nature 464(7291):993–998. https://doi.org/10.1038/nature08987. Erratum in: Nature. 2010 Jun 17;465(7300):966

Jiang T, Shi T, Zhang H, Hu J, Song Y, Wei J, Ren S, Zhou C (2019) Tumor neoantigens: from basic research to clinical applications. J Hematol Oncol 12(1):93. https://doi.org/10.1186/s13045-019-0787-5. PMID: 31492199; PMCID: PMC6731555

Karnes JH, Miller MA, White KD, Konvinse KC, Pavlos RK, Redwood AJ, Peter JG, Lehloenya R, Mallal SA, Phillips EJ (2019) Applications of immunopharmacogenomics: predicting, preventing, and understanding immune-mediated adverse drug reactions. Annu Rev Pharmacol Toxicol 59:463–486. https://doi.org/10.1146/annurev-pharmtox-010818-021818. Epub 2018 Aug 22. PMID: 30134124; PMCID: PMC6409210

Khan DA, Phillips EJ (2020) Pharmacogenomic biomarkers in allergy and immunology practice. J Allergy Clin Immunol 146(3):509–512. https://doi.org/10.1016/j.jaci.2020.05.050

Kiyotani K, Chan HT, Nakamura Y (2018) Immunopharmacogenomics towards personalized cancer immunotherapy targeting neoantigens. Cancer Sci 109(3):542–549. https://doi.org/10.1111/cas.13498

Kiyotani K, Toyoshima Y, Nakamura Y (2021) Immunogenomics in personalized cancer treatments. J Hum Genet 66(9):901–907. https://doi.org/10.1038/s10038-021-00950-w. Epub 2021 Jun 30. PMID: 34193979.

Krebs K, Milani L (2019) Translating pharmacogenomics into clinical decisions: do not let the perfect be the enemy of the good. Hum Genomics 13:39. https://doi.org/10.1186/s40246-019-0229-z

Levy KD, Wu RR, Goto D et al (2020) Translating pharmacogenetics from research to routine clinical practice—a survey of the IGNITE network. Transl Med Commun 5:7. https://doi.org/10.1186/s41231-020-00057-y

Mehta RS, Rezvani K (2016) Immune reconstitution post allogeneic transplant and the impact of immune recovery on the risk of infection. Virulence 7(8):901–916. https://doi.org/10.1080/21505594.2016.1208866

Miyagawa F, Asada H (2021) Current perspective regarding the immunopathogenesis of drug-induced hypersensitivity syndrome/drug reaction with eosinophilia and systemic symptoms (DIHS/DRESS). Int J Mol Sci 22(4):2147

Nakamura Y (2015) Immunopharmacogenomics. Springer. https://www.springer.com/gp/book/9784431557258

Peng M, Mo Y, Wang Y, Wu P, Zhang Y, Xiong F, Guo C, Wu X, Li Y, Li X, Li G, Xiong W, Zeng Z (2019) Neoantigen vaccine: an emerging tumor immunotherapy. Mol Cancer 18(1):128. https://doi.org/10.1186/s12943-019-1055-6. PMID: 31443694; PMCID: PMC6708248

Pirmohamed M (2011) Pharmacogenetics: past, present and future. Drug Discov Today 16(19–20):852–861. https://doi.org/10.1016/j.drudis.2011.08.006. Epub 2011 Aug 22. PMID: 21884816.

Pompeu YA, Stewart JD, Mallal S, Phillips E, Peters B, Ostrov DA (2012) The structural basis of HLA-associated drug hypersensitivity syndromes. Immunol Rev 250(1):158–166

Qi Q, Liu Y, Cheng Y, Glanville J, Zhang D, Lee JY, Olshen RA, Weyand CM, Boyd SD, Goronzy JJ (2014) Diversity and clonal selection in the human T-cell repertoire. Proc Natl Acad Sci U S A 111(36):13139–13144. https://doi.org/10.1073/pnas.1409155111. Epub 2014 Aug 25. PMID: 25157137; PMCID: PMC4246948

Ramos-Casals M, Brahmer JR, Callahan MK, Flores-Chávez A, Keegan N, Khamashta MA, Lambotte O, Mariette X, Prat A, Suárez-Almazor ME (2020) Immune-related adverse events of checkpoint inhibitors. Nat Rev Dis Primers 6(1):38. https://doi.org/10.1038/s41572-020-0160-6. PMID: 32382051

Scott SA (2011) Personalizing medicine with clinical pharmacogenetics. Genet Med 13(12):987–995. https://doi.org/10.1097/GIM.0b013e318238b38c

Segal AW (2019) Studies on patients establish Crohn's disease as a manifestation of impaired innate immunity. J Intern Med 286(4):373–388. https://doi.org/10.1111/joim.12945. Epub 2019

Shemesh CS, Hsu JC, Hosseini I, Shen BQ, Rotte A, Twomey P, Girish S, Wu B (2021) Personalized cancer vaccines: clinical landscape, challenges, and opportunities. Mol Ther 29(2):555–570. https://doi.org/10.1016/j.ymthe.2020.09.038. Epub 2020 Sep 30. PMID: 33038322; PMCID: PMC7854282

Shyr D, Liu Q (2013) Next generation sequencing in cancer research and clinical application. Biol Proced Online 15(1):4

Song R, Jia X, Zhao J, Du P, Zhang JA (2021) T cell receptor revision and immune repertoire changes in autoimmune diseases. Int Rev Immunol 41:517–533

Szekanecz Z, Meskó B, Poliska S et al (2013) Pharmacogenetics and pharmacogenomics in rheumatology. Immunol Res 56(2–3):325–333. https://doi.org/10.1007/s12026-013-8405-z

Tickotsky-Moskovitz N, Louzoun Y, Dvorkin S, Rotkopf A, Kuperman AA, Efroni S (2021) CDR3 and V genes show distinct reconstitution patterns in T cell repertoire post-allogeneic bone marrow transplantation. Immunogenetics 73(2):163–173. https://doi.org/10.1007/s00251-020-01200-7. Epub 2021 Jan 21. PMID: 33475766

Ventola CL (2011) Pharmacogenomics in clinical practice: reality and expectations. P T 36(7):412–450

Vijverberg SJ, Pieters T, Cornel MC (2010) Ethical and social issues in pharmacogenomics testing. Curr Pharm Des 16(2):245–252. https://doi.org/10.2174/138161210790112700. PMID: 20205666.

Yamamoto-Furusho JK (2017) Pharmacogenetics in inflammatory bowel disease: understanding treatment response and personalizing therapeutic strategies. Pharmgenomics Pers Med 10:197–204. https://doi.org/10.2147/PGPM.S109648

Yerly D, Pompeu YA, Schutte RJ, Eriksson KK, Strhyn A, Bracey AW, Buus S, Ostrov DA (2017) Structural elements recognized by Abacavir-induced T cells. Int J Mol Sci 18(7):1464

Yew PY, Alachkar H, Yamaguchi R, Kiyotani K, Fang H, Yap KL, Liu HT, Wickrema A, Artz A, van Besien K, Imoto S, Miyano S, Bishop MR, Stock W, Nakamura Y (2015) Quantitative characterization of T-cell repertoire in allogeneic hematopoietic stem cell transplant recipients. Bone Marrow Transplant 50(9):1227–1234

Zewde M, Kiyotani K, Park JH et al (2018) The era of immunogenomics/immunopharmacogenomics. J Hum Genet 63:865–875. https://doi.org/10.1038/s10038-018-0468-1

Zhang Z, Lu M, Qin Y, Gao W, Tao L, Su W, Zhong J (2021) Neoantigen: a new breakthrough in tumor immunotherapy. Front Immunol 12:672356. https://doi.org/10.3389/fimmu.2021.672356. PMID: 33936118; PMCID: PMC8085349

# Immunology Encounters Nano-technology: Recent Developments and Future Prospects of Nano-based Drug Delivery Systems

**Soumitra Mohanty**

**Abstract** Biomaterials possess a unique property that makes them suitable for different biomedical applications. The emergence of drug resistance strain demands reduced usage of antibiotics and urges search for an alternate treatment method to directly combat the bacterial infection or to formulate efficient drug delivery methods. Apart from antimicrobial activity, these materials also serve as potent anticancer agent. These formulations need to be biologically active with a potent ability to induce the immune response. However, on many occasions these particles are of not prime choice because of its high toxicity towards host cells and poor biocompatibility which restrict their application. On contrary, in recent time, several nano formulations have been developed which can possibly impact the human health and the environment. Hence, in this present review we provide a series of comprehensive methods of nanoparticle synthesis and its immune response alongside with its biomedical application.

**Keywords** Antimicrobial peptides · Biomaterial · Biomaterial scaffolds · Dendrimers · Hydrogels · Microporous annealed particle · Nanoencapsulation

## 1 Introduction

In recent times, emergence of several new diseases and multiple challenges with existing drugs demand new drug development with a better delivery method. Even if we consider only bacterial infections, rise of multiple drug resistance and

S. Mohanty (✉)
Biotechnology Research and Innovation Council-National Institute of Biomedical Genomics (BRIC-NIBMG), Kalyani, India

Department of Microbiology, Tumor and Cell Biology, Karolinska Institutet, Stockholm, Sweden
e-mail: sm14@nibmg.ac.in



A. Das Mohapatra, P. S. Sahu (eds.), *Advances in Immunology and Immuno-techniques*, Interdisciplinary Biotechnological Advances,
https://doi.org/10.1007/978-981-97-5508-0_8

extremely drug-resistant bacteria further worsens the situation. This is not just limited to microbial infections rather situation is quite alarming for other metabolic disorders like diabetes, cancer, cardiovascular and chronic kidney disorders.

Over the years, plant-based natural products contribute largely to the pharmaceutical industries. Natural products are mainly used for drug development because of several salient features like easy affordability, low toxicity, and better therapeutic potential (Patra et al. 2018). However, biocompatibility, non-specificity, adverse side effects, poor efficiency, and uncontrolled biodistribution are major concerns that need special attention (Kadam et al. 2012).

In recent times, nanomedicine is emerged as an efficient approach for disease diagnosis and therapy. Nanoparticles are mainly ranging from 1 to 100 nanometers (nm) of size. Being small in size, the penetration to desired deeper tissue with targeted delivery is achievable with proper bioavailability. Nanoparticulate drug carriers can easily bypass the complex blood–brain barrier and the tight epithelial junctions which normally impede drug delivery. A combination of therapeutic molecules with nanoparticles also offers a unique combination of targeted delivery to desired cells or tissues in the body with improved stability (Jahan et al. 2017). Particularly, nano-technology-based drug delivery is important when drug is of protein or nucleic acid in nature (Emerich and Thanos 2007). Targeted delivery will reduce the concentration of therapeutic agent which eventually reduces the usage of therapeutic agent resulted in minimal adverse effect. This will also benefit water in soluble therapeutic agents with improved bioavailability (Yetisgin et al. 2020).

In future, these novel methods of drug delivery will lead to more effective and targeted treatments and will influence new drug formulations for already existing drugs (Emerich and Thanos 2007). Here, we summarize different approach of nanoparticle methods and its potential role in immune modulation of host.

## 2 Different Mode of Drug Delivery Methods

Currently, several methods are used for the designing of different efficient drug delivery methods (Fig. 1). A detailed description about several synthesis methods is highlighted below.

## 3 Coated Metallic Nanoparticles

Most metallic nanoparticles with high concentration are toxic in nature. Reduction of toxicity can be achieved by coating nanoparticles with several polymeric carbohydrates like starch and chitosan. Metallic nanoparticles such as silver, titanium dioxide, iron oxide, and gold can effectively target against bacteria and host cells. For example, starch and chitosan-coated silver nanoparticles can effectively kill several human Gram-negative pathogens like *P. aeruginosa*, *S. typhi*, *S. flexneri*;

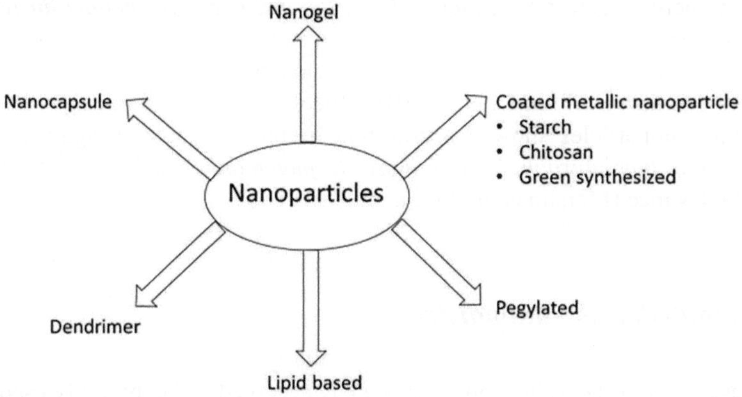


**Fig. 1** Different mode of nanoparticle synthesis and drug delivery methods

Gram-positive bacteria such as *S. aureus* and acid-fast bacteria, *M. smegmatis* (Jena et al. 2012; Mohanty et al. 2012). Similarly, starch-coated gold nanoparticles are known to have antibacterial activity against a various microorganism, including *E. coli*, *E. aerogenes*, *P. aeruginosa*, *S. epidermidis*, *E. durans*, and *S. bovis* (Pender et al. 2013). Similar antibacterial reports were also available for iron oxide (Xu et al. 2019) and titanium dioxide nanoparticles (Anaya-Esparza et al. 2020). This in turn highlights the wide range of application of antibacterial activity which might in future can lead to a new therapeutic approach with minimal chances of drug resistance.

Apart from antibacterial activity, these nanoparticles are also known to induce significant cytotoxicity against prostate (Morais et al. 2021) and colon (De Matteis et al. 2018) cancer cells.

In recent times, green synthesized nanoparticles have become popular to reduce the environmental toxicity. Thereafter, the focus has been shifted towards the green synthesis of nanoparticles exploiting bacterial, fungal biomasses and plant extracts. For example, biogenic nanoparticles synthesized using *A. macrophylla* and *Trichoderma* sp. biomasses showed high antibacterial activity against *M. smegmatis* and *M. marinum* (Mohanty et al. 2013).

## *3.1 PEGylated Nanoparticles*

Polyethylene glycol (PEG) is most commonly established polymer for the greater stability as a nano carrier. PEGylated conjugated drugs imposed reduced renal filtration, decreased uptake by the reticuloendothelial system, and diminished enzymatic degradation which also improve the half-life of conjugated drug (Knop et al. 2010). Both silver and gold nanoparticles can effectively conjugate in PEG. There are several reports of antibacterial and anticancer activity of PEGylated nanoparticles which are active against *S. epidermidis, S. aureus, and also methicillin-resistant S. aureus,* MRSA (Fan et al. 2019). PEGylated and silver nanoparticles loaded graphene oxide and ZnO nanoparticles for significant bactericidal activity against both *E. coli* and *S. aureus* (Pranjali et al. 2019; Zhao et al. 2017). Similarly, PEGylated ofloxacin nanoparticles showed a profound bactericidal activity against *E. coli, P. vulgaris, S. typhimurium, P. aeruginosa, K. pneumoniae*, and *S. aureus* of major clinical relevance (Marslin et al. 2015).

## *3.2 Lipid-Based Nanoparticles*

Micelles and lipid-based liposomes are widely accepted and commonly used methods for drug loading and delivery. The main component of the structure is vesicle which is similar to cell membrane.

Micelles are mostly composed of a hydrophilic head exposed out with a hydrophobic tail pointing to the micellar core, whereas liposomes are primarily made of amphiphilic lipids to form a structure identical as cell membranes. The large and complexity of outer and inner membrane allows relatively higher loading of antimicrobials for effective delivery to the targeted site (Puri et al. 2009).

By this targeted delivery, this will allow for better clearance of the invading pathogens without distressing the commensals and normal microbiota. Interestingly, the antimicrobial therapy like β-lactams, colistin, aminoglycosides, and glycopeptides for respiratory pathogens is quite challenging. This could be due to poor lung penetration (Ferreira et al. 2021).

Similarly, several aerosol-based antibiotics were approved because of controlled immunomodulatory activity with minimized systemic toxicity (Zarogoulidis et al. 2013). Likewise, liposomal aerosol formulations with antibiotics like amikacin, ciprofloxacin, and tobramycin found to be highly specific against pathogens including *P. aeruginosa* and *Mycobacterium* spp. (Bassetti et al. 2020).

Interestingly, in recent past a liposomal formulation combining colistin and ciprofloxacin as dry powder inhaler has been developed and tested against clinical isolates *P. aeruginosa,* in vitro (Wang et al. 2018). Similarly, reports are also present against chronic wounds such as in diabetes which are often difficult to treat (Lipsky and Hoey 2009). Similarly, several reports have shown liposomal formulations against skin infections focusing a significant role in skin regeneration and wound healing. Silver sulfadiazine encapsulated in liposomes is found to be competent enough in one dose to clear *P. aeruginosa* soft tissue infections (Price et al. 1990).

### 3.3 Dendrimers

Dendrimers are long polymers of nano meter range, with well-characterized, primarily composed of a single molecular size, and homogeneous structure comprising of symmetric branches forming around a small molecular core or a linear polymer core (Abbasi et al. 2014). In recent times, several biomedical applications are highlighted with dendritic polymers. Most likely dendrimers are used for their ability to perform a well-controlled and specific drug delivery ability.

Although in higher concentrations these molecules are toxic towards the cell, a series of cationic dendrimers derived from diethylene triamine lead to hyperbranched poly(amidoamine) (h-PAMAM) and simple one-pot and commercial synthesis method to synthesize different moles percent of methyl acrylate against *S. aureus*, *B. subtilis*, *E. coli*, *P. aeruginosa*, *methicillin-resistant S. aureus*, *C. albicans,* and *A. niger* (Labena et al. 2016).

Similarly, Boc-protected amino-functional bis-MPA dendrimers of generation one to five with trimethylopropane cores (TMP-Gx-NHBoc) were successfully synthesized from hydroxyl dendrimers through functionalization with boc-protected β-alanine and found to be active against *E. coli* (Stenström et al. 2017). The major advantage of these cationic dendrimer is its natural positively charged functional group which directly acts on the negatively charged membrane.

In line with that quaternary-ammonium compounds onto hyperbranched polyurea coatings found to be very active against *S. epidermidis* after adherence to bacterial surface with significant membrane damage of the bacteria (Asri et al. 2014). Therefore, these cationic hyperbranched polymers mimic conventional antimicrobial peptides and reduce the risk of drug resistance.

### 3.4 Nanocapsules

Nanoencapsulation is a technology used to deliver a drug molecule upon with encapsulation with biocompatible polymers like polyethylene glycol. For example, antibacterial essential oil in a nanocapsule found to be very targeted and efficient eradication of *S. aureus* (Ivanova et al. 2020). Similarly, nanoformulated antibiotics like penicillin and vancomycin have been effectively studied against Gram-negative bacteria (Fernandes et al. 2016; Fernandes et al. 2017). Likewise, a metal enzyme nanoaggregate using silver and active α-amylase observed to have a strong antibiofilm activity (Ferreres et al. 2018). It is also reported that lipid nanocapsules comprises of polyoxyl 15 hydroxystearate, triglycerides, soybean lecithin along with the drug chlorpromazine can effectively release the drug and have shown potent antibacterial activity against both Gram-positive *Staphylococcus aureus* and Gram-negative *E. coli*, *P. aeruginosa*, *K. pneumoniae,* and *A. baumannii (*Nehme et al. 2018*)*. Polymeric nanocapsules are also present with many advantages over conventional drugs to show their ability to display immediate and sustained drug release while keeping the integrity of the loaded drug. Notably, a chlorhexidine-loaded nanocapsule-based gel showed an immediate antibacterial effect within 30 s as tested against the skin flora (Nhung et al. 2007). Nanocapsule method might constitute a unique approach to deal with sensitive drugs which are less stable in body fluids.

### *3.5 Nanogels*

Nanogels are comprised of swollen polymer nanoparticles that can efficiently carry drugs to the site of infection. A major advantage of nanogel is the elasticity of nanogels that can be achieved with controlled drug delivery with very minimal side effects. Currently, delivery of a single drug using nanogel drug delivery systems has been established (Mahmoudian and Ganji 2017), which limits the application of drug. However, recently multiple drugs can be combined in nanogels for better activity which will eventually reduce the risk of drug resistance. Recently, hydrophobic antibiotic ciprofloxacin was embedded along with hydrophilic antibiotic novobiocin within a hybrid hydrogel. This combination resulted in homogenous distribution of antibiotics and found to be highly antibacterial against *E. coli* and *S. aureus (*Fan et al. 2021*)*. Similarly, hybrid nanogels prepared with varying lysozyme content carrying silver nanoparticle enhances the stabilization of silver nanoparticle with better bactericidal properties against *E. coli* and *S. aureus (*Coll Ferrer et al. 2014*)*. A lysine-based α-peptide/β-peptoid hybrid together with octenyl succinic anhydride-modified hyaluronic acid formulated into a nanogel has been shown potent activity against *P. aeruginosa* with improved cell selectivity (Kłodzińska et al. 2018). Apart from conventional drugs, nanogel containing silver nanoparticles showed a potent bactericidal activity against *E. coli* and *S. aureus*. This nanogel formulation was also found to heal the diabetic foot ulcer which also minimizes the risk of other infections in the wound site (Choi et al. 2013). Polysaccharide nanogels loaded with zinc ions also found to be active against target MRSA (Malzahn et al. 2014).

## 4 Immunological Responses to Nanoparticles

Immunological response is a key phenomenon that occurs in response to pathogenic microbes as well as drugs. The immune system differentiates between foreign and within and helps the host to eliminate toxic or allergenic substances. A detailed description of both innate and adaptive immune responses (Fig. 2) upon interaction with different nanoconjugates is highlighted as follows.

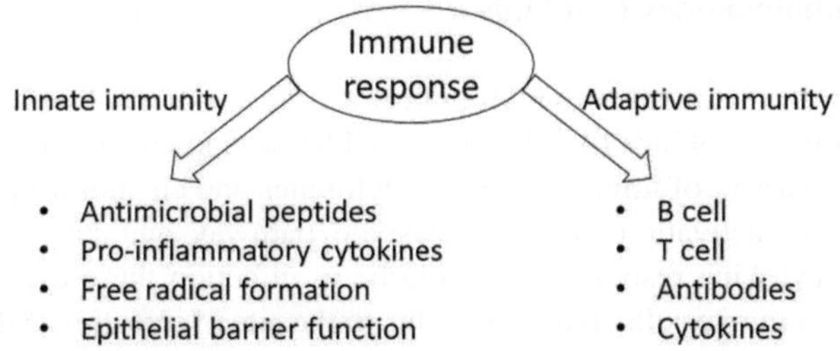


**Fig. 2** Different parameters involved in the immune response

# 5 Alternation in Innate Immune Response

## 5.1 *Endogenous Antimicrobial Peptides*

Upon encounters with pathogens, host cells that secret small peptides with strong antimicrobial activity are commonly known as antimicrobial peptides (AMPs). These AMPs are secreted by many different cell types such as epithelial cells, macrophages, and neutrophils. These peptides are cationic in nature and mainly act on negatively charged cell membrane. This led to the formation of pore on the surface of the bacteria. Apart from making pores, these AMPs are also known to inhibit DNA replication and protein synthesis. Some commonly available AMPs are cathelicidin, defensin family proteins like human β defensin 1, 2, and 3, RNase7 superfamily protein, RNase7 and S100 family protein, and psoriasin. The level of these endogenously expressed AMPs is crucial to control the invading pathogens. Therefore, drugs activating these peptides in a controllable range is an optimal option to control the infection. Recently, several nanoparticles are known to induce the host AMPs. In this approach, these nanoparticles can resolve the infection by two different mechanisms, one being the direct antimicrobial action and secondly by inducing these host AMPs. Recently, a report demonstrated that silver nanoparticles coated with ZnO, Ag@ZnO can induce the expression two antimicrobial peptides, human β defensin 2 and RNase7, which indicates that Ag@ZnO can also activate the skin innate immunity against bacterial infection (Majhi et al. 2021). As *S. aureus* is sensitive to hBD2 in vitro, (Midorikawa et al. 2003) RNase7 prevents skin colonization by *S. aureus* (Simanski et al. 2010). Similarly, silver nanoparticles overexpress psoriasin in epidermal keratinocytes (Kim et al. 2019). These findings can lead to future topical application options to combat skin infection. Similar approaches are now relevant to confirm in other infection model like urinary tract and lung infection.

### 5.1.1 Pro-inflammatory Cytokines

Like AMPs, pro-inflammatory cytokines play a vital role in innate immune defense. Pro-inflammatory cytokines like IL-1β, IL-6, TNF-α, and chemokine IL-8 released in a series of a variety of infections. These cytokines and chemokines are primarily mediated due to activation of inflammasome pathway and known as the pro-inflammatory cytokine response. Particularly, in infection these cytokines help in recruitment of immune cells like neutrophils to the site of infection. Silver nanoparticles can considerably induce pro-inflammatory cytokine gene expression of IL-1, IL-6, and TNF-α in human monocyte-derived macrophages, THP-1 and primary blood monocytes (Murphy et al. 2016). It is not just limited to silver nanoparticles rather all other nanoparticles such as titanium dioxide and zinc oxide nanoparticles also stimulate IL-8 in THP-1 monocyte-derived macrophages and human epithelial colorectal adenocarcinoma, Caco-2 cells (Tada-Oikawa et al. 2020). Non-coated and polyacrylic acid-coated iron oxide nanoparticles overexpressed IL-1β, IL-8, IL-6, TNF, and IL-10 in primary human blood cells (Couto et al. 2015). Similarly, copper and polystyrene also induce IL-1β in skin and in the intestine of zebrafish (Brun et al. 2018). This highlights the importance of nanoparticles in upregulation of pro-inflammatory cytokines.

### 5.1.2 Free Radical Formation

Free radical formation is also a host-driven phenomenon in response to infection. There are several types of radicals, but most known forms are from oxygen, and commonly known as reactive oxygen species like superoxide anion, peroxide, hydroxyl radical, and hydroxyl ion. However, there are also nitrogen reactive species known as nitric oxide, nitrous oxide, peroxynitrite, and nitroxyl anion. Free radicals are short lived but significantly induce the genotoxicity in the pathogens. In several nanoparticles like gold-silver nanocage, ZnO induces destruction of cell membrane and inhibits the enzymatic activity, DNA synthesis, and induction of apoptosis (Sirelkhatim et al. 2015; Wang et al. 2016).

### 5.1.3 Epithelial Barrier Function

Epithelial barrier is a major barrier function to prevent the entry of pathogen. Apart from regulating nutrient absorption, a strong epithelial barrier is good enough to protect the host from the infection. For example, during skin injury disruption of epithelial barrier allows the bacteria to reach deeper layer of tissues, where bacteria often change the morphology, survive and cause severe infection, and form biofilm within. Often antibiotics are difficult and are unable to reach and treatment is not possible. Although there are no extensive reports available for direct impact of nanoparticles on the tight junction proteins from the infection aspect, chitosan nanoparticles are known to open claudin expression in Caco-2 cells in certain concentration (Zhou et al. 2020). But this could be an important area to investigate in future. Interestingly, silver nanoparticles accelerate re-epithelialization and promote wound healing (Rigo et al. 2013). However, in other conditions like cancer, the role of nanoparticles is vital where partial disruption of tight junctions allows the uptake of therapeutic molecules with poor bioavailability; such observation allows a better drug delivery approach. As nanoparticles can differentially regulate the expression of tight junction proteins based on the nature of the nanoparticle, these molecules potentially can be designed as a therapeutic molecule.

## 5.2 *Alternation in Adaptive Immune Response by Different Nanoparticles*

B cells are crucial in the adaptive immune response with a very specific and well-regulated role. Selection process is regulated in the bone marrow. Migration of immature B cells to secondary lymph nodes where B cells are mainly activated. Upon activation, isotype switching and affinity maturation help later, faster, and stronger immune response (Wardemann et al. 2003).

Similarly, like B cells, T cells undertake highly explicit differentiation for antigen specificity. T cells are produced in bone marrow; however, early premature T cells are transported to the Thymus for further selective maturation to produce either CD4 or CD8 T cells. Upon positive selection, affinity towards the specific antigen occurred through MHC bindings with antigen presenting cells. CD4+ cells further differentiate into helper T cells can differentiate into Th1, Th2, Th9, Th17, Th22, Treg, and Tfh with different cytokine (Deng 2018) and gene expression profiles while CD8+ T cells are most responsible for the elimination of cells with non-self-peptide:MHC I complexes (Andersen et al. 2006).

### 5.2.1 Silver and Gold Nanoparticles

Apart from direct antibacterial activity, silver nanoparticles are believed to trigger immunological responses. It is also evident from the study where silver nanoparticles triggerd programmed cell death of inflammatory cells in a mouse model to study allergic contact dermatitis (185) and by activating free radicals, apoptosis in THP-1 monocytes (186). This process can trigger both innate and adaptive immune system with involvement of pro-inflammatory cytokine IL-1β. It is also important to note that silver nanoparticles not only activate cytokines but also are found to have a differential role in multiple cell types. For example, silver nanoparticles also significantly inhibit phytohemagglutinin (PHA)-induced IL-5, IFN-γ, and TNF-α in human PBMCs. Similarly, they inhibit both PHA and ConA-induced IL-2 production in human PBMCs. However, polyvinylpyrrolidone (PVP)-coated AgNPs significantly increased the generations of IL-6 and IL-8 but significantly decreased the release of IL-1ra from human PBMCs (Li et al. 2018). This differential regulation and immune response make silver nanoparticle an exciting candidate to design novel therapeutics to regulate the immune response. At the same time, there are few reports about gold nanoparticles, one study in dendritic cells highlighted inhibitory effect on the release of interleukin IL-12p70, a well-known activator for T cell (Villiers et al. 2010).

### 5.2.2 Iron Oxide and Titanium Dioxide

Iron oxide nanoparticles are found to have a mixed effect on different cells. Iron nanoparticles treated in a murine model showed both helper T cells and cytotoxic T cells in circulating blood, with higher expression of IL-2 and interferon, mostly secreted by T cells promoted an immunological activity (Chen et al. 2010). Iron oxide nanoparticles also found to have a negative effect on endothelial cells. It has been also shown that iron nanoparticles caused alteration in the aortic endothelial cells with increased levels of cell mortality and IL-8 secretion (Zhu et al. 2011).

Titanium dioxide nanoparticles are among some commonly used nanoparticles in therapeutics. However, long-term exposure to these nanoparticles can potentially induce spleen damage with alternation in inflammatory response (Sang et al. 2012). Similarly, an impaired chemotaxis of phagocytic macrophages was observed with decreased MHC-class II expression. However, there is also a report of low concentration of titanium dioxide that significantly induced the phagocytic activity of macrophages (Liu et al. 2010). There are also several other reports that state the involvement of cytokines like IL-2, IL-4, IL-6, IL-10, and IL-18, indicating immunomodulatory effect of titanium dioxide.

### 5.2.3 Hydrogels

In regenerative medicine, restoration of tissue function is a major event. In a subcutaneous implantation model, microporous annealed particle (MAP) gel showed that recruitment of myeloid cells within and around the scaffold are CD11b positive. Immune response is often triggered by the release of IL-33. It has also reported the presence of IL-33-expressing F4/80$^{+}$CD11b$^{+}$ macrophages around the implants (Griffin et al. 2021). Similarly, macrophages releases CXCL13, which are often known to prime the recruitment of B cell. Interestingly, colony stimulating factor-1 receptor (CSF1R) is overexpressed following the implantation of multiple biomaterials that are comprised of ceramic, polymer, and hydrogel (Doloff et al. 2017). From prosthetic implants, microparticles are also released and macrophages stimulated to produce IL-33 leading to sterile inflammation (Mishra et al. 2019). It is also evident from the studies that among the innate immune effector cells, notably macrophages are vital and form a fibrous capsule around synthetic implants (Chung et al. 2017). Biomaterial scaffolds induced a pro-regenerative response, characterized by an mTOR/Rictor-dependent T helper 2 pathway that guides interleukin-4-dependent macrophage polarization (Sadtler et al. 2016). It is also evident that injection of microparticles into the knees of mice replicates the type 2 immunity induction in intraperitoneal model (de Kouchkovsky et al. 2019).

## 6 Conclusion

We hereby conclude a series of different nanoparticle synthesis methods and its biomedical applications in different disease conditions with special focus on bacterial infections. A detailed immunological response involving both innate and trained immune activation further highlighted the immunogenicity property of these biomaterials with a strong potential in clinical applications. In future, these methods further need to be improved to serve as a better therapeutic agent against conventional antibiotics. This in turn will also minimize the risk of increased cases of drug resistance and will open the avenue for new drug development.

## Glossary

**Antimicrobial peptides** A class of small peptides that widely exist in nature and they have inhibitory effects against bacteria, fungi, parasites, and viruses.

**Biomaterial** A substance that has been engineered to interact with biological systems for a medical purpose.

**Biomaterial scaffolds** Three-dimensional porous, fibrous, or permeable biomaterials intended to permit transport of body liquids and gases, having a great application in tissue engineering.

**Dendrimers** Long polymers, with well-characterized, primarily composed of a single molecular size, and homogeneous structure comprising of symmetric branches forming around a small molecular core or a linear polymer core.

**Epithelial barrier** A cellular barrier (composed of epithelial cells, tight junctions, and a mucus layer) that regulates nutrient absorption as well as prevents the invasion of pathogenic bacteria in the host.

**Hydrogels** Three-dimensional network structures able to imbibe large amounts of water; they exist naturally in the form of polymer networks such as collagen or gelatin, or can be made synthetically.

**Inflammasome** Large cytosolic multiprotein complexes that assemble in response to infection which is responsible for the activation of inflammatory responses.

**Microporous annealed particle** An emerging class of porous biomaterials and a versatile platform for regenerative medicine.

**Nanoencapsulation** Technology used to deliver a drug molecule upon with encapsulation with biocompatible polymers like polyethylene glycol.

**Re-epithelialization** Resurfacing of a wound with new epithelium.

**Regenerative medicine** A branch of medicine that focuses on developing and applying new treatments to restore the function lost due to aging, disease, damage, or defects.

**Zebrafish** Native to India and South Asia; it is a popular aquarium fish, frequently sold under the trade name zebra danio; represents a promising model system for the discovery of novel biological pathways, pharmacological candidates.

## References

Abbasi E, Aval SF, Akbarzadeh A, Milani M, Nasrabadi HT, Joo SW, Hanifehpour Y, Nejati-Koshki K, Pashaei-Asl R (2014) Dendrimers: synthesis, applications, and properties. Nanoscale Res Lett 9:247. https://doi.org/10.1186/1556-276x-9-247

Anaya-Esparza LM, Villagrán-de la Mora Z, Ruvalcaba-Gómez JM, Romero-Toledo R, Sandoval-Contreras T, Aguilera-Aguirre S, Montalvo-González E, Pérez-Larios A (2020) Use of titanium dioxide (TiO2) nanoparticles as reinforcement agent of polysaccharide-based materials. Processes 8:1395

Andersen MH, Schrama D, Thor Straten P, Becker JC (2006) Cytotoxic T cells. J Invest Dermatol 126:32–41. https://doi.org/10.1038/sj.jid.5700001

Asri LATW, Crismaru M, Roest S, Chen Y, Ivashenko O, Rudolf P, Tiller JC, van der Mei HC, Loontjens TJA, Busscher HJ (2014) A shape-adaptive, antibacterial-coating of immobilized quaternary-ammonium compounds tethered on hyperbranched polyurea and its mechanism of action. Adv Funct Mater 24:346–355. https://doi.org/10.1002/adfm.201301686

Bassetti M, Vena A, Russo A, Peghin M (2020) Inhaled liposomal antimicrobial delivery in lung infections. Drugs 80:1309–1318. https://doi.org/10.1007/s40265-020-01359-z

Brun NR, Koch BEV, Varela M, Peijnenburg WJGM, Spaink HP, Vijver MG (2018) Nanoparticles induce dermal and intestinal innate immune system responses in zebrafish embryos. Environ Sci Nano 5:904–916. https://doi.org/10.1039/C8EN00002F

Chen BA, Jin N, Wang J, Ding J, Gao C, Cheng J, Xia G, Gao F, Zhou Y, Chen Y et al (2010) The effect of magnetic nanoparticles of Fe(3)O(4) on immune function in normal ICR mice. Int J Nanomedicine 5:593–599. https://doi.org/10.2147/ijn.s12162

Choi JB, Park JS, Khil MS, Gwon HJ, Lim YM, Jeong SI, Shin YM, Nho YC (2013) Characterization and antimicrobial property of poly(acrylic acid) nanogel containing silver particle prepared by electron beam. Int J Mol Sci 14:11011–11023. https://doi.org/10.3390/ijms140611011

Chung L, Maestas DR Jr, Housseau F, Elisseeff JH (2017) Key players in the immune response to biomaterial scaffolds for regenerative medicine. Adv Drug Deliv Rev 114:184–192. https://doi.org/10.1016/j.addr.2017.07.006

Coll Ferrer MC, Dastgheyb S, Hickok NJ, Eckmann DM, Composto RJ (2014) Designing nanogel carriers for antibacterial applications. Acta Biomater 10:2105–2111. https://doi.org/10.1016/j.actbio.2014.01.009

Couto D, Freitas M, Porto G, Lopez-Quintela MA, Rivas J, Freitas P, Carvalho F, Fernandes E (2015) Polyacrylic acid-coated and non-coated iron oxide nanoparticles induce cytokine activation in human blood cells through TAK1, p38 MAPK and JNK pro-inflammatory pathways. Arch Toxicol 89:1759–1769. https://doi.org/10.1007/s00204-014-1325-4

de Kouchkovsky DA, Ghosh S, Rothlin CV (2019) Induction of sterile type 2 inflammation. Nat Mater 18:193–194. https://doi.org/10.1038/s41563-019-0300-0

De Matteis V, Cascione M, Toma CC, Leporatti S (2018) Silver nanoparticles: synthetic routes, in vitro toxicity and theranostic applications for cancer disease. Nanomaterials (Basel) 8:319. https://doi.org/10.3390/nano8050319

Deng G (2018) Tumor-infiltrating regulatory T cells: origins and features. Am J Clin Exp Immunol 7:81–87

Doloff JC, Veiseh O, Vegas AJ, Tam HH, Farah S, Ma M, Li J, Bader A, Chiu A, Sadraei A et al (2017) Colony stimulating factor-1 receptor is a central component of the foreign body response to biomaterial implants in rodents and non-human primates. Nat Mater 16:671–680. https://doi.org/10.1038/nmat4866

Emerich DF, Thanos CG (2007) Targeted nanoparticle-based drug delivery and diagnosis. J Drug Target 15:163–183. https://doi.org/10.1080/10611860701231810

Fan Y, Pauer AC, Gonzales AA, Fenniri H (2019) Enhanced antibiotic activity of ampicillin conjugated to gold nanoparticles on PEGylated rosette nanotubes. Int J Nanomedicine 14:7281–7289. https://doi.org/10.2147/ijn.S209756

Fan Y, Lüchow M, Zhang Y, Lin J, Fortuin L, Mohanty S, Brauner A, Malkoch M (2021) Nanogel encapsulated hydrogels as advanced wound dressings for the controlled delivery of antibiotics. Adv Funct Mater 31:2006453. https://doi.org/10.1002/adfm.202006453

Fernandes MM, Ivanova K, Francesko A, Rivera D, Torrent-Burgués J, Gedanken A, Mendonza E, Tzanov T (2016) Escherichia coli and Pseudomonas aeruginosa eradication by nano-penicillin G. Nanomedicine 12:2061–2069. https://doi.org/10.1016/j.nano.2016.05.018

Fernandes MM, Ivanova K, Hoyo J, Pérez-Rafael S, Francesko A, Tzanov T (2017) Nanotransformation of vancomycin overcomes the intrinsic resistance of gram-negative bacteria. ACS Appl Mater Interfaces 9:15022–15030. https://doi.org/10.1021/acsami.7b00217

Ferreira M, Ogren M, Dias JNR, Silva M, Gil S, Tavares L, Aires-da-Silva F, Gaspar MM, Aguiar SI (2021) Liposomes as antibiotic delivery systems: a promising nanotechnological strategy against antimicrobial resistance. Molecules 26:2047. https://doi.org/10.3390/molecules26072047

Ferreres G, Bassegoda A, Hoyo J, Torrent-Burgués J, Tzanov T (2018) Metal-enzyme nanoaggregates eradicate both gram-positive and gram-negative bacteria and their biofilms. ACS Appl Mater Interfaces 10:40434–40442. https://doi.org/10.1021/acsami.8b14949

Griffin DR, Archang MM, Kuan CH, Weaver WM, Weinstein JS, Feng AC, Ruccia A, Sideris E, Ragkousis V, Koh J et al (2021) Activating an adaptive immune response from a hydrogel scaffold imparts regenerative wound healing. Nat Mater 20:560–569. https://doi.org/10.1038/s41563-020-00844-w

Ivanova K, Ivanova A, Ramon E, Hoyo J, Sanchez-Gomez S, Tzanov T (2020) Antibody-enabled antimicrobial Nanocapsules for selective elimination of Staphylococcus aureus. ACS Appl Mater Interfaces 12:35918–35927. https://doi.org/10.1021/acsami.0c09364

Jahan ST, Sadat SMA, Walliser M, Haddadi A (2017) Targeted therapeutic nanoparticles: an immense promise to fight against cancer. J Drug Deliv 2017:9090325. https://doi.org/10.1155/2017/9090325

Jena P, Mohanty S, Mallick R, Jacob B, Sonawane A (2012) Toxicity and antibacterial assessment of chitosan-coated silver nanoparticles on human pathogens and macrophage cells. Int J Nanomedicine 7:1805–1818. https://doi.org/10.2147/ijn.S28077

Kadam RS, Bourne DW, Kompella UB (2012) Nano-advantage in enhanced drug delivery with biodegradable nanoparticles: contribution of reduced clearance. Drug Metab Dispos 40:1380–1388. https://doi.org/10.1124/dmd.112.044925

Kim CM, Jeong SH, Lee H, Ryu HJ, Son SW (2019) Silver nanoparticles induce Egr-1-dependent psoriasin expression via the ERK and p38 pathways. Clin Exp Dermatol 44:390–396. https://doi.org/10.1111/ced.13758

Kłodzińska SN, Molchanova N, Franzyk H, Hansen PR, Damborg P, Nielsen HM (2018) Biopolymer nanogels improve antibacterial activity and safety profile of a novel lysine-based α-peptide/β-peptoid peptidomimetic. Eur J Pharm Biopharm 128:1–9. https://doi.org/10.1016/j.ejpb.2018.03.012

Knop K, Hoogenboom R, Fischer D, Schubert US (2010) Poly(ethylene glycol) in drug delivery: pros and cons as well as potential alternatives. Angew Chem Int Ed Engl 49:6288–6308. https://doi.org/10.1002/anie.200902672

Labena A, Kabel KI, Farag RK (2016) One-pot synthesize of dendritic hyperbranched PAMAM and assessment as a broad spectrum antimicrobial agent and anti-biofilm. Mater Sci Eng C Mater Biol Appl 58:1150–1159. https://doi.org/10.1016/j.msec.2015.09.042

Li WT, Wang LY, Chang HW, Yang WC, Lo C, Pang VF, Chen MH, Jeng CR (2018) Th2 cytokine bias induced by silver nanoparticles in peripheral blood mononuclear cells of common bottlenose dolphins (Tursiops truncatus). PeerJ 6:e5432. https://doi.org/10.7717/peerj.5432

Lipsky BA, Hoey C (2009) Topical antimicrobial therapy for treating chronic wounds. Clin Infect Dis 49:1541–1549. https://doi.org/10.1086/644732

Liu R, Zhang X, Pu Y, Yin L, Li Y, Zhang X, Liang G, Li X, Zhang J (2010) Small-sized titanium dioxide nanoparticles mediate immune toxicity in rat pulmonary alveolar macrophages in vivo. J Nanosci Nanotechnol 10:5161–5169. https://doi.org/10.1166/jnn.2010.2420

Mahmoudian M, Ganji F (2017) Vancomycin-loaded HPMC microparticles embedded within injectable thermosensitive chitosan hydrogels. Prog Biomater 6:49–56. https://doi.org/10.1007/s40204-017-0066-x

Majhi RK, Mohanty S, Khan MI, Mishra A, Brauner A (2021) Ag@ZnO nanoparticles induce antimicrobial peptides and promote migration and antibacterial activity of keratinocytes. ACS Infect Dis 7:2068–2072. https://doi.org/10.1021/acsinfecdis.0c00903

Malzahn K, Jamieson WD, Dröge M, Mailänder V, Jenkins ATA, Weiss CK, Landfester K (2014) Advanced dextran based nanogels for fighting Staphylococcus aureus infections by sustained zinc release. J Mater Chem B 2:2175–2183. https://doi.org/10.1039/c3tb21335h

Marslin G, Revina AM, Khandelwal VK, Balakumar K, Sheeba CJ, Franklin G (2015) PEGylated ofloxacin nanoparticles render strong antibacterial activity against many clinically important human pathogens. Colloids Surf B Biointerfaces 132:62–70. https://doi.org/10.1016/j.colsurfb.2015.04.050

Midorikawa K, Ouhara K, Komatsuzawa H, Kawai T, Yamada S, Fujiwara T, Yamazaki K, Sayama K, Taubman MA, Kurihara H et al (2003) Staphylococcus aureus susceptibility to innate antimicrobial peptides, beta-defensins and CAP18, expressed by human keratinocytes. Infect Immun 71:3730–3739. https://doi.org/10.1128/iai.71.7.3730-3739.2003

Mishra PK, Palma M, Buechel B, Moore J, Davra V, Chu N, Millman A, Hallab NJ, Kanneganti TD, Birge RB et al (2019) Sterile particle-induced inflammation is mediated by macrophages releasing IL-33 through a Bruton's tyrosine kinase-dependent pathway. Nat Mater 18:289–297. https://doi.org/10.1038/s41563-018-0271-6

Mohanty S, Mishra S, Jena P, Jacob B, Sarkar B, Sonawane A (2012) An investigation on the antibacterial, cytotoxic, and antibiofilm efficacy of starch-stabilized silver nanoparticles. Nanomedicine 8:916–924. https://doi.org/10.1016/j.nano.2011.11.007

Mohanty S, Jena P, Mehta R, Pati R, Banerjee B, Patil S, Sonawane A (2013) Cationic antimicrobial peptides and biogenic silver nanoparticles kill mycobacteria without eliciting DNA damage and cytotoxicity in mouse macrophages. Antimicrob Agents Chemother 57:3688–3698. https://doi.org/10.1128/aac.02475-12

Morais M, Machado V, Dias F, Palmeira C, Martins G, Fonseca M, Martins CSM, Teixeira AL, Prior JAV, Medeiros R (2021) Starch-capped AgNPs' as potential cytotoxic agents against prostate cancer cells. Nanomaterials (Basel) 11:256. https://doi.org/10.3390/nano11020256

Murphy A, Casey A, Byrne G, Chambers G, Howe O (2016) Silver nanoparticles induce pro-inflammatory gene expression and inflammasome activation in human monocytes. J Appl Toxicol 36:1311–1320. https://doi.org/10.1002/jat.3315

Nehme H, Saulnier P, Ramadan AA, Cassisa V, Guillet C, Eveillard M, Umerska A (2018) Antibacterial activity of antipsychotic agents, their association with lipid nanocapsules and its impact on the properties of the nanocarriers and on antibacterial activity. PLoS One 13:e0189950. https://doi.org/10.1371/journal.pone.0189950

Nhung DT, Freydiere AM, Constant H, Falson F, Pirot F (2007) Sustained antibacterial effect of a hand rub gel incorporating chlorhexdine-loaded nanocapsules (Nanochlorex). Int J Pharm 334:166–172. https://doi.org/10.1016/j.ijpharm.2006.10.017

Patra JK, Das G, Fraceto LF, Campos EVR, Rodriguez-Torres MDP, Acosta-Torres LS, Diaz-Torres LA, Grillo R, Swamy MK, Sharma S et al (2018) Nano based drug delivery systems: recent developments and future prospects. J Nanobiotechnol 16:71. https://doi.org/10.1186/s12951-018-0392-8

Pender DS, Vangala LM, Badwaik VD, Willis CB, Aguilar ZP, Sangoi TN, Paripelly R, Dakshinamurthy R (2013) Bactericidal activity of starch-encapsulated gold nanoparticles. Front Biosci (Landmark Ed) 18:993–1002. https://doi.org/10.2741/4158

Pranjali P, Meher MK, Raj R, Prasad N, Poluri KM, Kumar D, Guleria A (2019) Physicochemical and antibacterial properties of PEGylated zinc oxide nanoparticles dispersed in peritoneal dialysis fluid. ACS Omega 4:19255–19264. https://doi.org/10.1021/acsomega.9b02615

Price CI, Horton JW, Baxter CR (1990) Topical liposomal delivery of antibiotics in soft tissue infection. J Surg Res 49:174–178. https://doi.org/10.1016/0022-4804(90)90258-4

Puri A, Loomis K, Smith B, Lee JH, Yavlovich A, Heldman E, Blumenthal R (2009) Lipid-based nanoparticles as pharmaceutical drug carriers: from concepts to clinic. Crit Rev Ther Drug Carrier Syst 26:523–580. https://doi.org/10.1615/critrevtherdrugcarriersyst.v26.i6.10

Rigo C, Ferroni L, Tocco I, Roman M, Munivrana I, Gardin C, Cairns WR, Vindigni V, Azzena B, Barbante C et al (2013) Active silver nanoparticles for wound healing. Int J Mol Sci 14:4817–4840. https://doi.org/10.3390/ijms14034817

Sadtler K, Estrellas K, Allen BW, Wolf MT, Fan H, Tam AJ, Patel CH, Luber BS, Wang H, Wagner KR et al (2016) Developing a pro-regenerative biomaterial scaffold microenvironment requires T helper 2 cells. Science 352:366–370. https://doi.org/10.1126/science.aad9272

Sang X, Zheng L, Sun Q, Li N, Cui Y, Hu R, Gao G, Cheng Z, Cheng J, Gui S et al (2012) The chronic spleen injury of mice following long-term exposure to titanium dioxide nanoparticles. J Biomed Mater Res A 100:894–902. https://doi.org/10.1002/jbm.a.34024

Simanski M, Dressel S, Gläser R, Harder J (2010) RNase 7 protects healthy skin from Staphylococcus aureus colonization. J Invest Dermatol 130:2836–2838. https://doi.org/10.1038/jid.2010.217

Sirelkhatim A, Mahmud S, Seeni A, Kaus NHM, Ann LC, Bakhori SKM, Hasan H, Mohamad D (2015) Review on zinc oxide nanoparticles: antibacterial activity and toxicity mechanism. Nanomicro Lett 7:219–242. https://doi.org/10.1007/s40820-015-0040-x

Stenström P, Hjorth E, Zhang Y, Andrén OCJ, Guette-Marquet S, Schultzberg M, Malkoch M (2017) Synthesis and in vitro evaluation of monodisperse amino-functional polyester dendrimers with rapid degradability and antibacterial properties. Biomacromolecules 18:4323–4330. https://doi.org/10.1021/acs.biomac.7b01364

Tada-Oikawa S, Eguchi M, Yasuda M, Izuoka K, Ikegami A, Vranic S, Boland S, Tran L, Ichihara G, Ichihara S (2020) Functionalized surface-charged SiO(2) nanoparticles induce pro-inflammatory responses, but are not lethal to Caco-2 cells. Chem Res Toxicol 33:1226–1236. https://doi.org/10.1021/acs.chemrestox.9b00478

Villiers C, Freitas H, Couderc R, Villiers MB, Marche P (2010) Analysis of the toxicity of gold nano particles on the immune system: effect on dendritic cell functions. J Nanoparticle Res 12:55–60. https://doi.org/10.1007/s11051-009-9692-0

Wang Y, Wan J, Miron RJ, Zhao Y, Zhang Y (2016) Antibacterial properties and mechanisms of gold-silver nanocages. Nanoscale 8:11143–11152. https://doi.org/10.1039/c6nr01114d

Wang S, Yu S, Lin Y, Zou P, Chai G, Yu HH, Wickremasinghe H, Shetty N, Ling J, Li J et al (2018) Co-delivery of ciprofloxacin and colistin in liposomal formulations with enhanced in vitro antimicrobial activities against multidrug resistant Pseudomonas aeruginosa. Pharm Res 35:187. https://doi.org/10.1007/s11095-018-2464-8

Wardemann H, Yurasov S, Schaefer A, Young JW, Meffre E, Nussenzweig MC (2003) Predominant autoantibody production by early human B cell precursors. Science 301:1374–1377. https://doi.org/10.1126/science.1086907

Xu C, Akakuru OU, Zheng J, Wu A (2019) Applications of iron oxide-based magnetic nanoparticles in the diagnosis and treatment of bacterial infections. Front Bioeng Biotechnol 7:141. https://doi.org/10.3389/fbioe.2019.00141

Yetisgin AA, Cetinel S, Zuvin M, Kosar A, Kutlu O (2020) Therapeutic nanoparticles and their targeted delivery applications. Molecules 25:2193. https://doi.org/10.3390/molecules25092193

Zarogoulidis P, Kioumis I, Porpodis K, Spyratos D, Tsakiridis K, Huang H, Li Q, Turner JF, Browning R, Hohenforst-Schmidt W et al (2013) Clinical experimentation with aerosol antibiotics: current and future methods of administration. Drug Des Devel Ther 7:1115–1134. https://doi.org/10.2147/dddt.S51303

Zhao R, Lv M, Li Y, Sun M, Kong W, Wang L, Song S, Fan C, Jia L, Qiu S et al (2017) Stable nanocomposite based on PEGylated and silver nanoparticles loaded graphene oxide for long-term antibacterial activity. ACS Appl Mater Interfaces 9:15328–15341. https://doi.org/10.1021/acsami.7b03987

Zhou S, Deng H, Zhang Y, Wu P, He B, Dai W, Zhang H, Zhang Q, Zhao R, Wang X (2020) Thiolated nanoparticles overcome the mucus barrier and epithelial barrier for oral delivery of insulin. Mol Pharm 17:239–250. https://doi.org/10.1021/acs.molpharmaceut.9b00971

Zhu MT, Wang B, Wang Y, Yuan L, Wang HJ, Wang M, Ouyang H, Chai ZF, Feng WY, Zhao YL (2011) Endothelial dysfunction and inflammation induced by iron oxide nanoparticle exposure: risk factors for early atherosclerosis. Toxicol Lett 203:162–171. https://doi.org/10.1016/j.toxlet.2011.03.021

# Novel Drug Targets in Neuro-immune Disorders

**Bapi Gorain, Varnita Karmakar, Manisha Pandey, Pooja, Biswa Mohan Sahoo, Dhirodatta Senapati, and Subrat Kumar Bhattamisra**

**Abstract** Neuro-immune disorders are mediated through the immune system damaging the central and peripheral nervous systems. Several such disorders are diagnosed in different age groups and are mainly linked to autoimmune damage or infections. Persons with a compromised immune system, familial history, and environmental factors are the primary cause of these disorders. Based on the pattern of pathological lesions in the nervous system, neuro-immune disorders are either monofocal or multifocal deficits. Some examples of neuro-immune disorders are optic neuritis, neuromyelitis optica, multiple sclerosis, encephalomyelitis, etc. Currently, there are conventional approaches for treating these disorders where antimicrobials and psychotropic drugs such as antidepressants, antipsychotics, and psychostimulants are widely used. These agents benefit the patients by reducing infections and inflammations in the nervous system and relieving neurobehavioral symptoms. Research on various drug targets is in progress to find out better and safer therapeutics to control these disorders. Newer drug targets that control the inflammation and/or provide neuroprotection are taken into a priority to develop

B. Gorain · V. Karmakar
Department of Pharmaceutical Sciences, Birla Institute of Technology, Ranchi, Jharkhand, India

M. Pandey · Pooja
Department of Pharmaceutical Sciences, Central University of Haryana, Mahendragarh, Haryana, India

B. M. Sahoo
School of Pharmacy and Life Sciences, Centurion University of Technology & Management, Bhubaneswar, Odisha, India

D. Senapati
KIIT School of Pharmacy, KIIT (Deemed to be University), Bhubaneswar, Odisha, India

S. K. Bhattamisra (✉)
Department of Pharmacy, School of Health Sciences, Central University of South Bihar, Gaya, India
e-mail: subrat@cusb.ac.in



A. Das Mohapatra, P. S. Sahu (eds.), *Advances in Immunology and Immuno-techniques*, Interdisciplinary Biotechnological Advances,
https://doi.org/10.1007/978-981-97-5508-0_9

newer drugs for neuro-immune disorders. Such drug targets are the glutathione system, SIRT1 pathway, Th17 cells, neuropeptide Y, toll-like receptors, and tryptophan catabolites which are discussed in this book chapter.

**Keywords** Neuro-immune disorders · Antimicrobials · Psychotropic drugs · SIRT1 · Glutathione · Toll-like receptors · Th17 · Inflammation

# 1 Introduction

Neuro-immune disorders refer to immune-mediated neurological disorders in which the immune system mainly attacks the central and peripheral nervous systems. Autoimmune disorders affect any part of the nervous system that includes the central (spinal cord and brain), peripheral (neuromuscular junction and peripheral nerves), and autonomic nervous system (ANS) (Galetta and Bhattacharyya 2019). Several neuro-immune disorders such as neuromyelitis optica, optic neuritis, multiple sclerosis, transverse myelitis, encephalomyelitis, and spasticity are reported (Flanagan 2016). The characteristic symptoms of neuro-immune disorders are based on the nerves affected such as visual impairments in neuromyelitis optica and optic neuritis; muscle weakness, tremors, painful nerve-joint, loss of muscle coordination, locomotor functions and paralysis are common in neuromuscular nerve and some central autoimmune disorders; other symptoms of central autoimmune disorders are cognitive impairment, dementia, depression, seizures, convulsions, and behavioral changes (Greenberg et al. 2019).

Neuro-immune disorders are categorized into monofocal or multifocal neurological deficits depending on the pathological lesions pattern in the nervous system. The radiological diagnosis has characterized the lesions in three types, i.e., lesional (focal areas of hyperintensity), black holes (regional loss of tissue), and atrophy (loss of brain volume). Neuro-immune disorders are complex and multifactorial due to autoimmune-related neuronal demyelination, infection, degeneration, and paraneoplastic conditions (Ballatori et al. 2009). The neuronal infections are mostly due to the compromised immune system, environmental conditions, and hereditary. Biologically, the immune system is hyperactivated with excess autoantibody production, effector T-cell generation, and elevated phagocytic cell counts. Such changes cause inflammation of the blood–brain barrier (BBB) and activation of microglia, oligodendrocytes, and astrocytes. Both B and T cells play important roles in the development of different autoimmune inflammatory neuronal disorders. Several biomarkers are useful for clinical assessment, monitoring, and understanding the pathophysiological mechanisms with targeted treatments of neuro-immune disorders (Constantinescu et al. 1998). Technological advancement in diagnostics helps to diagnose and predict a specific neuro-immune condition precisely. Magnetic resonance imaging (MRI) in association with pathological and clinical findings is found promising in diagnosing specific disorders with precision. Positron emission tomography (PET) imaging has also been useful for assessing patients with neuro-immune disorders (Maher 2005).

Patient's immune system attacks their own cells in the brain and spinal cord causing inflammation and scarring in multiple sclerosis. Myelitis may be present in the state of acute or chronic progressive disorder. The acute type of transverse myelitis normally develops over a few days reaching the lowest point within 28 days (Graus et al. 2004). Whereas chronic myelopathy is the most common paraneoplastic myelopathy and evolves over weeks to months. Optic neuritis is a non-specific autoantibody-related optic nerve damage which leads to a painful partial loss of vision. It may associate with other types of autoantibodies and is common in patients with antibody-mediated CNS demyelination (Flanagan 2019). Neuromyelitis Optica is also referred to as Devic's disease. It is a severe autoimmune inflammatory condition that mainly affects the optic nerves and the spinal cord which leads to extensive inflammation and tissue damage. Encephalitis is an autoimmune condition in which specific antibodies or pathogenic immune cells attack the brain that causes confusion, seizures, and movement disorders. Spasticity occurs due to an abnormal increase in muscle tone secondary to any pathological conditions that affect motor tracts in the brain or spinal cord. It leads to body stiffness and spasms which cause problems with motor function (Sies 1999). Multifocal motor neuropathy is considered an important peripheral nervous system (PNS) immune-mediated disorder. The neuro-immune disease selectively involved in PNS includes chronic inflammatory demyelinating polyneuropathy (CIDP) that explicitly targets peripheral nerves (Kieseier et al. 2012). Such groups of autoimmune diseases are linked to high clinical disabilities (Leboyan et al. 2023).

Based on the research evidence, it could be said that there is a strong interconnection between infection, immune dysregulation, and neurological disorders (Obregon et al. 2012). There is a possibility of induction of neuroinflammatory responses in the brain because of instabilities of the CNS homeostasis due to associated disease conditions including neurodegenerative diseases, psychiatric disorders, ischemia, infection, trauma, and neuro-developmental disorders. It is well understood that the key cellular components associated with the neuro-immune system comprise glial cells and associated microglia, astrocytes, and oligodendrocytes (Zorrilla et al. 2001). Basically, the neurons do not always serve as the immune cells in this neuro-immune system; however, they are known to interact with the immune effector cells significantly with a positive contribution (McDaneld et al. 2010). Thereby, due to the rapid processing of the information and through coordination with the immune system and defense mechanism, invading threats could be eliminated. In this aspect, while combating any infection or tissue injury, the tightly integrated local and systemic reflexes are coordinated to maintain hemostasis (Ashraf et al. 2022). On the other hand, secretion of neuropeptides, for example, substance P, and neurotransmitters also attributes towards the modulatory release of cytokines and activation of mast cells (Mashaghi et al. 2016).

The correlation among the three key players is evident and reported through supporting the role of infection and immune dysregulation in the pathogenesis of various neurological disorders. The involvement of the immune system has been established in Alzheimer's disease and Parkinson's disease, the two most predominant neurodegenerative disorders (Alcocer-Gómez et al. 2017). For example, neurotoxins are produced by the activated monocytes and microglial cells, whereas amyloid-β takes the opportunity to activate them during Alzheimer's disease (Pellegrino and Bayer 2000). Alternatively, the progression of Parkinson's disease had shown the presence of α-synuclein-specific T cell reactivity (Lindestam Arlehamn et al. 2020); moreover, T cell epitopes were also recognized in Parkinson's disease patients (Peña et al. 2005).

Furthermore, several respiratory disorders, for example, chronic cough and asthma, can influence the neuro-immune system, where the release of immune molecules and positive feedback mechanisms results in a hypersensitized state of sensory neurons (Mössner and Lesch 1998). In this context, it would be pertinent to mention that the severe acute respiratory syndrome caused by COVID-19 (Gorain et al. 2020) is linked to peripheral and central neurological problems. Around 36.4–57.4% of the COVID-19 patients admitted to the hospital were reported with neurological issues (Oliveira et al. 2021). Concurrently, the pathogenesis of schizophrenia has also been linked to neuro-immune regulation because of the involvement of inflammatory pathways as evident from the neurobiological, neuroanatomical, genetic, and environmental study data (Smyth and Lawrie 2013). Although, evidence is supporting the linkage of several infections and neuropsychiatric disorders due to the involvement of the neuro-immune system. However, due to a lack of clinical research and available conflicting reports, the acceptance of these theories is still contradictory (Obregon et al. 2012). However, the above suggestions support the correlation between infection, immune dysregulation, and neurological disorders.

## 2 Conventional Drugs in Neuro-immune Disorders

### *2.1 Antimicrobials in Neuro-immune Disorders*

Antimicrobial substances and psychotropic drugs frequently have overlapping pharmacological effects and mechanisms of action. Many researchers revealed that many antimicrobials have neurological and direct psychotropic effects. These psychotropic drugs have significant immunomodulatory properties that indicate that the immune system appears to be a key interaction site for antimicrobials. These common pharmacological effects of psychotropic and antimicrobial may offer crucial insight into elucidating complexities of neuro-immune pathophysiology and can guide the development of potential new treatments. However, the therapeutic potential of antibiotics as psychotropic agents is not utilized yet (Kass and Shandera 2010). It is more clear that neuropsychiatric conditions are heterogeneous groups of illnesses with some shared pathologic mechanisms including dysregulated immune responses. It is evident that the adaptive and innate systems of the body are often affected by the microbes which lead to peripheral signal pathway activation with the involvement of different processes compared with the peripheral immune counterpart. Moreover, after microbial infection, there is the abrupt generation of IgG autoreactive antibodies that can penetrate BBB opening a new door for reactive neuro-immune modulation (Obregon et al. 2012). According to current findings, the effective treatment of neuro-immune-mediated disorders requires the combination of an antimicrobial agent with neuroprotective properties.

### 2.1.1 β-Lactams: Penicillin

The core molecular structural component of β-lactams is a cyclic amide ring that binds to the transpeptidase and inhibits the peptidoglycan synthesis, which is the main constituent of the bacterial cell wall. The disruption of peptidoglycan inhibits bacterial cell wall formation and thus prevents cell division (Thompson 2010). It was reported that β-lactams enhance glutamate transporter GLT1 expression and showed neuroprotective activity when used in motor neuron degeneration and ischemic injury (Rothstein et al. 2005). Penicillin and azithromycin (AZM) have been used to treat pediatric autoimmune neuropsychiatric disorders associated with streptococcal infections (PANDAS), with improvements seen in neuropsychiatric symptoms (Snider et al. 2005). A recent study revealed the potential of ceftriaxone to treat the major depressive disorder in a mouse model via enhancing glutamate uptake (Mineur et al. 2007). However, further investigations are needed to use β-lactams in neurogenerative disorders, depression, obsessive-compulsive disorder (OCD), and schizophrenia,.

### 2.1.2 Clavulanic Acid

A common combination of amoxicillin and the β-lactamase inhibitor clavulanate is used to combat antibiotic resistance. Clavulanate easily crosses the BBB and has shown anxiolytic effects in nonhuman primates and rodents (Kim et al. 2009). This action may result from increased dopamine release (Kost et al. 2011) or potential effects on glutamate transmission via inhibition of the N-acetyl-L-aspartyl-L--glutamate (NAAG) peptidase. Specifically, NAAG-peptidase inhibition by clavulanate may lower glutamate, improve NAAG stability, and ultimately increase metabotropic glutamate (mGlu)3 receptor (mGlu3R) activation. Animal studies suggest that known mGlu3R agonists have antidepressant, anxiolytic, and neuroprotective effects (Pałucha et al. 2004). Additionally, mGlu3R activation is linked to decreased excitotoxic glutamate-mediated neuronal damage and reduced glutamate release from the presynaptic neurons (Zhou et al. 2005). Furthermore, trophic factors appear to be released by glial cells following activation of the mGlu3R by NAAG (Bacich et al. 2005). Clavulanate increases the TGF-β release from glial cells, perhaps due to its capacity to inhibit NAAG peptidase (Thomas et al. 2001). It concludes that clavulanate exhibits noteworthy potential as an antidepressant and anxiolytic agent through the reduction in glutamate, increment in dopamine release, and trophic factors (Kost et al. 2011).

### 2.1.3 Tetracyclines

Tetracyclines are broad-spectrum antibiotics that prevent the t-RNA subunit association with ribosomal complex, thus inhibiting the translational machinery formation (Semenkov et al. 1982). Compared to the first generation of tetracyclines, second generations are widely used, for instance, the therapeutic uses of minocycline have expanded due to its high lipophilicity and longer half-lives in addition to its enhanced antimicrobial efficacy (Griffin et al. 2011). The immunomodulatory and psychotropic effects of minocycline have been demonstrated to be independent of the antibiotic's mode of action. These characteristics have been linked to the prevention of oxidative stress by reducing the expression of matrix metalloproteinase (MMP), COX2, and inducible nitric oxide synthase (iNOS) (Paribello et al. 2010). Minocycline reduces MMP9 levels, which are high in fragile X syndrome (FXS), and strengthens brain connections in FXS animal models. Minocycline showed a significant effect when MMPs play a significant role in disease pathology, such as FXS (Siller and Broadie 2011). Moreover, MMPs are also linked to axonal guidance, neurotransmission, synaptogenesis, synaptic behavioral learning, and plasticity (Ethell and Ethell 2007). Along with suppressing the viral load in the brain and reducing inflammatory markers' expression in the CNS, minocycline also decreases the severity of encephalitis. The therapeutic efficacy of minocycline in psychiatric disorders is well explained by its immunomodulatory potential and BBB permeability (Zink et al. 2005). However, further clinical trial is needed for safety and efficacy evaluation.

### 2.1.4 Linezolid

Linezolid belongs to the oxazolidinones class and has broad-spectrum antibacterial activity by preventing the synthesis of bacterial proteins. This inhibition is done through a novel mechanism in which the drug binds to the 50S rRNA subunit active site. Other than Linezolid, many agents from this class are under evaluation such as radezolid, torezolid, etc. These substances were reported as the initial selective-reversible MAO-A inhibitors. Oxazolidinones were originally formed as antidepressants, but they are still used as psychotropics except in the USA (Provost et al. 1992). Linezolid uses as a psychotropic based on its CNS penetration and capacity to inhibit the breakdown of monoamines (Mahesh et al. 2011). It also impacts immune cells which express monoamine receptors such as lymphocytes, microglia, and natural killer (Färber et al. 2005). Despite all these studies and evidence, linezolid's psychotropic effect in neuro-immune disorders by neuro-immune modulation has not been investigated yet.

### 2.1.5 Vancomycin

The antibiotic effect of vancomycin is carried out by preventing the synthesis of cell walls; however, its use is limited. On the other hand, Vancomycin use has been explored for the potential autism treatment (Finegold 2011). Sudden regression was reported in autistic children who are showing normal development. It was hypothesized that these might be due to the imbalance of intestinal bacteria. In an open-label clinical trial, it was observed that autistic symptoms were reduced in children with regressive autism when treated with oral vancomycin. Especially improvements in communication and behavior were evident in this trial though it was not observed at follow-up (Sandler et al. 2000).

### 2.1.6 Macrolides

The antibacterial activity of macrolides is due to their ability to hinder bacterial protein synthesis via inhibition of peptidyl transferase activity (Mazzei et al. 1993). In addition, macrolides have reduced the production of cytokines in preclinical and clinical studies which suggests their ability to immunomodulation besides their antimicrobial effects (Zarogoulidis et al. 2012). Although there is little information on the therapeutic use of macrolides in neuropsychiatric disorders is available, clarithromycin has been found to have neuropsychiatric side effects, which may indicate a potential role in CNS modulation. The futuristic application of macrolides in neuro-immune disorders needs further investigation (Prime and French 2001).

### 2.1.7 Azithromycin

Azithromycin (AZM) is a derivative of erythromycin and has a methyl-substituted nitrogen atom in the lactone ring (Echeverría-Esnal et al. 2021). This modification increases stability, improves tissue penetration, and broadens the antimicrobial spectrum. AZM has been demonstrated to inhibit the release of proinflammatory cytokines (interleukin-6 and interleukin-12) and increased the production of anti-inflammatory cytokines (IL-10) in macrophage cell lines (Murphy et al. 2008). AZM also inhibits iNOS-mediated nitric oxide synthesis, promotes apoptosis, and decreases nuclear transcription factors by reducing mRNA expression (Morikawa et al. 2002). Some studies show that AZM acts to downregulate NF-κB signalling and co-stimulatory molecules. After AZM treatment, toll-like receptor 4 (TLR4) and IL-12 levels are decreased. Recent clinical research on the role of TLR2 and TLR4 polymorphisms in susceptibility to group A streptococcal (GAS) infections suggested that the D299G and T399I mutations in TLR4 were linked to susceptibility to recurrent GAS infection (Liadaki et al. 2011). It was suggested that none of the antibiotics is significantly different from the other in terms of their ability to prevent GAS infection and reduce neuropsychiatric exacerbations. AZM is currently being researched as a potential treatment for childhood OCD with an acute onset and to determine how it affects immune function (Loof et al. 2008).

### 2.1.8 Methylene Blue

In the 1890s, the thiazine dye methylene blue was successfully used to treat malaria (Adjalley et al. 2011). Methylene blue also has significant effects on the CNS. The redox ability of methylene blue showed neuroprotective effects against amyloid aggregation, oxidative damage, mitochondrial dysregulation, and tau phosphorylation in neurodegenerative disorders. Such activity was linked with the modulation of glutathione reductase (Riedel et al. 2003; Verge et al. 2011). Studies offer intriguing proof in favor of the theory that neuroprotection via immunomodulation is an important mode of action for the psychotropic effects of methylene blue. Its oxidation/reduction characteristics have also been linked to the inhibition of monoamine oxidase (MAO), which was mediated by oxidization and a one-electron reductant at the active site of MAO-A (Ramsay et al. 2007). The use of methylene blue in the treatment of bipolar disorder was evident in several clinical trials, despite its association with serotonin syndrome (Naylor et al. 1987). The anxiolytic and antidepressant effects of methylene blue on rats have been demonstrated to be mediated by the regulation of the nitrergic system (Eroğlu and Çağlayan 1997).

### 2.1.9 D-Cycloserine

D-cycloserine is a second-line bacteriostatic agent that shows antimicrobial action by inhibiting alanine racemase and D-alanine ligase. D-alanine is one of the important components of the bacterial cell wall (Thompson 2010). D-cycloserine has been evaluated especially for treating anxiety disorders, including OCD, depression, post-traumatic stress disorder, etc. (Otto et al. 2007). These psychotropic effects attribute its agonist activity to the glycine site of N-methyl-D-aspartate (NMDA) receptors in the CNS. It is well known as a cognitive enhancer and has also been investigated for its effectiveness as an adjunctive treatment for Parkinson's, schizophrenia, and Alzheimer's (Jones et al. 2002). Effects on neuro-immune activity could be anticipated that NMDA receptors are expressed in various immune cells of the periphery and CNS (Müller and Schwarz 2006). Despite these reports of psychotropic effects that seem to be caused by immune modulation, a precise immunomodulatory mechanism still hasn't been identified.

### 2.1.10 Isoniazid Family

Isoniazid was the first mycobactericidal drug taken by oral route. Isoniazid becomes bacteriocidal when it is converted by mycobacterium katG catalase-peroxidase, which produces isonicotinic radicals (Assefa et al. 2022). It exhibited mood-enhancing properties through MAOI activity. This coincidental finding sparked the creation of the entire MAOI antidepressant class of drugs. Iproniazid, an isoniazid relative, also has MAOI properties similar to those of isoniazid. This agent seems to increase T- and B-cell populations under conditions of phorbol esters-induced proliferation, in contrast to other studies where isoniazid reduced IL-1 and IL-2 signalling in immune cells (Kucharz and Sterakowski 1990).

## 2.2 *Psychotropic Agents in Neuro-immune Disorders*

Immune system-related complications leading to neurological and psychiatric illness are regarded as among the most challenging central nervous system disorders, affecting approximately 30 million people globally. Although the aetiology of neurological illnesses is complex, numerous studies have supported the immune system's participation in developing these diseases (Ashraf et al. 2022). The immunomodulatory effects of psychotropic medications, particularly when used in conjunction with infection-related neuro-immune disorders, are among the therapeutic approaches to treat these disorders. Utilization of these drugs in the treatment of the neuropsychiatric components of infection is beneficial; however, it comes with aversive adverse effects on the immune system. Moreover, in-depth research is required to hunt the potential drug candidates to control infection-related neuro-immune disorders. An additional investigation could provide a useful basis for evaluating prospective psychoactive drugs to enhance the outcomes of treating various neurological conditions. Expanded studies with new approaches for treating illnesses that have both an infectious and a neuropsychiatric component, as well as a better knowledge of the role immunological dysregulation may contribute to neuropsychiatric disorders.

### 2.2.1 Antidepressants

Depression is frequently linked to innate immune system activation. For depressed symptoms that are concurrently associated with inflammatory diseases, immunomodulatory medications may be effective (McDaneld et al. 2010). Novel immune therapeutics can produce antidepressant effects in depressed patients indicating that antidepressant therapy has been linked to the immune system (Di Rosso et al. 2016). In depression, serotonin (5-HT) plays an important role (Mössner and Lesch 1998). Serotonin has been confirmed to have important modulatory effects. Numerous studies have demonstrated that 5-HT has the ability to control a variety of immune functions by increasing the proliferation of mitogen-stimulated lymphocytes, encouraging the activation of natural killer cells, supporting the function of macrophage/dendritic accessory cells, and helping to trigger the delayed-type hypersensitivity response. Depression has been linked to changes in lymphocyte proliferation. A higher CD4+/CD8+ ratio and leukocytosis with lymphopenia are two additional immunological alterations that are apparent (Zorrilla et al. 2001; Pellegrino and Bayer 2000). Additionally, a higher proportion of natural killer (NK) cells and an increase in the CD3+/CD8+ ratio without change in CD3+CD4+ lymphocytes have also been linked to serious depression in some studies (Pavón et al. 2006). Drugs such as selective serotonin reuptake inhibitors (SSRIs) modify immune function via a unique independent mechanism and a serotonin-dependent pathway. Serotonergic neurons in the central nervous system have immunomodulatory effects through their control of neuroendocrine signals. These drugs also directly affect T lymphocytes, influencing their proliferation in two ways depending on the level of cellular activation (Pellegrino and Bayer 2000; Di Rosso et al. 2016). For instance, it has been demonstrated that fluoxetine and mirtazapine increase the expression of the serotonin transporter on the surface of T lymphocytes. This may improve serotonin reuptake, lowering its concentration in the environment of lymphocytes and, as a result, reducing the lymphocytes' capacity to proliferate (Szałach et al. 2019).

The neurotransmitters level in the synaptic cleft is affected by drugs like selective serotonin reuptake inhibitors (SSRI)/serotonin and noradrenaline reuptake inhibitors (SNRI) and tricyclic antidepressants (TCA), as well as antiepileptic drugs (e.g., valproic acid, lamotrigine) and antipsychotics (including olanzapine, amisulpride, quetiapine) used for the pharmacotherapy of depression (Peña et al. 2005; Szałach et al. 2019). In a study conducted by Alcocer-Gómez et al., the mononuclear cells extracted from patients receiving antidepressants from the TCA (amitriptyline, imipramine), SNRI (desvenlafaxine, venlafaxine), and SSRI group (paroxetine), and patients treated with mianserin or agomelatine scrutinized the gene expression encoding NLRP3 inflammasome elements in mononuclear cells. All of these drugs not only significantly lowered gene expression but also inflammasome activity, which reduced IL-1 and IL-18 plasma concentrations (Alcocer-Gómez et al. 2017). All the above findings may be attributed to the fact that the management of depression may be effective due to the alteration of the immune response. In this context, polyunsaturated fatty acids, cytokine inhibitors, nonsteroidal anti-inflammatory drugs (NSAIDs), or curcumin have also been used in studies for depression (Menon and Sudheer 2007).

### 2.2.2 Antipsychotics

Similar to antidepressants, antipsychotic drugs may cause immunomodulatory effects. It has been demonstrated that antipsychotics affect immunological function, particularly cytokine expression and T-cell activation. Clozapine and olanzapine, two antipsychotic medicines that are frequently associated with proinflammatory effects (on TNF-, IL-6, and CRP), are also those that are widely acknowledged to have the highest therapeutic efficacy (Tourjman et al. 2012). Mantere et al. evaluated the immunological alterations and potential immunomodulatory consequences in first-episode psychosis (FEP) by contrasting FEP patients ($n = 67$) with matched controls ($n = 38$), all of whom were between the ages of 18 and 40. Risperidone was linked to a statistically significant upregulation of 11 immune systems genes, such as those for cytokines and its receptors (IL1R1, IL1R2, SPP1), proteins associated with apoptosis (FAS), pattern recognition molecules (dectin-1/CLEC7A, TLR1, 2 and 6), and other molecules having the immune stimulant functions (BDKRB1, IGF1R, CR1). Thus, risperidone has potent immunomodulatory characteristics that primarily alter the innate immune response in FEP patients (Mantere et al. 2019). The impact of antipsychotic drugs on immune-compromised peripheral blood mononuclear cell (PBMC) cultures collected from blood samples of 12 patients affected with schizophrenia first-episode was demonstrated in a different study by Al-Amin et al. The levels of IL-10 and IL-4 were dramatically raised by haloperidol, quetiapine, and clozapine according to the results. Antipsychotic medication affected the immune role by increasing anti-inflammatory cytokines (IL-4 and IL-10) levels and reducing proinflammatory cytokines IFN-γ levels (Al-Amin et al. 2013).

Deviations in inflammatory indicators have a variety of effects on patients. C-reactive protein (CRP), one of the most prevalent inflammation markers, raises the action of microglial cells in the CNS. Particularly in schizophrenia patients, the rise in CRP and initiation of microglia trigger the release of IL6 and TGF-β, enhancing neuroinflammatory interactions (Fernandes et al. 2016). Supporting this fact, a study reported by Kim et al. evaluated the impact of several antipsychotics on healthy volunteer whole blood cell cultures' in vitro C-reactive protein (CRP) production. The test revealed that the levels of CRP in the blood culture samples were dramatically raised by amisulpride, haloperidol, and chlorpromazine. Such studies suggest that antipsychotics have immunomodulatory effects (Kim and Kim 2019).

### 2.2.3 Psychostimulants

Psychostimulants generally manage psychological conditions including attention-deficit hyperactivity disorder (ADHD). It is a common neurodevelopmental condition with a childhood origin that is characterized by persistent inattention, hyperactivity, and impulsivity (Obregon et al. 2012). Amphetamines, including the prodrug lisdexamfetamine dimesylate, and methylphenidate are the main stimulants used in pharmacologic treatment (Cortese 2020). These illnesses have been treated with medications that are known to target this neurotransmitter network, in particular antagonists of adrenergic receptors. These medications have substantial anti-inflammatory effects on inflammatory mediators, which are mediated by their enhancement of noradrenaline activity (Obregon et al. 2012). This suggests a significant association of involvement psychostimulants as immune modulators in the brain and ADHD.

Stimulant use is also associated with immune activation in HIV-positive persons. Methamphetamine, cocaine, and crack are sympathomimetic stimulants that can promptly cause various physiological and psychological changes, including hyperthermia, increased blood pressure, decreased appetite, sleeplessness, increased sex drive, and feelings of pleasure (Carrico et al. 2008; Irwin et al. 2007). Despite the strong effects of stimulants, relatively few studies have looked at how immunomodulatory they are (Kopnisky et al. 2007). As a result, stimulants may have direct and indirect effects on increased immunological activation. An experiment involving heroin injectors found that cocaine usage was independently linked to greater levels of neopterin, even after adjusting for HIV serostatus, providing more evidence of the possible direct effects of stimulants on immunological activation (Fuchs et al. 1987). These findings offer some early evidence that stimulant use among HIV-positive people receiving ART may directly enhance HIV viral load and encourage immunological activation (Carrico et al. 2008).

## 3 New Drug Targets in Neuro-immune Disorders

### *3.1 Glutathione in Neuro-immune Disorders*

Glutathione (GSH) is a tripeptide and is chemically known as L-γ-glutamyl-L-cysteinyl glycine. GSH is obtained from glycine, glutamate, and cysteine (Meister and Mary 1983). GSH is mainly synthesized in a two-step process catalyzed by L-glutamate L-cysteine γ-ligase or γ-glutamyl-L-cysteine ligase (γGLCL) or γ-glutamylcysteine synthase and glutathione synthase (GLS) (Fig. 1). In the first step, the enzyme γGLCL forms a peptide bond between the γ-carboxyl of glutamate and the amino group of cysteine. In the second step, the dipeptide is linked with glycine in the presence of GLS. In both cases, the reaction is driven by the energy produced by the hydrolysis of ATP (Forman et al. 2009). Synthesis of GSH occurs intracellularly whereas its degradation occurs extracellularly on the surface of cells that express the enzyme γ-glutamyl transpeptidase or γ-glutamyltransferase (GGT) (Taylor et al. 1996).

Glutathione is present in both oxidized (GSSG) and reduced (GSH) form (Fig. 2). The reduced GSH is the major form under physiological conditions which is approximately 10–100 folds higher than the oxidized form (Zhang and Forman 2009).

GSH plays a vital role in antioxidant defense mechanisms by directly reacting with free radicals such as reactive oxygen species (ROS) or reactive nitrogen species (RNS) and as an essential cofactor for glutathione peroxidases and glutathione S-transferases. Further, GSH exerts action in association with its dependent enzymes called the glutathione system (O'Donovan and Fernandes 2000). In the brain, GSH showed neuroprotective activities due to antioxidant, neuronal modulation, and survival. It has been reported that the presence of thiol (–SH) moiety in GSH is essential for exhibiting antioxidant activity in the scavenging of free radical species (Fig. 3). GSH production is significantly depleted in mitochondrial dysfunction

Step-I: γ-L-Glutamate + L-Cysteine + ATP ⟶ γ-L-Glutamyl-L-Cysteine + ADP + Pi

Step-II: γ-L-Glutamyl-L-Cysteine + Glycine + ATP ⟶ GSH + ADP + Pi

**Fig. 1** Synthesis of glutathione (Author's drawing using chem draw (original figure))

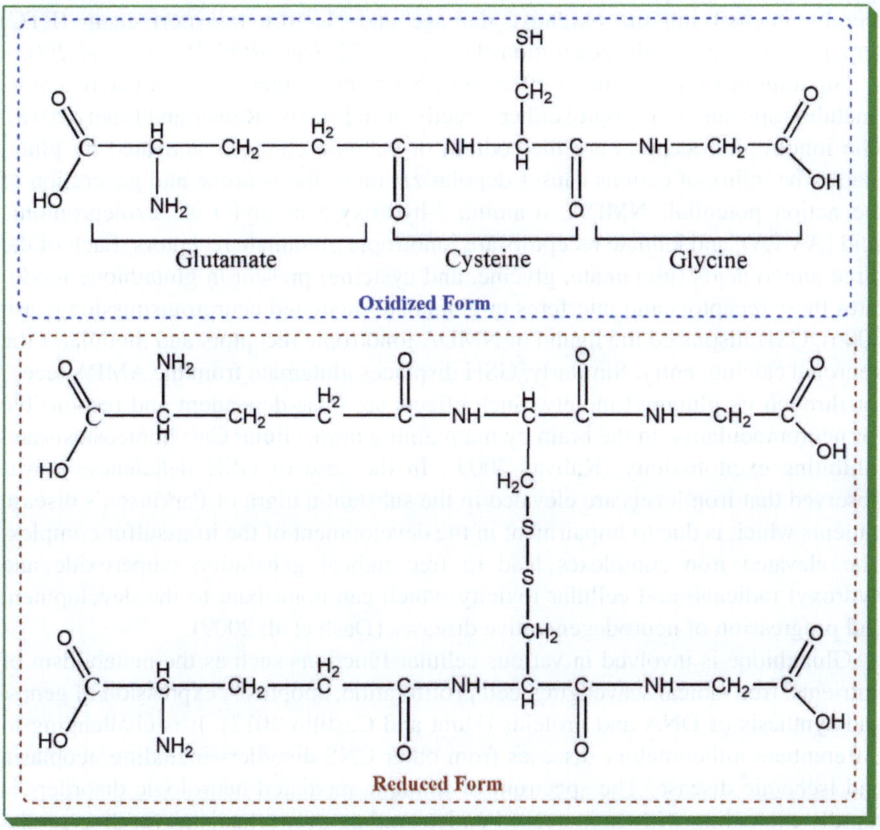


**Fig. 2** Structures of oxidized (GSSG) and reduced (GSH) form of glutathione (Author's drawing using chem draw (original figure))

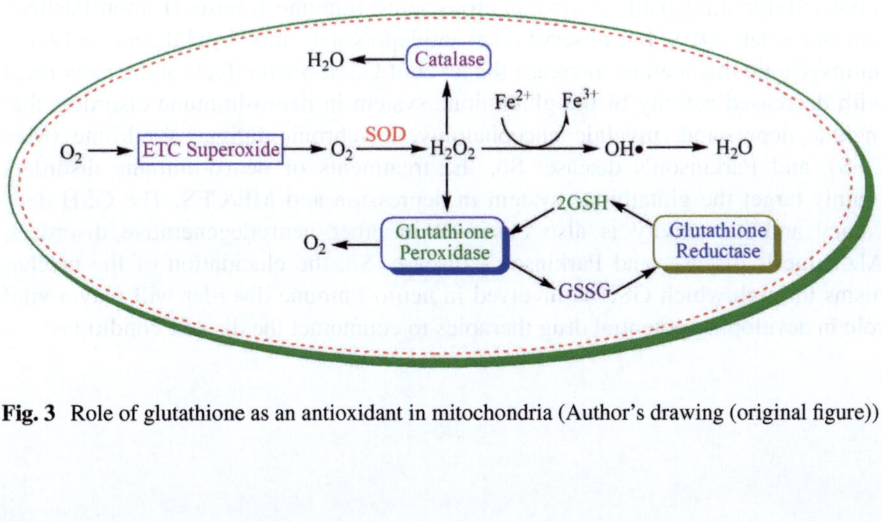


**Fig. 3** Role of glutathione as an antioxidant in mitochondria (Author's drawing (original figure))

thereby intensifying the oxidative damage and electron transport chain (ETC) impairment leading to the generation of excessive RNS and ROS (Pastore et al. 2003).

Glutathione exerts its major effects in CNS through altering glutamate receptors (metabotropic and ionotropic) either directly or indirectly (Reiner and Levitz 2018). The ionotropic receptors are the sodium or calcium channels activated by glutamate. The influx of cations causes depolarization of the neurone and generation of the action potential. NMDA, α-amino-3-hydroxy-5-methyl-4-isoxazolepropionic acid (AMPA), and kainate receptors are ionotropic glutamate receptors. Each of the three amino acids (glutamate, glycine, and cysteine) present in glutathione modulates these receptors and interferes in glutamate-mediated neurotransmission (Lash 2006). GSH displaced the ligand of NMDA ionotropic receptors and modulates the neuronal calcium entry. Similarly, GSH displaces glutamate from the AMPA receptor through its glutamyl moiety. Such effects are dose-dependent and responsible for neuromodulation in the brain by maintaining intracellular $Ca^{2+}$ homeostasis and inhibiting excitotoxicity (Kalivas 2009). In the case of GSH deficiency, it was observed that iron levels are elevated in the substantia nigra of Parkinson's disease patients which is due to impairment in the development of the iron-sulfur complex. The elevated iron complexes lead to free radical generation (superoxide and hydroxyl radicals) and cellular toxicity which can contribute to the development and progression of neurodegenerative diseases (Dash et al. 2009).

Glutathione is involved in various cellular functions such as the metabolism of nutrients, free radical scavenging, cell proliferation, apoptosis, expression of genes, and synthesis of DNA and proteins (Hunt and Castillo 2012). It is challenging to differentiate inflammatory diseases from other CNS disorders including neoplasia and ischemic disease. The spectrum of immune-mediated neurologic disorders is rapidly expanding with an increased understanding of mechanisms for diverse disease conditions that can lead to the central and peripheral nervous system. Future directions for this study mainly focus on developing more targeted treatments for neuro-immune disorders. Future studies also highlight the vital role of immune-mediated signalling pathways in numerous neuro-immune diseases (Limon-Pacheco and Gonsebatt 2010). It is observed that antidepressants, mood stabilizers, and some antipsychotic medications increase the level of GSH. Such effects are corroborated with decreased activity of the glutathione system in neuro-immune disorders that include depression, myalgic encephalomyelitis/chronic fatigue syndrome (ME/CFS), and Parkinson's disease. So, the treatments of neuro-immune disorders mainly target the glutathione system in depression and ME/CFS. The GSH deficiency and its activity is also observed in other neurodegenerative disorders, Alzheimer's disease, and Parkinson's disease. So, the elucidation of the mechanisms through which GSH is involved in neuro-immune disorder will play a vital role in developing potential drug therapies to counteract the disease conditions.

## 3.2 *SIRT1 Pathway in Neuro-immune Disorders*

Sirtuins are the nicotinamide adenine dinucleotide (NAD+)-dependent enzymes that belong to the histone deacetylase class (Michan and Sinclair 2007). There are four classes (Class I to Class IV) of human sirtuins are reported. Class I SIRTs are SIRT1, SIRT2, and SIRT3. Class II for SIRT4 and Class III and IV for SIRT5, respectively. SIRT6 and SIRT7 belong to class IV sirtuins (Frye 2000). The first sirtuin to be recognized and extensively investigated among all sirtuins is SIRT1. Sirtuins have NAD+ binding domains, a catalytic domain, and variable flanking COOH− and $NH_2$− terminal sections in their structural makeup (Yamamoto et al. 2007). The cellular location, catalytic activity, and function of sirtuins are influenced by the amino acid sequences present in different domains. Both neurons and glial cells highly express SIRT1 during embryonic stages and the expressions gradually decline throughout development (Quintas et al. 2012). Age and pathological alterations appear to have an impact on the sirtuins' expression levels. The activity of SIRT1 deacetylase is downregulated in the aging brain and several neurodegenerative diseases (Tang 2017). Several studies have suggested that SIRT1 may control the immunological response in neural tissues (Mendes et al. 2017). Finding the signalling axis may be a valuable starting point for developing a therapy for treating neuroinflammation and neurodegenerative diseases.

SIRT1 could deacetylate various nonhistone substrates, including p53, nuclear factor kappa B (NF-κB), hypoxia-inducible factor-1a (HIF-1), HIF-2, and signal transducer and activator of transcription 3 (STAT3). The innate immune system which includes various cells (astrocytes, microglia, mast cells, natural killer (NK) cells, and oligodendrocytes) serves as the body's first line of defense against disease, and it is essential for maintaining brain homeostasis (Jiao and Gong 2020). By stimulating microglial BV2 cells with lipopolysaccharides, the SIRT1 activator reduces the secretion of interleukin 6 (IL-6) and tumor necrosis factor (TNF-α) whereas the SIRT1 inhibitor increases the release of inflammatory mediators (Ye et al. 2013). SIRT1 inhibition increases the TNF-α and IL-1 gene expression and ROS formation in astrocytes (Cheng et al. 2014). Newly identified possible regulatory pathways for SIRT1 in developing neuro-immune disorders may include the NF-κB, TLR, and mitogen-activated protein kinase (MAPK) signal pathway (Fig. 4).

### 3.2.1 Functional Relation Between SIRT1 and NF-κB Pathway

The expression of numerous inflammatory genes is controlled by the NF-κB family of transcription factors (DiDonato et al. 2012). The NF-κB complex p65/p50 mediates the canonical NF-κB pathway and SIRT1 has shown the capacity to deacetylate the p65 subunit of the complex (Yeung et al. 2004). In SIRT1 knockout mice, NF-κB is hyperacetylated and proinflammatory cytokines are produced in greater amounts (Schug et al. 2010). Resveratrol, which acts as an activator of SIRT1, deacetylates

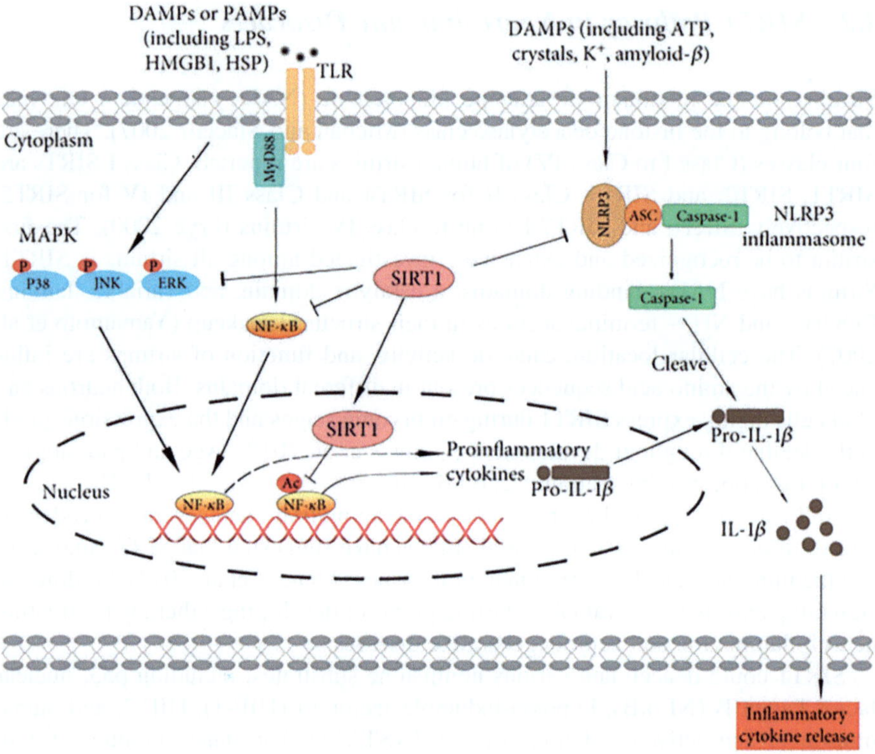


**Fig. 4** Mechanism of SIRT1 in neuroinflammation (Jiao and Gong 2020) (This is an open-access article distributed under the Creative Commons Attribution License, which permits unrestricted use, distribution, and reproduction in any medium, provided the original work is properly cited)

the p65 subunit of the NF-κB complex, and inhibits the TNF-induced inflammatory responses in 3T3 fibroblast cells (Zhu et al. 2011). Accumulating studies indicate that the decrease in SIRT1 level activates the NF-κB pathway that leads to the neuroinflammation progression. Aging mice with microglial SIRT1 loss exhibit memory impairments and higher IL-1 production (Cho et al. 2015). Resveratrol also lowers inflammatory cytokines in neural cells by inhibiting NF-κB transcriptional activity. The NF-κB pathway may also be indirectly regulated by SIRT1 through other targets, such as PPAR$\alpha$ (Planavila et al. 2010). Thus, SIRT1 is a potential target for suppressing neuroinflammation through the NF-κB signalling pathway that includes p65 subunit deacetylation and NF-κB transcriptional inhibition.

### 3.2.2 Functional Relation Between SIRT1 and TLRs Pathway

TLRs are the membrane receptors that are expressed widely in the immune cells and neurones of the brain (Kumar 2019). TLRs are stimulated by various external signals, including lipopolysaccharides, peptidoglycan, heat shock proteins (HSP) and high-mobility group protein B1(HMGB1). TLRs activation leads to the recruitment of various downstream effector proteins, such as myeloid differentiation factor-88 (MyD88), TIR-domain-containing adaptor protein (TIRAP), TNFR-associated factor 6 (TRAF6), and subsequent regulation of NF-κB activation. Finally activated TLR promotes the expression of inflammatory cytokines (Kay et al. 2014). Furthermore, several studies have shown that SIRT1 controls neuroinflammation by inhibiting the TLR pathway. Resveratrol decreases the level of cytokines IL-1β and TNF-α and inhibits the TLR2- MyD88-NF-κB signal pathway. Inhibition of both the mediator and signalling pathway improves spatial reference memory (Qi et al. 2018). In trigeminal neuralgia, resveratrol inhibits the expression of IL-1, TNF-α, MMP-9, and MMP-2 via downregulating the TLR-4/NF-κB signalling pathway. Reducing the expression level of cytokinins and MMP lead to improving the pathological condition of the disease (Yin et al. 2019). Activation of the SIRT1 pathway has been found to reduce neurological behavior impairment via the TLR4-MyD88-NF-κB pathway and interfere with the action of the proinflammatory cytokine in an experimental subarachnoid hemorrhage model (Zhang et al. 2016).

### 3.2.3 Functional Relation Between SIRT1 and MAPK Pathway

MAPK pathway linked to the innate immune response has received substantial attention recently. MAPK subfamily members such as Jun N-terminal kinase (JNK), p38 kinase, and extracellular signal-regulated kinases (ERK) are stimulated through the cell surface-located pattern recognition receptors (PRRs) (Arthur and Ley 2013). Pathogenesis of many CNS disorders has been directly linked with the MAPK pathway (Kim and Choi 2015). Specifically, the p38 MAPK pathway controls the SIRT1-regulated neuroinflammation. SIRT1 activation by resveratrol inhibits the IL-1β mediated astrocyte activation via inhibition of the MAPK signalling pathway (Li et al. 2017). Resveratrol also reduces the level of IL-1β and TNF-α in cytotoxic-BV2 microglial cells through inhibition of MAPK and NF-κB signalling pathways (Zhang et al. 2015). In contrast, the SIRT1 inhibitor “sirtinol” enhances the neuronal apoptosis process in traumatic brain injury (TBI) through p38 MAPK pathway activation (Yang et al. 2017).

## 3.3 Th17 in Neuro-immune Disorders

CD4 effector T cells or Th cells are categorized into four subsets based on their signalling pathways and cytokine production (Luckheeram et al. 2012). T helper 1 or Th-1 cells are known to produce IL10, TNF-β, IFN-γ, and lymphotoxin, whereas Th-2 cells induce production of different interleukins (e.g., IL4, IL5, IL10, IL13, IL21, IL31). On the other hand, Th-3 is known to produce TGF-β, IL35, and IL10. Th-17, the newest Th cell, is associated with the production of IL17, IL21, IL26, and chemokine ligand (CCL)-20. This IL17, by the presence of other signalling molecules, contributes to the pathogenesis of several autoimmune and inflammatory neuro-immune disorders (Vojdani and Lambert 2011; Vojdani and Erde 2006a, b).

Figure 5 depicted that naive CD4 cells are triggered by TGF-β and different ILs to differentiate to Th17. Upon activation, the released cytokines from Th17 cells start making and facilitating proinflammatory roles in the body system. Migration of this interleukin in the respiratory tract stimulated the accumulation and activation of the inflammatory cells, resulting in allergic inflammation or asthma (Ge et al. 2020; Zenobia and Hajishengallis 2015). Therefore, it is well understood that infiltration of these T helper cells would contribute to inflammatory diseases. This inflammatory response in different tissues occurs by the effect of different members of the IL17 family (e.g., IL17, IL17-A, and IL17-F). These members possess

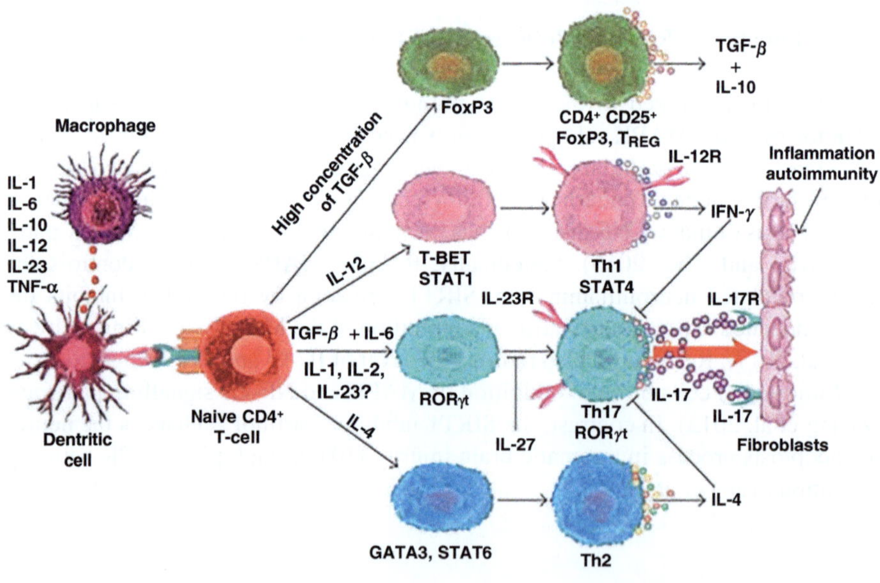


**Fig. 5** Activation of different T cells from naive CD4 cells and production of different cytokines along with regulation of Th17 cells by Th1 and Th2 cytokines (Vojdani and Lambert 2011) (This is an open-access article distributed under the Creative Commons Attribution License, which permits unrestricted use, distribution, and reproduction in any medium, provided the original work is properly cited)

proinflammatory responses on different cellular targets such as osteocytes, monocytes or macrophages, keratinocytes, fibroblasts, epithelial and endothelial cells (Ge et al. 2020; Zenobia and Hajishengallis 2015; Tesmer et al. 2008). Considering its inflammatory role in the CNS, microglia cells produce IL1 and IL23, which induces the production of IL17 in an autocrine manner in the microglial cells. This IL17 contributes to neuro-immune disorders.

On the other hand, activation of MMP3 by IL17 is responsible for recruiting neutrophils at the inflammatory site. Activated neutrophils then subsequently activate gelatinases and proteases, thereby resulting in the enhancement of the blood–brain barrier permeability. Concurrently, there will be damage to the axon and myelin (Vojdani and Lambert 2011; Zenobia and Hajishengallis 2015). Eventually, Th17 or IL17 could be targeted by the use of novel chemicals or by the use of chemicals from natural sources which could downregulate the activation of the particular T cells and production of IL17.

## 3.4 *Neuropeptide Y in Neuro-immune Disorders*

Neuropeptide Y (NPY), a neuroendocrine peptide consisting of 36 amino acid residues, has been established to regulate several functions of the CNS including stress coping, immune function, metabolism of the body, blood pressure, and digestion (Thorsell and Mathé 2017). This neuropeptide is widely distributed in the CNS and peripheral nervous system. More precisely, NPY is distributed in the thalamus, hypothalamus, cerebral cortex, cerebellum, and brainstem with the highest concentration in the hippocampus area of the CNS, which is formed by the GABAergic interneurons (Li et al. 2019). The action of this neuropeptide exerts by binding to the specific receptors and activation to modulate memory, learning, epilepsy, feeding, and endocrine secretions (Kovac and Walker 2013; O'Loughlin et al. 2014). These receptors are G-protein coupled receptors and five such receptors were already cloned (e.g., Y1, Y2, Y4, Y5, and Y6) (Diaz-delCastillo et al. 2018). These receptors are distributed in different regions of the brain, such as Y1 and Y2, responsible for memory functions and found in the thalamus, hypothalamus, amygdala, hippocampus, and cerebral cortex; similarly, Y4 was found to be expressed in the paraventricular nucleus of the hypothalamus, nucleus tractus solitarii, preoptic area, and area postrema. Alternatively, Y5 would be found in limbic brain regions, cingulate cortex, hippocampus, and thalamic and hypothalamic nuclei. Although the existence of Y6 has already been established in some mammals, it is usually recognized as non-functional, or its physiological roles are yet to be explored (Gao et al. 2017).

These neuropeptide receptors are known to modulate two different types of neuro-immune disorders, multiple sclerosis and Guillain-Barré syndrome, two autoimmune diseases related to the central and peripheral nervous system. Multiple sclerosis is characterized by axonal loss, extensive demyelination, multifocal inflammation in the brain and gliosis (Lazibat et al. 2018). Microglial cells in multiple sclerosis cause nerve damage due to the immune response (Zrzavy et al. 2017), where activation of Y1 receptors is shown to inhibit activation of microglial cells including inhibition of IL-1β production and nitric oxide with subsequent signal transduction through NF-κB-iNOS (Sørensen et al. 2012). Furthermore, deactivating neuropeptide receptors in multiple sclerosis could also decrease the production of cytokines, nitric oxide, and glutamate; also there will be inhibition of phagocytosis activity of the microglia by acting on the p38 signal system (Ferreira et al. 2012). A recent study demonstrated in an experimental animal model of multiple sclerosis, i.e., an inflammatory autoimmune disease of CNS experimental autoimmune encephalomyelitis (EAE), that there is a suppression of NPY (Chen et al. 2019; Schmitz et al. 2017). Furthermore, the administration of NPY could dose-dependently decrease the symptoms of the disease in experimental animals and improve the diseased symptoms (Schmitz et al. 2017).

There is damage to myelin sheaths and axons with inflammatory infiltration during the acute attack of Guillain-Barré syndrome (Li et al. 2019). Animal experiments have established the relationship between several cytokines and immune cells with the disease. Due to the release of different cytokines by the common CD4 T cells [Th1 (increase in IL-1β, IL-6, TNF-α, and IFN-γ production), Th2, Th17 (increase in IL-17A) and decrease in Treg cells], there will be an induction of immune response leading to Guillain-Barré syndrome. Th2 response is related to recovery from the diseased condition where there will be an increase in the level of IL-4 with the decrease in IFN-γ (Li et al. 2019; Zhang et al. 2013).

There are specific animal models already established to replicate human disease conditions, which could be used to modulate the diseased condition by the application of NPY. This could be used to correlate to immune cell trafficking, production or inhibition of cytokines, differentiation of T helper cells, phagocytosis, the activity of natural killer cells and generation of ROS with the responses of neuro-immune disorders.

## *3.5 Toll-like Receptors in Neuro-immune Disorders*

TLRs are also known to initiate immune response following recognition of damage-associated molecular patterns and pathogen-associated molecular patterns (Heinz et al. 2020; Li et al. 2021a), where activation of these receptors could be achieved by viral and bacterial nucleic acids, lipoproteins, lipopolysaccharides, and flagellin (Anthoney et al. 2018). These receptors are transmembrane receptors (Type 1) with a molecular weight of around 50–150 kDa. The extracellular leucine-rich repeats recognize damage-associated molecular patterns and pathogen-associated molecular patterns (Li et al. 2021a). The intracellular domain of this receptor has homology with IL-1R; thus, it is known as the N-terminal cytoplasmic Toll/IL-1 receptor (TIR). Due to involvement in the secretion of inflammatory mediators, they are participating in the pathogenesis of neuro-immune disorders where modulation of such activities has been shown to alter symptoms in preclinical studies (Li et al. 2021a; Zheng et al. 2019).

These TLRs consist of five TIR domain-containing adaptors. Among, MyD88-independent TIR domain-containing adaptor-inducing IFN-β and MyD88-dependent signalling pathways are known to be involved in the production of proinflammatory cytokines. Other domains of the receptors include TRIF-related adaptor molecule, TIR domain-containing adaptor protein, and sterile α- and armadillo-motif-containing protein (Li et al. 2021a; O'Neill 2008).

So far, 13 members of the TLRs have been recognized in mammals, out of which TLR1-10 are found in humans. Based on their localization, a few of them are found on the cell surface (TLR-1, -2, -5, -6, and -10) and others are found on the intracellular endosomal membrane (TLR-3, -7, -8, -9) (Li et al. 2021b). However, there are controversies on the location of TLR-4, which may be expressed in intracellular endosomal membranes or outer cell membranes (Li et al. 2021a). The extracellular domain of these receptors is activated by the specific patterns or ligands as discussed, where signalling of almost all the receptors corresponds to myD88-dependent pathways except for TLR-3. Upon binding of ligands at the leucine-rich repeats, it follows the formation of dimers, either homo- or hetero-, such as TLR-4-TLR-4/TLR-9-TLR-9/TLR7-TLR-8/TLR-2TLR-6/TLR-2-TLR-1 which activates IL-1-receptor-associated kinase (IRAK) complex, MyD88 or TIR domain-containing adapter protein (TIRAP). Such complex then stimulates TNF receptor-associated factor 6 (TRAF-6) to form transforming growth factor-β-activated kinase 1 (TAK1), TAK1 and 2 binding protein (TAB1 and 2). This activates NF-κB and protein kinases [mitogen-activated protein kinases (MAPK) including JNK and ERK]. In the next step, these will be producing proinflammatory cytokines (Fig. 6) (Li et al. 2021a).

Currently available therapeutic, natalizumab, is known to prevent infiltration of leukocytes in the brain through inhibition α4 integrin to improve the autoimmune response. TLR-3 agonists could also reverse immune surveillance of the CNS in EAE by inhibiting the α4 integrin receptors from producing a slight reduction in the progressive risk of multifocal encephalopathy (Hussain et al. 2018; Polman et al. 2006). This condition would further assist in reducing the risk of progressive multifocal encephalopathy, whereas treatment with immunosuppression of the autoimmune disease increases the risk (Hussain et al. 2018). Another agent that received regulatory approval for the treatment of multiple sclerosis is fingolimod, which modulates the sphingosine 1-phosphate receptors and thereby alters lymphocyte transport. Simultaneously, this agent is also known to decrease lymphocytes in circulation by decreasing the outflow of lymphocytes from the lymph glands (Huwiler and Zangemeister-Wittke 2018). Furthermore, the inhibitory effect of TLR-4 decreases the production of chemokines, CCL2, C-X-C motif chemokine ligand (CXCL)10, and CXCL5. This role has also been observed with fingolimod (O'Sullivan et al. 2018). Although the molecular mechanism of the disease state has already been established and several agents are identified that would improve the conditions, the poor prognosis for the disease and the severe toxicities to this disease need urgent focus.

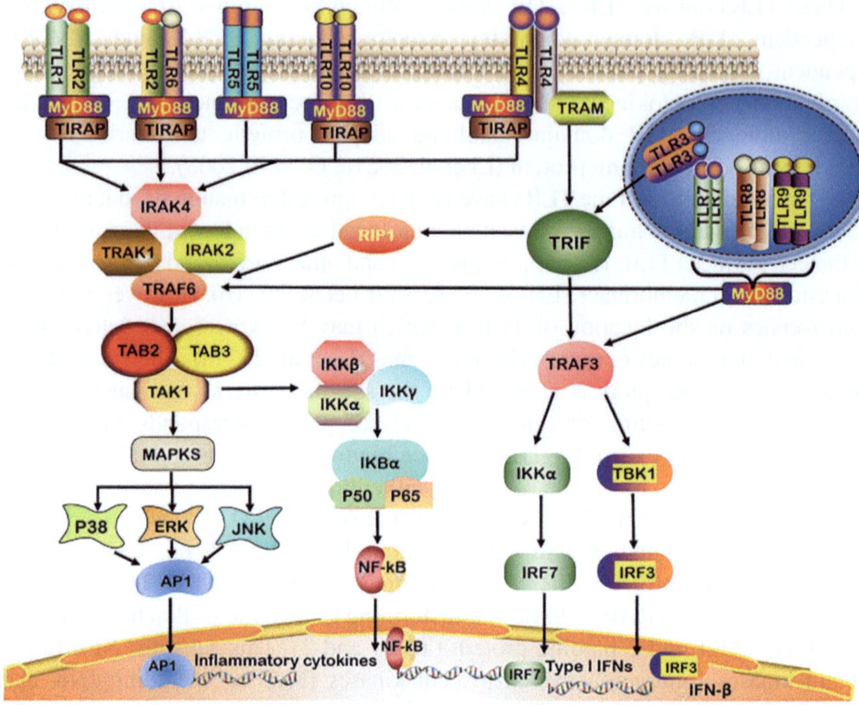


**Fig. 6** Toll-like receptor signalling pathways (Li et al. 2021a) (This is an open-access article distributed under the terms of the Creative Commons Attribution License (CC BY). The use, distribution, or reproduction in other forums is permitted, provided the original author(s) and the copyright owner(s) are credited and that the original publication in this journal is cited, in accordance with accepted academic practice. No use, distribution, or reproduction is permitted which does not comply with these terms)

## 3.6 *Tryptophan Catabolites in Neuro-immune Disorders*

It is essential to maintain the perfect functioning of the brain-gut axis for the maintenance of the gastrointestinal tract (GIT) and CNS homeostasis (Morris et al. 2016a). Dysbiosis is a condition that alters the normal microbial condition of the GIT, which might be caused by a wide variety of factors including genetic or dietary factors, alteration in communication between antigen-presenting cells in the mucosal immune system and gut commensals, and increased intestinal permeability. This might be due to stress, either physical or psychological, food habits in an individual, etc. (Bailey et al. 2011; Chassaing et al. 2015). Whatever the factors might be to alter such alteration to the axis, increased permeability of the GIT with translocation of the bacteria or its components (e.g., lipopolysaccharides, LPS) into the systemic circulation (Morris et al. 2016b). Simultaneously, the production of microbial metabolites, tryptophan catabolites which are neuroactive, plays an important role in the pathogenesis of neuro-immune diseases (Maes and Rief 2012). The increase in proinflammatory cytokines stimulates the production of bacterial indoleamine 2,3-dioxygenase (IDO), which is playing a role in the catabolism of tryptophan, thereby the level of tryptophan in the circulation drops with a simultaneous increase in tryptophan catabolites (Maes and Rief 2012). These catabolites are associated with the production of IL-22 from Th-22 cells, modulation of innate lymphoid cells, and differentiation of Tregs and Th17 by interacting at aryl hydrocarbon receptors. Thereby, these tryptophan catabolites are known to alter adaptive and innate immune responses (Mjösberg et al. 2012).

The upregulation of aryl hydrocarbon receptors on the dendritic cells happens through the accumulation of kynurenine, which is due to IDO expression on the cells. In addition, the production of IL-22 by the activated and naïve T cells leads to the development of Tregs (Morris et al. 2016a; Mezrich et al. 2010). Alternatively, short-chain fatty acids, and microbial metabolites, are also recognized as potential members of neuroinflammatory conditions and neuro-immune disorders (Morris et al. 2016a). Considering progress on the molecular aspects of the diseased condition by the role of tryptophan catabolites and short-chain fatty acids, recent treatment progresses are at the preclinical stages. Several hurdles to cross to bring them effectively without any toxic manifestations.

## 4 Conclusion

With the combining theories of molecular and network pharmacology, structural biology, and transcriptomics, the cellular phenomenon of neuro-immune consequences has been understood. Consequently, different novel targets have been identified to modulate expressions of diverse markers in the nervous systems during neuro-immune disorders. Apart from conventional therapeutics, several components have also been identified from different sources to aim at selected targets that could be established from fundamental research to translational medicine in the future. However, before considering their efficacy in managing neuro-immune disorders, the consequences of safety and toxicities should be ruled over.

## Glossary

**Attention-deficit hyperactivity disorder** A neurobehavioral problem that results in lacking the focus, attention, and the inability to control impulsive behavior/thought in childhood.

**Chronic inflammatory demyelinating polyneuropathy** A neurological condition where the progressive loss of sense and muscle strength of arms and legs develops due to damage to the myelin sheath of nerves.

**Encephalomyelitis** A neurological condition that resulted due to inflammation in the brain and spinal cord.

**Fragile X syndrome** A neurological condition caused by the mutation of Fragile X Messenger Ribonucleoprotein 1 resulting in the brain developmental problem.

**Guillain-Barré syndrome** An autoimmune neurological condition where the immune system attacks the peripheral nervous system.

**Immune dysregulation** Any proposed or confirmed breakdown or maladaptive change in the molecular control of immune system processes.

**Immunomodulation** Alteration, either increase or decrease, in the immune system of the body.

**Infection** Invasion and growth of microorganisms in the body system.

**Multiple sclerosis** An autoimmune disease where the immune system damages the myelin sheath of the nerve in the brain and spinal cord.

**Neurodegenerative disease** A neurological condition that results in progressive loss/damage of neurons in the central/peripheral nervous system.

**Neuroinflammation** Inflammation in the nervous system.

**Neurological disorder** Disorders of the central and/or peripheral nervous systems.

**Neuromyelitis Optica** An autoimmune disease where the immune system damages the spinal cord and optic nerve.

**Obsessive-compulsive disorder** A psychological condition where a person develops uncontrolled and repetitive thoughts or behavior.

**Optic neuritis** A condition where inflammation and swelling damage the optic nerve.

**Psychotropic agents** Therapeutic agents or other compounds affect the brain functioning, thus, resulting in alteration in mood, thoughts, feelings, awareness, or behavior.

**Spasticity** A neurological condition of excessive muscle stiffness and rigidity due to damage/disruption of the brain and spinal cord.

**Transverse myelitis** A rare disease caused by spinal cord inflammation.

## References

Adjalley SH, Johnston GL, Li T, Eastman RT, Ekland EH, Eappen AG et al (2011) Quantitative assessment of Plasmodium falciparum sexual development reveals potent transmissionblocking activity by methylene blue. Proc Natl Acad Sci U S A 108(47):E1214–E1223

Al-Amin MM, Uddin MMN, Reza HM (2013) Effects of antipsychotics on the inflammatory response system of patients with schizophrenia in peripheral blood mononuclear cell cultures. Clin Psychopharmacol Neurosci 11(3):144–151

Alcocer-Gómez E, Casas-Barquero N, Williams MR, Romero-Guillena SL, Cañadas-Lozano D, Bullón P et al (2017) Antidepressants induce autophagy dependent-NLRP3-inflammasome inhibition in major depressive disorder. Pharmacol Res 121:114–121

Anthoney N, Foldi I, Hidalgo A (2018) Toll and toll-like receptor signalling in development. Development 145(9):dev156018

Arthur JSC, Ley SC (2013) Mitogen-activated protein kinases in innate immunity. Nat Rev Immunol 13(9):679–692

Ashraf H, Solla P, Sechi LA (2022) Current advancement of immunomodulatory drugs as potential pharmacotherapies for autoimmunity based neurological diseases. Pharmaceuticals 15(9):1077

Assefa DG, Zeleke ED, Bekele D, Ejigu DA, Molla W, Woldesenbet TT et al (2022) Isoniazid preventive therapy for prevention of tuberculosis among people living with HIV in Ethiopia: a systematic review of implementation and impacts. Int J Environ Res Public Health 20(1):621

Bacich DJ, Wozniak KM, Lu XCM, O'Keefe DS, Callizot N, Heston WDW et al (2005) Mice lacking glutamate carboxypeptidase II are protected from peripheral neuropathy and ischemic brain injury. J Neurochem 95(2):314–323

Bailey MT, Dowd SE, Galley JD, Hufnagle AR, Allen RG, Lyte M (2011) Exposure to a social stressor alters the structure of the intestinal microbiota: implications for stressor-induced immunomodulation. Brain Behav Immun 25(3):397–407

Ballatori N, Krance SM, Notenboom S, Shi S, Tieu K, Hammond CL (2009) Glutathione dysregulation and the etiology and progression of human diseases. Biol Chem 390(3):191–214

Carrico AW, Johnson MO, Morin SF, Remien RH, Riley ED, Hecht FM et al (2008) Stimulant use is associated with immune activation and depleted tryptophan among HIV-positive persons on anti-retroviral therapy. Brain Behav Immun 22(8):1257–1262

Chassaing B, Koren O, Goodrich JK, Poole AC, Srinivasan S, Ley RE et al (2015) Dietary emulsifiers impact the mouse gut microbiota promoting colitis and metabolic syndrome. Nature 519(7541):92–96

Chen Y, Zhang ZX, Zheng LP, Wang L, Liu YF, Yin WY et al (2019) The adenosine A2A receptor antagonist SCH58261 reduces macrophage/microglia activation and protects against experimental autoimmune encephalomyelitis in mice. Neurochem Int 129:104490

Cheng Y, Takeuchi H, Sonobe Y, Jin S, Wang Y, Horiuchi H et al (2014) Sirtuin 1 attenuates oxidative stress via upregulation of superoxide dismutase 2 and catalase in astrocytes. J Neuroimmunol 269(1–2):38–43

Cho SH, Chen JA, Sayed F, Ward ME, Gao F, Nguyen TA et al (2015) SIRT1 deficiency in microglia contributes to cognitive decline in aging and neurodegeneration via epigenetic regulation of IL-1β. J Neurosci 35(2):807–818

Constantinescu CS, Milliard B, Fujioka T, Bhopale MK, Calida D, Rostami AM (1998) Pathogenesis of neuroimmunologic diseases. Immunol Res 17(1–2):217–227

Cortese S (2020) Pharmacologic treatment of attention deficit-hyperactivity disorder. N Engl J Med 383(11):1050–1056

Dash MB, Douglas CL, Vyazovskiy V, Cirelli C, Tononi G (2009) Long-term homeostasis of extracellular glutamate in the rat cerebral cortex across sleep and waking states. J Neurosci 29(3):620–629

Di Rosso ME, Palumbo ML, Genaro AM (2016) Immunomodulatory effects of fluoxetine: a new potential pharmacological action for a classic antidepressant drug? Pharmacol Res 109:101–107

Diaz-delCastillo M, Woldbye DPD, Heegaard AM (2018) Neuropeptide Y and its involvement in chronic pain. Neuroscience 387:162–169

DiDonato JA, Mercurio F, Karin M (2012) NF-κB and the link between inflammation and cancer. Immunol Rev 246(1):379–400

Echeverría-Esnal D, Martin-Ontiyuelo C, Navarrete-Rouco ME, De-Antonio Cuscó M, Ferrández O, Horcajada JP et al (2021) Azithromycin in the treatment of COVID-19: a review. Expert Rev Anti Infect Ther 19(2):147–163

Eroğlu L, Çağlayan B (1997) Anxiolytic and antidepressant properties of methylene blue in animal models. Pharmacol Res 36(5):381–385

Ethell IM, Ethell DW (2007) Matrix metalloproteinases in brain development and remodeling: synaptic functions and targets. J Neurosci Res 85:2813–2823

Färber K, Pannasch U, Kettenmann H (2005) Dopamine and noradrenaline control distinct functions in rodent microglial cells. Mol Cell Neurosci 29(1):128–138

Fernandes BS, Steiner J, Bernstein HG, Dodd S, Pasco JA, Dean OM et al (2016) C-reactive protein is increased in schizophrenia but is not altered by antipsychotics: meta-analysis and implications. Mol Psychiatry 21(4):554–564

Ferreira R, Santos T, Cortes L, Cochaud S, Agasse F, Silva AP et al (2012) Neuropeptide Y inhibits interleukin-1 beta-induced microglia motility. J Neurochem 120(1):93–105

Finegold SM (2011) State of the art; microbiology in health and disease. Intestinal bacterial flora in autism. Anaerobe 17(6):367–378

Flanagan EP (2016) Autoimmune myelopathies. In: Pittock SJ, Vincent A (eds) Handbook of clinical neurology. Elsevier, pp 327–351

Flanagan EP (2019) Neuromyelitis optica spectrum disorder and other non–multiple sclerosis central nervous system inflammatory diseases. Continuum 25(3):815–844

Forman HJ, Zhang H, Rinna A (2009) Glutathione: overview of its protective roles, measurement, and biosynthesis. Mol Asp Med 30(1–2):1–12

Frye RA (2000) Phylogenetic classification of prokaryotic and eukaryotic Sir2-like proteins. Biochem Biophys Res Commun 273(2):793–798

Fuchs D, Hausen A, Reibnegger G, Schönitzer D, Unterweger B, Blecha HG et al (1987) Immune status of drug abusers. Cancer Detect Prev Suppl 1:535–541

Galetta KM, Bhattacharyya S (2019) Multiple sclerosis and autoimmune neurology of the central nervous system. Med Clin North Am 103(2):325–336

Gao S, Zhang J, He C, Meng F, Bu G, Zhu G et al (2017) Molecular characterization of neuropeptide Y (NPY) receptors (Y1, Y4 and Y6) and investigation of the tissue expression of their ligands (NPY, PYY and PP) in chickens. Gen Comp Endocrinol 240:46–60

Ge Y, Huang M, Yao YM (2020) Biology of interleukin-17 and its pathophysiological significance in sepsis. Front Immunol 11:1558

Gorain B, Choudhury H, Molugulu N, Athawale RB, Kesharwani P (2020) Fighting strategies against the novel coronavirus pandemic: impact on global economy. Front Public Health 8:800

Graus F, Delattre JY, Antoine JE, Dalmau J, Giometto B, Grisold W et al (2004) Recommended diagnostic criteria for paraneoplastic neurological syndromes. J Neurol Neurosurg Psychiatry 75(8):1135–1140

Greenberg BM, Krishnan C, Harder L (2019) New onset transverse myelitis diagnostic accuracy and patient experiences. Mult Scler Relat Disord 30:42–44

Griffin MO, Ceballos G, Villarreal FJ (2011) Tetracycline compounds with non-antimicrobial organ protective properties: possible mechanisms of action. Pharmacol Res 63(2):102–107

Heinz LX, Lee JE, Kapoor U, Kartnig F, Sedlyarov V, Papakostas K et al (2020) TASL is the SLC15A4-associated adaptor for IRF5 activation by TLR7–9. Nature 581(7808):316–322

Hunt DL, Castillo PE (2012) Synaptic plasticity of NMDA receptors: mechanisms and functional implications. Curr Opin Neurobiol 22(3):496–508

Hussain RZ, Cravens PC, Doelger R, Dentel B, Herndon E, Loof N et al (2018) TLR3 agonism re-establishes CNS immune competence during α4-integrin deficiency. Ann Clin Transl Neurol 5(12):1543–1561

Huwiler A, Zangemeister-Wittke U (2018) The sphingosine 1-phosphate receptor modulator fingolimod as a therapeutic agent: recent findings and new perspectives. Pharmacol Ther 185:34–49

Irwin MR, Olmos L, Wang M, Valladares EM, Motivala SJ, Fong T et al (2007) Cocaine dependence and acute cocaine induce decreases of monocyte proinflammatory cytokine expression across the diurnal period: autonomic mechanisms. J Pharmacol Exp Ther 320(2):507–515

Jiao F, Gong Z (2020) The beneficial roles of SIRT1 in neuroinflammation-related diseases. Oxid Med Cell Longev 2020:6782872

Jones R, Laake K, Øksengård AR (2002) D-cycloserine for Alzheimer's disease. Cochrane Database Syst Rev 2002(2):3153

Kalivas PW (2009) The glutamate homeostasis hypothesis of addiction. Nat Rev Neurosci 10(8):561–572

Kass JS, Shandera WX (2010) Nervous system effects of antituberculosis therapy. CNS Drugs 24(8):665–667

Kay E, Scotland RS, Whiteford JR (2014) Toll-like receptors: role in inflammation and therapeutic potential. Biofactors 40(3):284–294

Kieseier BC, Lehmann HC, Zu Hörste GM (2012) Autoimmune diseases of the peripheral nervous system. Autoimmun Rev 11(3):191–195

Kim EK, Choi EJ (2015) Compromised MAPK signaling in human diseases: an update. Arch Toxicol 89(6):867–882
Kim EJ, Kim YK (2019) Immunomodulatory effects of antipsychotic drugs in whole blood cell cultures from healthy subjects. Curr Psychiatr Res Rev 15(4):261–266
Kim DJ, King JA, Zuccarelli L, Ferris CF, Koppel GA, Snowdon CT et al (2009) Clavulanic acid: a competitive inhibitor of beta-lactamases with novel anxiolytic-like activity and minimal side effects. Pharmacol Biochem Behav 93(2):112–120
Kopnisky KL, Bao J, Lin YW (2007) Neurobiology of HIV, psychiatric and substance abuse comorbidity research: workshop report. Brain Behav Immun 21(4):428–441
Kost GC, Selvaraj S, Lee YB, Kim DJ, Ahn CH, Singh BB (2011) Clavulanic acid increases dopamine release in neuronal cells through a mechanism involving enhanced vesicle trafficking. Neurosci Lett 504(2):170–175
Kovac S, Walker MC (2013) Neuropeptides in epilepsy. Neuropeptides 47(6):467–475
Kucharz EJ, Sterakowski SJ (1990) Studies on immunomodulatory properties of isoniazid: I. Effect of isoniazid on mitogen- and anti-CD3 antibody-induced proliferation of human peripheral blood mononuclear cells and T cells. J Hyg Epidemiol Microbiol Immunol 34(1):99–105
Kumar V (2019) Toll-like receptors in the pathogenesis of neuroinflammation. J Neuroimmunol 332:16–30
Lash LH (2006) Mitochondrial glutathione transport: physiological, pathological and toxicological implications. Chem Biol Interact 163(1–2):54–67
Lazibat I, Majdak MR, Županić S (2018) Multiple sclerosis: new aspects of immunopathogenesis. Acta Clin Croat 57(2):352–361
Leboyan A, Esselin F, Bascou A, Duflos C, Ion I, Charif M et al (2023) Immune-mediated diseases involving central and peripheral nervous systems. Eur J Neurol 30(2):490–500
Li D, Liu N, Zhao HH, Zhang X, Kawano H, Liu L et al (2017) Interactions between Sirt1 and MAPKs regulate astrocyte activation induced by brain injury in vitro and in vivo. J Neuroinflammation 14(1):67
Li C, Wu X, Liu S, Zhao Y, Zhu J, Liu K (2019) Roles of neuropeptide Y in neurodegenerative and neuroimmune diseases. Front Neurosci 13:869
Li H, Liu S, Han J, Li S, Gao X, Wang M et al (2021a) Role of toll-like receptors in neuroimmune diseases: therapeutic targets and problems. Front Immunol 12:777606
Li L, Acioglu C, Heary RF, Elkabes S (2021b) Role of astroglial toll-like receptors (TLRs) in central nervous system infections, injury and neurodegenerative diseases. Brain Behav Immun 91:740–755
Liadaki K, Petinaki E, Skoulakis C, Tsirevelou P, Klapsa D, Germenis AE et al (2011) Toll-like receptor 4 gene (TLR4), but not TLR2, polymorphisms modify the risk of tonsillar disease due to streptococcus pyogenes and haemophilus influenzae. Clin Vaccine Immunol 18(2):217–222
Limon-Pacheco JH, Gonsebatt ME (2010) The glutathione system and its regulation by Neurohormone melatonin in the central nervous system. Cent Nerv Syst Agents Med Chem 10(4):287–297
Lindestam Arlehamn CS, Dhanwani R, Pham J, Kuan R, Frazier A, Rezende Dutra J et al (2020) α-Synuclein-specific T cell reactivity is associated with preclinical and early Parkinson's disease. Nat Commun 11:1875
Loof TG, Goldmann O, Medina E (2008) Immune recognition of streptococcus pyogenes by dendritic cells. Infect Immun 76(6):2785–2792
Luckheeram RV, Zhou R, Verma AD, Xia B (2012) CD4+T cells: differentiation and functions. Clin Dev Immunol 2012:925135
Maes M, Rief W (2012) Diagnostic classifications in depression and somatization should include biomarkers, such as disorders in the tryptophan catabolite (TRYCAT) pathway. Psychiatry Res 196(2–3):243–249
Maher P (2005) The effects of stress and aging on glutathione metabolism. Ageing Res Rev 4(2):288–314

Mahesh R, Jindal A, Gautam B, Bhatt S, Pandey D (2011) Evaluation of anti-depressant-like activity of linezolid, an oxazolidinone class derivative—An investigation using behavioral tests battery of depression. Biochem Biophys Res Commun 409(4):723–726
Mantere O, Trontti K, García-González J, Balcells I, Saarnio S, Mäntylä T et al (2019) Immunomodulatory effects of antipsychotic treatment on gene expression in first-episode psychosis. J Psychiatr Res 109:18–26
Mashaghi A, Marmalidou A, Tehrani M, Grace PM, Pothoulakis C, Dana R (2016) Neuropeptide substance P and the immune response. Cell Mol Life Sci 73(22):4249–4264
Mazzei T, Mini E, Novelli A, Periti P (1993) Chemistry and mode of action of macrolides. J Antimicrob Chemother 31:C1–C9
McDaneld LM, Fields JD, Bourdette DN, Bhardwaj A (2010) Immunomodulatory therapies in neurologic critical care. Neurocrit Care 12(1):132–143
Meister A, Mary EA (1983) Glutathione. Annu Rev Biochem 52:711–760
Mendes KL, de Farias Lelis D, Santos SHS (2017) Nuclear sirtuins and inflammatory signaling pathways. Cytokine Growth Factor Rev 38:98–105
Menon VP, Sudheer AR (2007) Antioxidant and anti-inflammatory properties of curcumin. Adv Exp Med Biol 595:105–125
Mezrich JD, Fechner JH, Zhang X, Johnson BP, Burlingham WJ, Bradfield CA (2010) An interaction between kynurenine and the aryl hydrocarbon receptor can generate regulatory T cells. J Immunol 185(6):3190–3198
Michan S, Sinclair D (2007) Sirtuins in mammals: insights into their biological function. Biochem J 404(1):1–13
Mineur YS, Picciotto MR, Sanacora G (2007) Antidepressant-like effects of ceftriaxone in male C57BL/6J mice. Biol Psychiatry 61(2):250–252
Mjösberg J, Bernink J, Peters C, Spits H (2012) Transcriptional control of innate lymphoid cells. Eur J Immunol 42(8):1916–1923
Morikawa K, Torii I, Morikawa S (2002) Modulation of IgD and CD20 by ligation of CD5 on tonsillar B cells. Scand J Immunol 55(1):44–52
Morris G, Berk M, Carvalho A, Caso JR, Sanz Y, Walder K et al (2016a) The role of the microbial metabolites including tryptophan catabolites and short chain fatty acids in the pathophysiology of immune-inflammatory and neuroimmune disease. Mol Neurobiol 54(6):4432–4451
Morris G, Berk M, Galecki P, Walder K, Maes M (2016b) The neuro-immune pathophysiology of central and peripheral fatigue in systemic immune-inflammatory and neuro-immune diseases. Mol Neurobiol 53(2):1195–1219
Mössner R, Lesch KP (1998) Role of serotonin in the immune system and in neuroimmune interactions. Brain Behav Immun 12(4):249–271
Müller N, Schwarz MJ (2006) Neuroimmune-endocrine crosstalk in schizophrenia and mood disorders. Expert Rev Neurother 6(7):1017–1038
Murphy BS, Sundareshan V, Cory TJ, Hayes D, Anstead MI, Feola DJ (2008) Azithromycin alters macrophage phenotype. J Antimicrob Chemother 61(3):554–560
Naylor GJ, Smith AHW, Connelly P (1987) A controlled trial of methylene blue in severe depressive illness. Biol Psychiatry 22(5):657–659
O'Donovan DJ, Fernandes CJ (2000) Mitochondrial glutathione and oxidative stress: implications for pulmonary oxygen toxicity in premature infants. Mol Genet Metab 71(1–2):352–358
O'Loughlin EK, Pakan JMP, McDermott KW, Yilmazer-Hanke D (2014) Expression of neuropeptide Y1 receptors in the amygdala and hippocampus and anxiety-like behavior associated with Ammon's horn sclerosis following intrahippocampal kainate injection in C57BL/6J mice. Epilepsy Behav 37:175–183
O'Neill LAJ (2008) When signaling pathways collide: positive and negative regulation of toll-like receptor signal transduction. Immunity 29(1):12–20
O'Sullivan SA, O'Sullivan C, Healy LM, Dev KK, Sheridan GK (2018) Sphingosine 1-phosphate receptors regulate TLR4-induced CXCL5 release from astrocytes and microglia. J Neurochem 144(6):736–747

Obregon D, Parker-Athill EC, Tan J, Murphy T (2012) Psychotropic effects of antimicrobials and immune modulation by psychotropics: implications for neuroimmune disorders. Neuropsychiatry 2(4):331–343

Oliveira V, Seabra M, Rodrigues R, Carvalho V, Mendes M, Pereira D et al (2021) Neuro-COVID frequency and short-term outcome in the northern Portuguese population. Eur J Neurol 28(10):3360–3368

Otto MW, Basden SL, Leyro TM, McHugh RK, Hofmann SG (2007) Clinical perspectives on the combination of D-cycloserine and cognitive-behavioral therapy for the treatment of anxiety disorders. CNS Spectr 12(1):59–61

Pałucha A, Tatarczyńska E, Brański P, Szewczyk B, Wierońska JM, Kłak K et al (2004) Group III mGlu receptor agonists produce anxiolytic- and antidepressant-like effects after central administration in rats. Neuropharmacology 46(2):151–159

Paribello C, Tao L, Folino A, Berry-Kravis E, Tranfaglia M, Ethell IM et al (2010) Open-label add-on treatment trial of minocycline in fragile X syndrome. BMC Neurol 10:91

Pastore A, Federici G, Bertini E, Piemonte F (2003) Analysis of glutathione: implication in redox and detoxification. Clin Chim Acta 333(1):19–39

Pavón L, Sandoval-López G, Eugenia Hernández M, Loría F, Estrada I, Pérez M et al (2006) Th2 cytokine response in major depressive disorder patients before treatment. J Neuroimmunol 172(1–2):156–165

Pellegrino TC, Bayer BM (2000) Specific serotonin reuptake inhibitor-induced decreases in lymphocyte activity require endogenous serotonin release. Neuroimmunomodulation 8(4):179–187

Peña S, Baccichet E, Urbina M, Carreira I, Lima L (2005) Effect of mirtazapine treatment on serotonin transporter in blood peripheral lymphocytes of major depression patients. Int Immunopharmacol 5(6):1069–1076

Planavila A, Iglesias R, Giralt M, Villarroya F (2010) Sirt1 acts in association with PPAR to protect the heart from hypertrophy, metabolic dysregulation, and inflammation. Cardiovasc Res 90(2):276–284

Polman CH, O'connor PW, Havrdova E, Hutchinson M, Kappos L, Miller DH et al (2006) A randomized, placebo-controlled trial of natalizumab for relapsing multiple sclerosis. N Engl J Med 354(9):899–910

Prime K, French P (2001) Neuropsychiatric reaction induced by clarithromycin in a patient on highly active antiretroviral therapy (HAART). Sex Transm Infect 77(4):297–298

Provost J-C, Funck-Brentano C, Rovei V, D'Estanque J, Ego D, Jaillon P (1992) Pharmacokinetic and pharmacodynamic interaction between toloxatone, a new reversible monoamine oxidase—A inhibitor, and oral tyramine in healthy subjects. Clin Pharmacol Ther 52(4):384–393

Qi B, Shi C, Meng J, Xu S, Liu J (2018) Resveratrol alleviates ethanol-induced neuroinflammation in vivo and in vitro: involvement of TLR2-MyD88-NF-κB pathway. Int J Biochem Cell Biol 103:56–64

Quintas A, de Solís AJ, Díez-Guerra FJ, Carrascosa JM, Bogónez E (2012) Age-associated decrease of SIRT1 expression in rat hippocampus. Exp Gerontol 47(2):198–201

Ramsay RR, Dunford C, Gillman PK (2007) Methylene blue and serotonin toxicity: inhibition of monoamine oxidase A (MAO A) confirms a theoretical prediction. Br J Pharmacol 152(6):946–951

Reiner A, Levitz J (2018) Glutamatergic signaling in the central nervous system: ionotropic and metabotropic receptors in concert. Neuron 98(6):1080–1098

Riedel W, Lang U, Oetjen U, Schlapp U, Shibata M (2003) Inhibition of oxygen radical formation by methylene blue, aspirin, or α-lipoic acid, prevents bacterial-lipopolysaccharide-induced fever. Mol Cell Biochem 247(1–2):83–94

Rothstein JD, Patel S, Regan MR, Haenggeli C, Huang YH, Bergles DE et al (2005) β-Lactam antibiotics offer neuroprotection by increasing glutamate transporter expression. Nature 433(7021):73–77

Sandler RH, Finegold SM, Bolte ER, Buchanan CP, Maxwell AP, Väisänen ML et al (2000) Short-term benefit from oral vancomycin treatment of regressive-onset autism. J Child Neurol 15(7):429–435
Schmitz J, Zakrzewicz A, Wilker S, Kuncová J, Hecker A, Grau V et al (2017) Non-neuronal neuropeptide Y and its receptors during acute rejection of rat pulmonary allografts. Transpl Immunol 43–44:49–53
Schug TT, Xu Q, Gao H, Peres-da-Silva A, Draper DW, Fessler MB et al (2010) Myeloid deletion of SIRT1 induces inflammatory signaling in response to environmental stress. Mol Cell Biol 30(19):4712–4721
Semenkov YP, Makarov EM, Makhno VI, Kirillov SV (1982) Kinetic aspects of tetracycline action on the acceptor (A) site of Escherichia coli ribosomes. FEBS Lett 144(1):125–129
Sies H (1999) Glutathione and its role in cellular functions. Free Radic Biol Med 27(9–10):916–921
Siller SS, Broadie K (2011) Neural circuit architecture defects in a Drosophila model of fragile X syndrome are alleviated by minocycline treatment and genetic removal of matrix metalloproteinase. Dis Model Mech 4(5):673–685
Smyth AM, Lawrie SM (2013) The neuroimmunology of schizophrenia. Clin Psychopharmacol Neurosci 11(3):107–117
Snider LA, Lougee L, Slattery M, Grant P, Swedo SE (2005) Antibiotic prophylaxis with azithromycin or penicillin for childhood-onset neuropsychiatric disorders. Biol Psychiatry 57(7):788–792
Sørensen G, Jensen M, Weikop P, Dencker D, Christiansen SH, Loland CJ et al (2012) Neuropeptide Y Y5 receptor antagonism attenuates cocaine-induced effects in mice. Psychopharmacology 222(4):565–577
Szałach ŁP, Lisowska KA, Cubała WJ (2019) The influence of antidepressants on the immune system. Arch Immunol Ther Exp 67(3):139–151
Tang BL (2017) Sirtuins as modifiers of Parkinson's disease pathology. J Neurosci Res 95(4):930–942
Taylor CG, Nagy LE, Bray TM (1996) Nutritional and hormonal regulation of glutathione homeostasis. In: Stadtman ER, Chock PB (eds) Current topics in cellular regulations. Academic, pp 189–208
Tesmer LA, Lundy SK, Sarkar S, Fox DA (2008) Th17 cells in human disease. Immunol Rev 223(1):87–113
Thomas AG, Liu W, Olkowski JL, Tang Z, Lin Q, Lu XCM et al (2001) Neuroprotection mediated by glutamate carboxypeptidase II (NAALADase) inhibition requires TGF-β. Eur J Pharmacol 430(1):33–40
Thompson CJ (2010) Goodman and Gilman's the pharmacological basis of therapeutics. Focus Altern Complement Ther 7(2):207
Thorsell A, Mathé AA (2017) Neuropeptide Y in alcohol addiction and affective disorders. Front Endocrinol (Lausanne) 8:178
Tourjman V, Koué MÈ, Kouassi E, Potvin S, Tourjman V, Koué MÈ et al (2012) In vivo immunomodulatory effects of antipsychotics on inflammatory mediators: a review. Adv Biosci Biotechnol 3(4):551–565
Verge B, Alonso Y, Valero J, Miralles C, Vilella E, Martorell L (2011) Mitochondrial DNA (mtDNA) and schizophrenia. Eur Psychiatry 26(1):45–56
Vojdani A, Erde J (2006a) Regulatory T cells, a potent immunoregulatory target for CAM researchers: the ultimate antagonist (I). Evid Based Complement Alternat Med 3(1):25–30
Vojdani A, Erde J (2006b) Regulatory T cells, a potent immunoregulatory target for CAM researchers: modulating allergic and infectious disease pathology (II). Evid Based Complement Alternat Med 3(2):209–215
Vojdani A, Lambert J (2011) The role of Th17 in neuroimmune disorders: target for CAM therapy. Part I. Evid Based Complement Alternat Med 2011:927294
Yamamoto H, Schoonjans K, Auwerx J (2007) Sirtuin functions in health and disease. Mol Endocrinol 21(8):1745–1755

Yang H, Gu ZT, Li L, Maegele M, Zhou BY, Li F et al (2017) SIRT1 plays a neuroprotective role in traumatic brain injury in rats via inhibiting the p38 MAPK pathway. Acta Pharmacol Sin 38(2):168–181

Ye J, Liu Z, Wei J, Lu L, Huang Y, Luo L et al (2013) Protective effect of SIRT1 on toxicity of microglial-derived factors induced by LPS to PC12 cells via the p53-caspase-3-dependent apoptotic pathway. Neurosci Lett 553:72–77

Yeung F, Hoberg JE, Ramsey CS, Keller MD, Jones DR, Frye RA et al (2004) Modulation of NF-κB-dependent transcription and cell survival by the SIRT1 deacetylase. EMBO J 23(12):2369–2380

Yin Y, Guo R, Shao Y, Ge M, Miao C, Cao L et al (2019) Pretreatment with resveratrol ameliorate trigeminal neuralgia by suppressing matrix metalloproteinase-9/2 in trigeminal ganglion. Int Immunopharmacol 72:339–347

Zarogoulidis P, Papanas N, Kioumis I, Chatzaki E, Maltezos E, Zarogoulidis K (2012) Macrolides: from in vitro anti-inflammatory and immunomodulatory properties to clinical practice in respiratory diseases. Eur J Clin Pharmacol 68(5):479–503

Zenobia C, Hajishengallis G (2015) Basic biology and role of interleukin-17 in immunity and inflammation. Periodontol 2000 69(1):142–159

Zhang H, Forman HJ (2009) Redox regulation of γ-glutamyl transpeptidase. Am J Respir Cell Mol Biol 41(5):509–515

Zhang HL, Zheng XY, Zhu J (2013) Th1/Th2/Th17/Treg cytokines in Guillain–Barré syndrome and experimental autoimmune neuritis. Cytokine Growth Factor Rev 24(5):443–453

Zhang Q, Yuan L, Zhang Q, Gao Y, Liu G, Xiu M et al (2015) Resveratrol attenuates hypoxia-induced neurotoxicity through inhibiting microglial activation. Int Immunopharmacol 28(1):578–587

Zhang XS, Li W, Wu Q, Wu LY, Ye ZN, Liu JP et al (2016) Resveratrol attenuates acute inflammatory injury in experimental subarachnoid hemorrhage in rats via inhibition of TLR4 pathway. Int J Mol Sci 17(8):1331

Zheng C, Chen J, Chu F, Zhu J, Jin T (2019) Inflammatory role of TLR-MyD88 signaling in multiple sclerosis. Front Mol Neurosci 12:314

Zhou J, Neale JH, Pomper MG, Kozikowski AP (2005) NAAG peptidase inhibitors and their potential for diagnosis and therapy. Nat Rev Drug Discov 4:1015–1026

Zhu X, Liu Q, Wang M, Liang M, Yang X, Xu X et al (2011) Activation of Sirt1 by resveratrol inhibits TNF-α induced inflammation in fibroblasts. PLoS One 6(11):e27081

Zink MC, Uhrlaub J, DeWitt J, Voelker T, Bullock B, Mankowski J et al (2005) Neuroprotective and anti-human immunodeficiency virus activity of minocycline. JAMA 293(16):2003–2011

Zorrilla EP, Luborsky L, McKay JR, Rosenthal R, Houldin A, Tax A et al (2001) The relationship of depression and stressors to immunological assays: a meta-analytic review. Brain Behav Immun 15(3):199–226

Zrzavy T, Hametner S, Wimmer I, Butovsky O, Weiner HL, Lassmann H (2017) Loss of "homeostatic" microglia and patterns of their activation in active multiple sclerosis. Brain 140(7):1900–1913

# The Role of Host Genetics in the Susceptibility, Severity, and Immune Response to SARS-CoV-2

**Arpit Kumar Shrivastava and Swaminathan Ravichandran**

**Abstract** SARS-CoV-2 (severe acute respiratory syndrome coronavirus 2) is the causative agent of COVID-19 disease. This virus emerged in China and spread quickly worldwide, responsible for nearly 596 million infections and 6.4 million deaths globally. The different variants of concern (VOCs) of SARS-CoV-2 identified in different parts of the world are alarming for serious public health concerns. The VOCs place constraints on utilizing monoclonal antibodies to treat SARS-CoV-2 infection. The SARS-CoV-2 virus has a spherical envelope and single-stranded positive-sense RNA. It primarily encodes four structural proteins (S, E, M, and N) and numerous non-structural proteins (NSPs). Numerous vaccines that target SARS-CoV-2 various SARS-CoV-2 sites have recently been developed, and more are in works. The concerns regarding the newly emerging virus strains are still very high. The latest scientific information on disease susceptibility suggests that multiple genetic factors are associated with COVID-9 disease severity. The genetic architect of an individual contributes to the susceptibility and response to viral infection. The different VOCs contain various mutations and many of which are localized to the S protein. Mostly these mutations are responsible for the changes in viral behavior and pathogenesis. Several host genetic factors that are associated with illness susceptibility, severity, and immune response to SARS-CoV-2 have been covered in this book chapter.

**Keywords** Allelic polymorphisms · Corona · CoV-2 · DNA methylation · Epigenetic inflammation · Pandemic SARS · RT-qPCR · Thromboembolic complications

A. K. Shrivastava (✉)
Atal Bihari Vajpayee Government Medical College, Vidisha, India

Office of International Relations, Indian Institute of Technology Indore, Indore, India

S. Ravichandran
Department of Pathology, School of Medicine, Medical University of the Americas, Nevis, St. Kitts & Nevis


A. Das Mohapatra, P. S. Sahu (eds.), *Advances in Immunology and Immuno-techniques*, Interdisciplinary Biotechnological Advances,
https://doi.org/10.1007/978-981-97-5508-0_10

## 1 Introduction

The Coronavirus disease 2019 (COVID-19) was declared a pandemic by the World Health Organization (WHO) and was caused by severe acute respiratory syndrome coronavirus 2 (SARS-CoV-2). The first outbreak of SARS-CoV-2 was documented in Wuhan, China, in December 2019. According to WHO (https://data.who.int/dashboards/covid19/cases?n=o), as of 29 August, 2024 total of 775,917,102 confirmed cases and 7,058,381 deaths have been documented worldwide.

Coronaviruses come under the Coronavirine subfamily, which contains four genera Alpha, Beta, Gamma, and Delta coronaviruses. Alpha and Beta coronavirus are mainly responsible for mammalian infection while Gamma and Delta coronavirus primarily infect birds (Corman et al. 2018). The name "Corona" comes from the crownlike shape of the spike protein; seven known Coronavirus species commonly cause infection in humans (Alipoor et al. 2021) out of these seven Severe Acute Respiratory Syndrome Coronavirus-1 (SARS-CoV-1), the Middle East respiratory syndrome (MERS), and the Severe Acute Respiratory Syndrome Coronavirus-2 (SARS-Cov-2) are Beta coronaviruses that emerged in the last two decades as human pathogens and linked to serious diseases.

The COVID-19 disease susceptibility, resistance, and severity depend on several extrinsic and intrinsic factors including viral, host, genetics, microbiome, metabolome, environmental, blood group, vitamins, and others. The emerging scientific information on disease susceptibility suggests that multiple genetic factors are linked with COVID-19 disease severity (Fig. 1); important elements of genetic factors include proteins involved in the viral life cycle and host defensive mechanisms (Brest et al. 2020; Casanova et al. 2020). The different immunological responses including delayed IgM responses and greater S protein IgG titers have been seen in mild and severe COVID-19 illness, especially in patients who were not in intensive

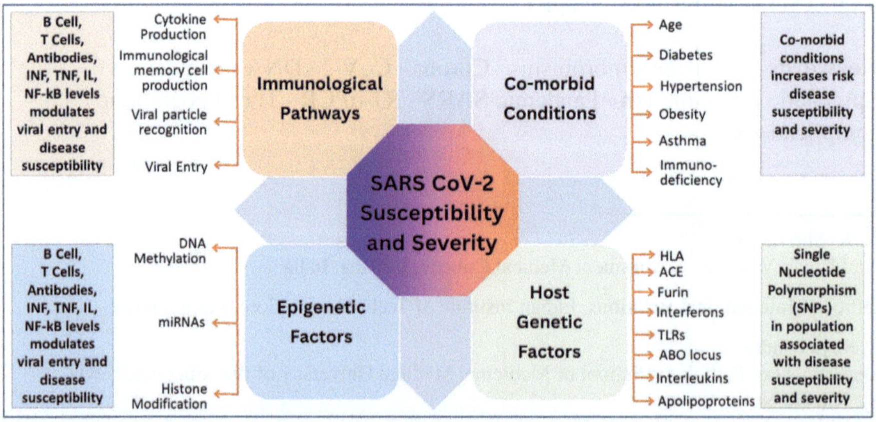


**Fig. 1** Various host genetic factors associated with SARS CoV-2 infection susceptibility and severity

care units (Shen et al. 2020; Wang et al. 2021). In particular, proteins implicated in coronavirus-cell entrance and fusion, like Angiotensin I converting enzyme 2 (ACE2), transmembrane protease, serine 2 (TMPRSS2), and CD26, are thought to be linked to an increased risk of COVID-19, according to a recent study (Kaltoum 2021).

During this global health crisis, it became extremely important to determine the correlation between genetic factors of the individual and the susceptibility towards disease progression to better understand the individual's risk for SARS-CoV-2. Genome-wide association studies (GWAS) offered a substantial understanding of the genetic basis of various infectious diseases. The genetics of the host and the virus must be studied in order to comprehend the pathophysiology of SARS-CoV-2. These studies are also required to interpret why COVID-19 manifests differently among individuals and for the development of novel antiviral therapeutics and vaccines for SARS-CoV-2.

## 2 Clinical and Epidemiological Features of SARS-CoV-2

A unique class of viruses known as coronaviruses affects a variety of animal species as well as humans, causing moderate to severe respiratory diseases. Human coronaviruses (CoV) were initially recognized as pathogens in the middle of the 1960s. From the beginning of twenty-first century there are three lethal zoonotic coronaviruses have emerged as a major global health concern and caused epidemics and pandemics in the last two decades: in November 2002, the SARS pandemic was brought on by the severe acute respiratory syndrome coronavirus (SARS-CoV-1). The Middle East Respiratory Syndrome (MERS) coronavirus (MERS-CoV), which causes MERS, was originally discovered in Jeddah, Saudi Arabia, in September 2012. The continuing COVID-19 pandemic is caused by a brand-new coronavirus called SARS-CoV-2, which first surfaced in Wuhan, China, towards the end of 2019. Many interactions between animals, people, and the environment frequently result in the emergence of novel viral diseases with high epidemic-pandemic potential (Coker et al. 2011).

The COVID-19 crisis profoundly changed our lives, COVID-19 is causing remarkable prolonged implications for public health, human well-being, and quality of life. Earlier it was suggested that in December 2019 very first SARS-CoV-2 transmission in Wuhan was linked with human see food wholesale market (Abebe et al. 2020). Although genetic evidence directed that SARS-CoV-2 is a natural virus that seems to originate in animals, but there is no conclusive evidence that when and where the virus first entered humans. However, SARS-CoV-2 shared 79.6% of the genetic sequence with SARS-CoV-1 and shared 96% of the genome with a bat CoV, according to reports (Zhou et al. 2020).

Transmission of SARS-CoV-2 either direct or indirect (sneezing, coughing, through eye and nasal mucous membranes) and apart from that blood, fecal-oral routes, urine, and breastfeeding are some of the other possible routes for

SARS-CoV-2 transmission (Halaji et al. 2020). The most common method of spreading SARS-CoV-2 is by airborne transmission, which is extremely virulent (Xu et al. 2020a). The unique virological features of this virus are responsible for its high transmissibility. SARS-CoV-2 is mostly spread from an infected individual through the air in the form of droplets and aerosol particles. The asymptomatic carriers can be fast spread SARS-CoV-2 because they unknowingly transmit the virus to others; according to one study, asymptomatic illness may be responsible for at least 50% of community transmission (Chang et al. 2020; Johansson et al. 2021).

All the age groups of the human population are susceptible to COVID-19 infection. Age can be a strong risk factor for severe illness, health-related problems, and mortality (Guan et al. 2020). SARS-CoV-2 expected to take about 5–14 days to incubate inside the human body (Lauer et al. 2020). The majority of individuals presented signs of diseases after an incubation period; from the onset of the disease illness, it took median time of 8 days to develop dyspnea and pneumonia (Wu et al. 2020). The most typical symptoms of SARS-CoV-2 infection include fever, exhaustion, and a dry cough (Chen et al. 2020a). Other less common symptoms involve the sore throat, headache, chest pain, diarrhea, anorexia, nausea, and vomiting (Huang et al. 2020). Olfactory and taste disorders were also observed in COVID-19 infected individuals (Villarreal et al. 2021). Lymphopenia found to be a common immunological abnormalities in majority of severe COVID-19 patients (Jafarzadeh et al. 2021).

The United States witnessed three simultaneous epidemiological waves of SARS-CoV-2. Emerging variants of concern (VOCs) in SARS-CoV-2 limit the use of monoclonal antibodies to treat COVID-19. The Alpha (B.1.1.7), Beta (B.1.351), Gamma (P.1), Kappa, and Delta (B.1.617.1 and B.1.617.2) are major VOCs that have been identified between December 2020 and May 2021. Advanced structural protein mutations and alterations in the other non-structural proteins in VOCs can cause the escape of authorized vaccines to some extent (Almubaid and Al-Mubaid 2021; Jiang et al. 2022). Delta Strain which was first identified in India replaced all other SARS-CoV-2 variants in terms of infectivity one study suggests an R0 of 5.08, well superior to the R0 of the ancestral strain of 2.79; suggesting lineage B.1.617.2 was responsible for a higher transmission resulting in the second wave of COVID-19 in India (Liu and Rocklöv 2021; Cherian et al. 2021; Pascarella et al. 2021). All-inclusive communities are concerned about the second and third waves of COVID-19. Recently, a new variant of concern was first identified on 11th November 2021 in Botswana and later many parts of South Africa is known as Omicron (B.1.1.529) (https://www.who.int/en/activities/tracking-SARS-CoV-2-variants). This new variant had undergone significant mutations when compared to its previous variants. The Omicron (B.1.1.529) might have improved transmissibility and immune escape. This novel variant may re-infect people who are prior infected with other SARS-CoV-2 variants (Kumar et al. 2021). The Omicron variant may have high infectivity, but it causes less severe symptoms than previous variants and is probably capable of escaping immunity (Ren et al. 2022). The genetic mutations in SARS-CoV-2 genome is the most significant causes of numerous waves. The virus gets longer to mutate and find ways to avoid or deceive antibodies if vaccination takes place slowly. New SARS-CoV-2 variants and mutations in circulating strains need to constantly identify and analyze to develop better diagnostics and therapeutics against SARS-CoV-2.

## 3 SARS-CoV-2: General Overview

Coronaviruses contain the largest genomes of among all RNA viruses ranging in between 25 and 31 kilobases. COVID-19 disease involved multiple genetic and environmental factors for disease severity suggesting COVID-19 as a multigenic and multifactorial disease. The SARS-CoV-2 is enveloped, positive-stranded RNA (+ssRNA) virus; its genome consists of 15 ORFs encoding 29 proteins (Al-Qaaneh et al. 2021). The whole genome sequence of SARS-CoV-2 suggested that it shows high sequence similarity (96.2%) with bat SARS-related coronavirus but showed a low level of similarity with SARS-CoV (about 79%) or MERS-CoV (about 50%) (Lu et al. 2020; Paraskevis et al. 2020; Gralinski and Menachery 2020). Recent phylogenetic investigation revealed that SARS-CoV-2 differs from SARS-CoV in that it has a unique evolution and a proteolytically sensitive activation loop, which may increase the pathogenicity and transmission potential of SARS-CoV-2 (Jaimes et al. 2020).

SARS-CoV-2 mainly consists of Spike (S), Envelop (E), Membrane (M), and Neucleocapsid (N) structural proteins. In contrast to other CoVs, SARS-CoV-2 possess glycoprotein acetyl esterase and hemagglutination; among all structural protein M protein is possibly the highest effective which provide shape to the viral structure (Kumar et al. 2021). The S protein plays a role in the recognition of transmembrane receptors and promotes viral entry into the host cell. In the pathophysiology of COVID-19, glycoprotein S plays a significant role (Zhang et al. 2020a). The first step in the infection process of SARS-CoV-2 in human cells is the binding of protein S to ACE2 (Wrapp et al. 2020). The host specificity and infectiousness of the CoVs are significantly influenced by the S proteins (Letko et al. 2020). Instantaneously, the N protein attaches to viral RNA and is crucial for the assembly of viral components that enclose viral RNA (Parameshwar et al. 2020). The Viral E protein helps to release viral particles from the infected cells (Ashour et al. 2020). Unlike other coronaviruses, SARS-CoV-2 has a unique furin cleavage site at S1/S2 border ("RPPA" sequence) which makes it pathogenic (Ratre et al. 2021).

Many non-structural proteins (NSPs) involving replicases and proteases are encoded by the SARS-CoV genome, the 5′ Orf1a/Orf1ab encodes a polyprotein that is autoproteolytically processed at 16 NSPs [NSP1-NSP16]. NSP1 interacts with the host's DNA replication route, whereas NSP5, NSP8, and NSP13 connect with the host's gene expression and epigenetic regulatory pathways. Vesicle trafficking is associated with NSP6, NSP7, NSP10, NSP13, and NSP15. The NSP13 replicates its RNA genome while interacting with the host signaling network (Chen et al. 2020b). The N and E proteins and specific accessory proteins are involved in stopping the initial innate immune response during SARS-CoV-2 replication. It has also been observed that SARS-CoV-2 mutation in NSP2 and NSP3 contributes significantly to SARS-CoV-2 differentiation and virulence (Angeletti et al. 2020). Drug targeting heavily depends on protein architectures. The understanding of the insight mechanism of virus interaction with human proteins and the immune mechanism against the SARS-CoV-2 infection is crucial to classifying genes that control illness, susceptibility, and severity in COVID-19 disease.

Diagnostic tools are crucial in the battle against COVID-19 in order to promptly implement controls to halt SARS-CoV-2 transmission through case identification, contact tracking, and isolation of positive cases (Udugama et al. 2020). The availability and accessibility of viral genomic sequences allowed us to use molecular techniques to identify SARS-CoV-2. Numerous diagnostic assays have been commercialized for the accurate detection of SARS-CoV-2 and several others are actively being developed and validated in numerous laboratories. Nucleic acid-based detection methods are a crucial technique for the diagnosis of SARS-CoV-2. Specifically, RT-qPCR is a gold standard technique for virus detection only because of prompt identification and a high degree of sensitivity and specificity. The clinical management of COVID-19 includes antiviral drugs, immunomodulating therapy, and advanced organ support, whereas critically ill patients may require intensive care with ventilatory strategies, careful fluid monitoring, and prone ventilation. There are currently no effective methods to cure it. However, continuous efforts are still being made by medical professionals to make effective and affordable therapies to combat this global pandemic.

## 4 Immunological Responses to SARS-CoV-2

Two key elements make up a defensive immunological response towards SARS-CoV-2 infection; first rapid innate immunity including granulocytes, monocytes, and macrophages among other cells of the innate immune system that activates upon infection and resolved after the clearance of infection, and second is specific antiviral acquired immunity with T and B cells which is being triggered right after the encounter of SARS-CoV-2 (Schultze and Aschenbrenner 2021; Sette and Crotty 2021).

Our immune system's initial line of defense, innate immunity, is crucial in combating this new coronavirus by activating several signaling pathways to create interferons, pro-inflammatory cytokines, and chemokines, as well as to start the adaptive immunological response. SARS-CoV-2 affects the innate immune system by preventing type I interferon from controlling viral multiplication and by interacting with macrophages and dendritic cells. Virus infected cells produced and release IFN-I and III during the interplay with SARS-CoV-2; further several studies reported that cytokines including IL-6, IL-1β, IL-12, IL-1, TNF-α are responsible to generate hyperinflammatory response against SARS-CoV-2 infection (Quarleri and Delpino 2021; Choudhary et al. 2021). Both active and severe SARS-CoV-2 patients had extremely low levels of circulating NK and NKT cells, whereas critical COVID-19 patients reported a cytokine storm under severe circumstances (Chen et al. 2020a; Masselli et al. 2020; Sebastian-Valverde and Pasinetti 2020). In SARS-CoV-2 infection, IFNα/β levels vary among susceptible individuals having comorbid conditions like obesity or diabetes and non-susceptible individuals (Feuillet et al. 2021). In COVID-19 positive individuals, an excessive innate immune response can result in a cytokine storm, inflammation, and vascular disruption (Beacon et al. 2020). The post-mortem investigations of individuals who had died from SARS-CoV-2 infection showed a large density of mast cell progenitor cells and an enhanced concentration of perivascular and septal mast cells (da Silva Motta Junior et al. 2020). The increased neutrophil and reduced lymphocyte counts are related to the severity and mortality of COVID-19 (Sapkal et al. 2020; Reusch et al. 2021). It was found that individuals with severe illnesses had greater amounts of chemokines, chemokine receptors, neutrophils, and pro-inflammatory macrophages at the site of infection (Chua et al. 2020; Liao et al. 2020).

Human complement system involved more than 30 serum proteins. The majority of these proteins have protease activity, allowing one complement protein to sequentially activate another (Merle et al. 2015). Activation of the complement proteins can be initiated by lectin pathway and alternative pathway right after infection of SARS-CoV-2. Recent findings suggested that the complement activation is increased during the progression of COVID-19 (Cugno et al. 2021; Lin et al. 2021). Complement activation is shown in infected patients as cell surface C3 fragment deposition on circulating neutrophils (Fang et al. 2020; Akk et al. 2019; Maxwell et al. 2021).

Immunological memory formation is a dynamic process that occurs over time at various body sites and involves the generation of memory B- and T-cell populations. People over the age of 65 and those with coexisting illnesses are more susceptible to COVID-19, which increases disease pathology and prolongs infection (Wang et al. 2020a). A recent study showed that the peripheral T lymphocytes (specifically innate-like T and various CD8+ T-cell subsets) were notably low in critical SARS-CoV-2 infected patients in comparison to healthy individuals. Patients with critical COVID-19 have considerably increased plasma levels of pro-inflammatory cytokines and chemokines (IL-1, IL-2, IL-6, IL-7, IL-10, G-CSF, CXCL-10, CCL2, CCL3, and TNF-), which are frequently coupled with impaired T-cell and antibody response (Chen et al. 2020a; Wen et al. 2020; Qin et al. 2020). Additionally, severe cases of COVID-19 were shown to have high amounts of megakaryocytes and erythroid cells (Bernardes et al. 2020). A new study on anti-SARS-CoV-2 B-cell memory responses shows that, even 6 months after infection, memory responses are still robust and widen neutralization even as serum antibody titers fall (Gaebler et al. 2021). Furthermore, it has been shown that the IgA response elicited by Mucosa-Associated Invariant T (MAIT) cells might play a significant role in an efficient and protective response against SARS-CoV-2 (De Sanctis et al. 2021).

## 5 Association of Different Genetic and Epigenetic Factors with SARS-CoV-2 Severity

Numerous studies have examined coronavirus replication, cell cycle modulation, pathogenesis, innate immunity, and inflammatory response from a basic science perspective; however, in order to understand the significance of host genetic determinants in illness susceptibility and clinical development, only a few studies have looked at host genetic variation in humans. Research indicates that ACE2, TMPRSS2, CtsL, and CtsB genes are crucial for SARS-CoV-2 entry (Darbani 2020). SARS-CoV2 severity is correlated with several immune component variations, cellular proteases, and additional genetic and epigenetic variables.

### *5.1 Human Leukocyte Antigens (HLA) Locus*

The human MHC genes encode a group of proteins known as the human leukocyte antigen (HLA), which play a key role in the genetic diversity and presentation of antigens. The HLA region has more than 240 genes and is divided into three classes (Class I, II, and III) (Dendrou et al. 2018). T lymphocytes get antigenic peptides from HLA Classes I and II molecules, which helps the immune system discriminate between self and non-self. HLA peptides with higher binding specificities to the SARS-CoV-2 peptides will result in a better T-cell response to viral infection. HLA is a highly polymorphic genetic locus in the human genome that enables individuals to handle infections with all the various pathogens. It has been acknowledged that the HLA region of the human genome is significant for both disease risk and resistance (Dendrou et al. 2018). It has been discovered that HLA gene polymorphisms are associated with several infectious disorders, including infections brought on by RNA viruses. Most of the HLA alleles present in the human population fall into the three- to the six-allele range, and their distribution varies around the globe.

The existence of various allelic variations in people from various racial and ethnic backgrounds is well documented (Van Dyke et al. 2009; Maiers et al. 2007). In case of both mild and severe COVID-19 infection, different immunological responses, such as delayed IgM and IgG responses, have been observed and patients who are not in intensive care units have greater S protein IgG titers, which may be related to HLA allelic variance (Shen et al. 2020; Sun et al. 2020; Ovsyannikova et al. 2020).

The HLA profiles of different people might be hard to compare. Those with severe SARS-CoV-2 infections and others with moderate or no symptoms of a disease have different HLA profiles, which makes comparison and analysis challenging. In Covid-19 patients, the frequencies of the HLA-C*07:29 and B*15:27 alleles were much higher, and HLA-B*15:03 had the best ability to display the highly conserved SARS-CoV-2 peptides that are shared by other common human coronaviruses (Nguyen et al. 2020). The severity of disease may fluctuate depending on the activation of the T-cell antiviral response by different HLA genotypes. A recent study revealed that the HLA*A*11:01, B*51:01, and C*14:02 alleles are significantly predisposed to the worse COVID-19 outcome (Wang et al. 2020b). However, further in silico investigations revealed that HLA-A*02:01 might be associated with an increased risk for COVID-19 infection, and this allele is less able to display SARS-CoV-2 antigens than other HLA variations (Tomita et al. 2020). These findings will surely help in understanding the fundamental processes of protective innate and adaptive immunity and may lead to the construction of genetic markers.

## 5.2 Angiotensin-Converting Enzymes (ACE)

Two homologous genes, angiotensin-converting enzyme 1 (ACE1) and angiotensin-converting enzyme 2 (ACE2), regulate the physiological homeostasis of the renin-angiotensin system (RAS). The primary RAS regulator ACE2 is in charge of cleaving and generating peptides from the Angiotensin I and Angiotensin II. One of the crucial regulatory elements for chromatin modification and transcription factors is the ACE2 gene, which regulates epigenetic and hormonal regulation of expression levels. Several studies documented that the balance of ACE/ACE2 expression level is critical for Covid-19 progression and therapeutic interventions (Gemmati et al. 2020; Sriram and Insel 2020; Tseng et al. 2020). ACE2 and ACE genotype may alter the clinical consequences of disease as evidenced by the imbalance and dysregulation of ACE/ACE2 expression, which may enhance tissue damage, proliferation, thrombosis, fibrosis, or inflammation (Delanghe et al. 2020; Zheng and Cao 2020). Therefore, investigating the genetic variants of this ACE gene aids in our understanding of why some people are more severely affected by SARS-CoV-2 than others.

ACE2 receptor interaction allows SARS-CoV-2 to enter target cells (Hoffmann et al. 2020). The majority of human tissues express ACE2 at varying amounts and after SARS-CoV-2 infection, type II alveolar lung cells, epithelial cells of esophagus, myocardial cells of the heart, proximal tubule cells of the kidney, and small intestinal cells all showed tissue-specific expression of ACE2 (Li et al. 2020a; Hu et al. 2021). The SARS-CoV-2 infection downregulates expression of ACE2 which further facilitate inflammation at the site of infection (Verdecchia et al. 2020).

Asian people were shown to have higher ACE levels, which were detrimental. In addition, it was discovered that the expression of ACE2 in several tissues was influenced by age, gender, body mass index, and ethnicity (Chen et al. 2018). The higher death rates in the European population may be related to ethnic variations in ACE I/D polymorphism allele frequency, as evidenced by the higher recurrence of ACE DD genotype in the European community compared to the Asian population (Yamamoto et al. 2020). According to other studies, East Asians (allele frequencies of 11 eQTLs: 0.73–0.99) expressed more ACE2 than Europeans (allele frequencies: 0.44–0.65), indicating that different groups may be more or less susceptible to the SARS-CoV-2 virus (Cao et al. 2020). A detailed analysis of a significant genomic dataset revealed that nine ACE2 variants were related to increased susceptibility, whereas 17 variants were related to decreased S protein binding affinity and defense against SARS-CoV-2 infection (Stawiski et al. 2020). The K26R and I468V variations may affect the S protein's ability to attach to certain molecules, according to a research that used molecular dynamic simulations. Non-Finnish Europeans are more likely to have the K26R version modified than East Asians, who are more likely to have the I468V variant (Li et al. 2020b).

The NCBI database estimates that there are close to 18,000 single-nucleotide variants of human ACE2. Investigation of genetic predisposition during SARS-CoV-2 infection and disease progression requires comparative investigation of the ACE2 in individuals of various community settings. A recent in vitro study found that ACE2 polymorphisms had a negligible effect on SARS-CoV-2 entrance into ACE2-dependent cells (Hashizume et al. 2021). To comprehend the significance of ACE2 genotypes in Covid-19 illness progression and severity, additional functional genotype analysis of critical patients will be necessary. Experimental model analysis shows ACE2 expression is very low in the lung, mainly limited to type II alveolar cells. Whereas SARS-CoV-2 stimulates secretion of interferons (IFNs), inflammatory cytokines can increase the expression ACE2 which will be internalized later for SARS-CoV-2 entry and potentiate infection (Scialo et al. 2020). Children are substantially less vulnerable to COVID-19 infection due to elevated levels of sACE2 (soluble ACE2), which has a protective effect by covering the spike protein and preventing SARS-CoV-2 from binding to ACE2 (Ciaglia et al. 2020). Plasma ACE2 levels are a useful indicator of COVID-19 outcomes in hospitalized patients because, in particular, patients with greater plasma levels of ACE2 expression have worse outcomes. Patients with high blood pressure or cardiac disease have higher ACE2 plasma levels, which explains why they are the main targets for severe COVID-19 manifestation (Kragstrup et al. 2021). Angiotensin II excess is reduced as a result of therapeutic sACE2, and viral entry is blocked as a secondary benefit (Rahman et al. 2021). Soluble recombinant ACE2 protein can contain the spread of coronavirus by limiting its cell membrane attachment, entry, and replication (Batlle et al. 2020). In addition to taking soluble recombinant ACE2, patients with COVID-19 should continue taking angiotensin-converting enzyme inhibitors (ACEIs) and angiotensin-receptor blockers (ARBs), which block the action of the renin-angiotensin system. Additionally, we need to learn more about the ACE2 polymorphisms that may have adverse effects (Grover and Oberoi 2021).

## *5.3 Furin*

When a coronavirus infection occurs, particularly SARS-CoV-2, the fusion machinery of viral glycoproteins is primarily activated by proteolytic cleavage; the spike proteins activation entailed several cleavage events (Hulswit et al. 2016). The ubiquitous expression of furin, which effectively cleaves polybasic or multibasic sites, occurs in the Golgi apparatus of all cells and is crucial for pathogenicity of viruses

(Thomas 2002). The multibasic S1/S2 site of SARS-CoV-2 spike protein is cleaved by host cell protease furin (Hoffmann et al. 2020; Tang et al. 2021). Furin enhances both SARS-CoV-2 infectivity and cell-cell dissemination, suggesting that furin inhibitors may reduce but not stop viral transmission. The SARS-CoV-2 furin cleavage site plays a crucial role in SARS-CoV-2 infection and pathogenesis (Johnson et al. 2021; Papa et al. 2021).

Among SARS-CoV-2-positive patients, one Italian study found the existence of intronic and missense germline furin mutations with very low allelic frequencies (Latini et al. 2020). Researchers have proposed that a single noncoding regulatory variation at the Furin locus may affect SARS-CoV-2 infection after studying CRISPR-edited Furin rs4702 (GG) in lung cells and neurons (Dobrindt et al. 2021). A significant factor in SARS-CoV-2 transmission is the furin cleavage site. The previous structural study has demonstrated that furin cleavage makes it easier to bind a higher percentage of the S protein to the human ACE2 receptor, which may have facilitated emergence of SARS-CoV-2 in humans (Peacock et al. 2021; Wrobel et al. 2020).

## 5.4 *Interferons*

Interferon signaling at the time of various viral infections is the first line of defense against the pathogenic agent. Interferons are a specialized class of cytokines that activates the host defense mechanism by signaling cascade which regulates the activation of many genes. Three district types of interferons (Type I, Type II, and Type III) medicated antiviral response to combat viral infections (Acosta et al. 2020). Type I/III interferon production and the accompanying signaling cascade are both necessary to limit coronavirus infection in the early stages; given that SARS-CoV-2 is susceptible to IFN pre-treatment, IFN therapy may be a possibility for COVID-19 treatment (Yuen et al. 2020; Lei et al. 2020). SARS-CoV and MERS-CoV infections are characterized by a type I IFN response; type I IFNs bind to the type I interferon receptor (IFNAR), producing significant, protective levels of IFN through a positive feedback loop mechanism (Channappanavar et al. 2016; Channappanavar and Perlman 2017). The subunits IFNAR1 and IFNAR2 make up IFNAR. The IFN-$\alpha/\beta$-induced antiviral immunity reduces in the deficiency of IFNARs (Duncan et al. 2015). There are two distinct versions IFNAR1 and IFNAR2 were found in individuals hospitalized for SARS-CoV-2-induced pneumonia (Zhang et al. 2020b). Types I and II Interferon-inducible transmembrane (IFITM) proteins are minuscule homologous proteins that provide the IFN response. Age-dependently, more severe disease has been associated with homozygosity for the C allele of rs12252 in the IFITM3 gene in the COVID-19 sickness (Zhang et al. 2020c). A study using primary human airway epithelia (HAE) and immortalized cell lines revealed that primary HAE and model cells elicit a substantial increase of type I and III IFNs and melanoma differentiation-associated protein 5 (MDA-5) in lung cells (Rebendenne et al. 2021). The inappropriate and dysregulated cytokine response might contribute towards immune-mediated pathology against COVID-19.

## 5.5 *Toll-like Receptors (TLRs)*

TLRs, also referred to as pattern recognition receptors (PRRs), are a class of innate immune receptor molecules (Kawai and Akira 2010). These receptors detect the molecular patterns connected to pathogens and alert the host immune system. Many immune cells, such as macrophages, natural killer cells, dendritic cells (DCs), T cells, and B cells, express TLRs (Angelopoulou et al. 2020). Type I IFNs and inflammatory cytokines are produced when TLRs trigger immunological signaling cascades. TLR3, which recognizes virus-derived double-stranded (ds) sRNA, is the most ubiquitous and extensively expressed TLR. TLR7 and TLR8 recognize virus-derived single-stranded (ss) RNA. TLRs may be a possible target in the development of a vaccine against SARS-CoV-2 and for managing infection in the early stages of the disease (Khanmohammadi and Rezaei 2021). The extrinsic coagulation pathway, leukocyte infiltration, inflammatory signaling in endothelial cells, and neutrophil granule release may all be triggered by COVID-19-mediated TLR activation (Biswas and Khan 2020). Eight genetic loci have been studied to determine how autosomal-recessive and autosomal-dominant deficits are caused by important illness outcome variants in Covid-19 patients; type I IFN response controlled by TLR3 and IRF7 is linked to potentially fatal consequences (Zhang et al. 2020c). According to the immunoinformatic technique, the genome of SARS-CoV-2 contains more ssRNA pieces than TLR7/8 can recognize. Acute lung injury is the outcome of SARS-CoV-2-induced innate immune hyperactivation, which significantly increases the inflammatory response upon TLR7/8 recognition (Moreno-Eutimio et al. 2020).

## 5.6 *ABO Locus*

Infection risk was shown to be higher for blood group A while lower for blood group O, but this does not affect the course of the disease and is not associated with mortality (Goel et al. 2021; Solmaz and Araç 2021). According to accumulating evidence, ABO blood type may be implicated in the immunopathogenesis of SARS-CoV-2 infection. The underlying mechanism of how blood types influence the SARS-CoV-2 induced disease outcomes is still uncertain. One opinion is that the ABO blood type influences glycosyltransferase activity and the risk of venous thromboembolism, which is prevalent in severe Covid 19 cases and can have significant implications (Ibrahim-Kosta et al. 2020; Porfidia et al. 2020).

### 5.7 Interleukins

The significant levels of circulating inflammatory cytokines and increased T-cell and monocyte migration into the lungs are seen in people with severe SARS-CoV-2 infections (Xu et al. 2020b). Viral infection clearance, attenuated adaptive immune response, and sustained virologic response were related to low or high producing IL-6 polymorphisms (Barrett et al. 2001; Bogdanović et al. 2016; Nattermann et al. 2007). Inhibitors of interleukin (IL)-1 or IL-6 have been utilized to treat this patient population, but their relative efficacy has not been established; inhibition of IL-1 is directly linked to a large drop in mortality in COVID-19 hospitalized patients (Cavalli et al. 2021). Decreased plasma IL-3 levels in SARS-CoV-2 infected people are associated with worsening of the disease, a rise in viral load, and an increase in mortality (Bénard et al. 2021). Additional diagnostic markers, particularly in severely ill COVID-19 patients, can also predict the course of a disease, including ferritin, cardiac troponin I, D-dimer, CRP, and IL-6 (Velavan et al. 2021).

### 5.8 Apolipoproteins

The polymorphic gene encoding the tiny protein known as ApoE is named apolipoprotein E (APOE). Alzheimer's disease, cardiovascular disease, atherosclerosis, and shorter life expectancy are all associated with the ApoE4 allele (Smith 2000). The APOE promotes the entrance and infectiousness of SARS-CoV-2; the RBD of SARS-CoV-2 binds to ApoE, and SARS-CoV-2 could enter the cell via ApoE's metabolic pathway (Yin et al. 2021). The SARS-CoV-2 entrance sites are sensitive and dependent on cholesterol. The severity of COVID-19 may be due to ApoE4, as seen by the greater size and increased nuclear fragmentation of astrocytes after SARS-CoV-2 infection (Wang et al. 2021; Wang et al. 2020c).

Previous research suggested that APOE modifies proinflammatory pathways in a genotype-dependent manner (Kasparian et al. 2020). In comparison to apolipoprotein E3 homozygotes, apolipoprotein E4 alleles showed a two- to three-fold greater risk of SARS-CoV-2 infection or COVID-19 death (Hilser et al. 2021). According to one study, APOE-4 genetic variants may increase the clinical severity of SARS-CoV-2 infection (Del Ser et al. 2021).

### 5.9 Epigenetic Factors

Epigenetic processes are reversible, dynamic, and impacted by both internal and external stimuli. SARS-CoV-2 also interacts with multiple epigenetic factors similar to other numerous viruses and affects several epigenetic processes. The investigation of the SARS-CoV-2 interactome provide proofs that the viral proteins molecules (E, nsp5, nsp8, and nsp13) interact with regulators of gene expression and epigenetic processes, potentially disrupting the programming of gene expression (Beacon et al. 2020). DNA methylation occurs mostly at CpGs (five methyl cytosine, 5mC) sites and is involved in a variety of biological activities, including control of gene expression, chromatin architecture, and X chromosome inactivation (Adalsteinsson and Ferguson-Smith 2014; Jeltsch et al. 2018). Age, smoking, and other illness states, including hypoxia, are some of the significant factors that affect DNA methylation patterns (Camuzi et al. 2019; Reid et al. 2017; Nwanaji-Enwerem and Colicino 2020). Changes in DNA methylation patterns may result from the silent hypoxemia seen in some COVID-19 patients (Ottestad et al. 2020).

Targeting the epigenetic machinery could be a viable option for treating viral infections because epigenetic regulation of transcription is a fundamental host cell response to viral infection (Salgado-Albarrán et al. 2021). Several proteins involved in epigenetic response are possible therapeutic candidates for COVID-19 disease.

## 6 Association of Significant Lab Factors with SARS-CoV-2 Severity

Significant laboratory evaluations are helpful for the treating physicians to assess the disease progression and halt its severity before it involves the majority of the lung, liver, and renal tissue leading to multi-organ damage. Understanding immunopathology is very important to prevent severe pulmonary complications and multi-organ dysfunction syndrome (MODS), especially among COVID-19 patients who need to get treated in ICU. Severe and non-severe COVID-19 cases were significantly different from one another in terms of the Complete Blood Count test, evaluation of liver and renal function results, identification of appropriate markers for inflammation/infection, and serum electrolytes.

### *6.1 Inflammatory/Infection Markers*

Since COVID-19 severe presentation is due to an uncontrolled inflammatory response and we need more research on understanding its immunopathology. Inflammation and infection markers like C-reactive protein, erythrocyte sedimentation rate, and lactate dehydrogenase levels were related to severe COVID-19 patients even though it is not specific markers and their high levels signify the severity of inflammation in COVID-19 patients. The most spoken immunological problem of the decade with COVID-19 patients is cytokine storm. High titers of serum pro-inflammatory cytokines in COVID-19 patients are considered as eye of the Cytokine storm/Hurricane (Anka et al. 2021). The higher probability of a poor prognosis was linked to the elevated ratio of IL-6/LY. Additionally, the early clinical identification of high-risk patients and risk stratification are both possible uses for IL-6/LY (Yang et al. 2021). Earlier research was mentioning that no strong association between severe COVID-19 patients and levels of their ESR and IL-6. But later studies prove the strong association between severe cases and IL-6 levels ESR levels (Yang et al. 2021; Hoseinyazdi et al. 2021).

## *6.2 Coagulation Studies*

At the beginning of the pandemic (COVID-19), we lost many lives due to thromboembolic complications. The imbalance created with coagulation and anti-coagulation in severe cases of COVID-19 leads to many thromboembolic complications (including myocardial infarction and stroke). SARS-CoV-2 targets the small blood vessel endothelial cells and increases the expression of ACE2, then ACE2 gets internalized and used as a tool for virus entry. All the changes lead to a reduction of the number of ACE2 expressions on endothelial cells which favors the thrombotic and inflammatory processes. Ultimately reduced ACE2 expression leads to a reduction in des-Arg bradykinin (DABK) into inactive peptides causing bradykinin receptors (BKRs) activation and increased pro-thrombotic signaling. Additionally, SARS-CoV-2 activates the complement cascade through the lectin, alternative, and classical pathways, resulting in the production of the membrane attack complex (MAC) on target cells. Therefore, SARS-CoV-2 targets and activates endothelial cells, increasing the expression of P-selectin, von Willebrand factor (vWf), and fibrinogen, which causes platelet-platelet binding, fibrin and clot formation, which will eventually progress to systemic thrombosis, and disseminated intravascular coagulation (Perico et al. 2021).

Coagulation profiles (Fibrinogen levels, PT (Prothrombin time), and D-dimer) were positively related to the COVID-19 severity (Ghahramani et al. 2020). D-dimer is a small protein fragment, a fibrin degradation product, normally no significant levels have been seen in the blood. After many blood clots have been fibrinolysed, especially the patients who have been progressing to develop DIC (Disseminated intravascular coagulation), fibrin protein D fragments joined by a cross-link will be present in the blood and considered as positive D-dimer test. D-dimer protein increased about four-fold is an indicator of poor prognosis in hospitalized COVID-19 patients (Asakura and Ogawa 2021). As part of a regular preventative treatment, low-molecular-weight heparin is advised for COVID-19 patients, particularly for those with aberrant D-Dimer levels (Tomo et al. 2021).

## *6.3 Significant Liver Function Test (LFT)*

Abnormal LFTs in severe COVID-19 patients can be due to many reasons including drug-induced liver injury has to be ruled out before confirming that hepatocytes are infected with SARS-CoV-2. Especially after the use of acetaminophen, ritonavir, and remdesivir medications in COVID-19 cases which are potentially hepatotoxic and cause severe damage to the hepatocytes (Bertolini et al. 2020). Isolated abnormal LFTs in severe COVID-19 patients are not directly related to the mortality of the patients. Abnormal serum glutamic oxaloacetic transaminase (SGOT) is likely due to inflammation-induced direct hepatocyte injury caused by the SARS-CoV-2 virus. Abnormal SGOT has a very strong association with severe COVID-19 cases (Leulseged et al. 2021). Alanine, aspartate transaminase, and alkaline phosphatase values were higher among critically ill Covid-19 patients admitted to ICU (Araya et al. 2022). All the COVID-19-related damages to the liver are temporary and could return back to normalcy at the end of 1-year follow-up (Liao et al. 2022).

## *6.4 Significant Renal Function Tests (RFT)*

Abnormal renal function tests (high serum creatinine/blood urea nitrogen levels) are a substantial risk factor for death in patients admitted to Intensive Care Unit (ICU). SARS-CoV-2 is commonly found in the glomerulus (Podocytes, Endothelial cells) and proximal tubular cells. Glomerular filtration is believed to be impaired as a result of SARS-CoV-2-induced podocyte dysfunction, which results in proteinuria and hemorrhage. Endothelial cell involvement of SARS-CoV-2 alters the glomerular capillary hemostasis and induces fibrin thrombi formation. The proximal tubular cells, which contain high amounts of ACE2 at the apical brush border, are the most often affected tissue type by SARS-CoV-2. A patient who got hospitalized with severe COVID-19 illness needs to be assessed for acute kidney injury in order to identify the patients who are at high risk for mortality and morbidity (La Porta et al. 2022). To evaluate the long-term effects of AKI in individuals who have recovered from COVID-19, extensive follow-up for potential renal disease is necessary (Gu et al. 2022). In severely ill COVID-19 cases with renal involvement, dialysis is an important tool for management (Perico et al. 2021).

## *6.5 Electrolyte Imbalance*

Electrolyte imbalance specifically Na and K levels were found to be significant predictors of immunopathological outcomes of COVID-19 disease. The reason for huge variation in Na and K levels can be due to the renal damage and decreased activity of the ACE2 receptors. This electrolyte imbalance can cause inappropriate body response to stress leading to more complications and death (Leulseged et al. 2021). The reason behind high serum potassium and the development of acute kidney injury may likely be due to the cytokine storm-induced tumor lysis syndrome (De Carvalho et al. 2021).

## 7 Conclusion

Although viral and host genetic differences are expected to impact the severity of an illness and the immune response against sickness, the processes behind the population's variety in COVID-19 susceptibility and illness presentation remain little understood. The large diversity of clinical manifestations suggests that host genetics may possibly play a significant role in the variations in host–pathogen relationships. The development of contemporary scientific techniques has made it possible to conduct genome-wide association studies (GWAS) and gene-specific research, which has substantial implications for understanding the genetic basis of various infectious diseases. These studies determine genetic susceptibility to infections by identifying associated genetic loci and allelic polymorphisms. Several consortiums (COVID-19 Host Genetics Initiative, COVID Human Genetic Effort) are being created to further explore the impact of genetics on SARS-CoV-2 infection and immune response. Many notable discoveries have been published for both viral and host genetic variations. These investigations make it easier to locate and assess whether newly emerging virus strains are related to various clinical disease outcomes or to people who carry particular genes for those particular disease conditions. The susceptibility, severity, and general clinical consequences of COVID-19 may be significantly influenced by host genetic variation. Along with host genetic factors understanding the laboratory features of severe COVID-19 patients help us to get more insight into the severity of the problem and communicating that with scientific society will improve the overall prognosis in high-risk patients around the globe. More research should be focusing on integrated immunopathology behind COVID-19 and that should modify our current immunotherapeutic approaches towards COVID-19 treatment. A greater knowledge of the host genetic and epigenetic factors, as well as the viral interactions of SARS-CoV-2, is necessary to develop novel treatments and improved prognostic tools. To completely comprehend the immunological response to SARS-CoV-2, large-scale genome-wide association analyses and biological validation studies using larger cohorts of convalescent individuals are necessary.

## Glossary

**Allelic polymorphisms** Allelic polymorphisms refer to the genetic variation that occurs within a population where multiple forms of a gene (alleles) exist at a specific locus (position) in the DNA sequence.

**DNA methylation** DNA methylation is a chemical modification of DNA molecules that involves the addition of a methyl group (CH3) to the cytosine nucleotide in the DNA molecule.

**Epigenetic factors** The term "epigenetic factors" describes alterations or changes in DNA's structure or function that can impact gene expression without changing the underlying DNA sequence.

**Hemostasis** Hemostasis refers to the physiological process by which bleeding is stopped or prevented from occurring.

**Immunopathology** The study of immune responses to illnesses, especially how immune responses may contribute to the etiology or development of diseases, is known as immunopathology.

**Immunotherapeutic** Immunotherapy, commonly referred to as immune-based therapy or immunotherapeutic, is a discipline of medicine that makes use of the immune system to treat conditions including cancer, autoimmune disorders, and infectious illnesses.

**Inflammation** Inflammation is a complex biological response that occurs in the body as a defense mechanism in response to injury, infection, or irritation.

**Microbiome** The microbiome refers to the vast and diverse community of microorganisms, including bacteria, viruses, fungi, and other microbes, that reside in and on the human body.

**Pandemic** A pandemic is an outbreak of a disease that occurs over a wide geographic area and affects an exceptionally high proportion of the population. The term is often used to describe the spread of infectious diseases, such as viruses, that can easily transmit from person to person.

**Prognostic tools** Prognostic tools are instruments used in medicine or healthcare to forecast the probable progression or result of a disease or condition in a specific patient. These technologies can give healthcare professionals useful data that can be used to influence treatment choices, determine prognosis, and improve patient care.

**Proteinuria** Proteinuria is a condition characterized by the presence of excess protein in the urine.

**RT-qPCR** RT-qPCR stands for Reverse Transcription Quantitative Polymerase Chain Reaction. It is a laboratory technique used to detect and quantify the amount of specific RNA molecules in a sample.

**Thromboembolic complications** Thromboembolic refers to the development of blood clots (thrombi) that can move through the circulation and obstruct blood vessel, causing organ damage and even death.

**Transcription** Transcription is the process by which genetic information encoded in DNA is converted into an RNA molecule. It is a fundamental cellular process that occurs in the nucleus of eukaryotic cells or in the cytoplasm of prokaryotic cells.

## References

Abebe EC, Dejenie TA, Shiferaw MY, Malik T (2020) The newly emerged COVID-19 disease: a systemic review. Virol J 17:96

Acosta PL, Byrne AB, Hijano DR, Talarico LB (2020) Human type I interferon antiviral effects in respiratory and reemerging viral infections. J Immunol Res 2020:1372494

Adalsteinsson BT, Ferguson-Smith AC (2014) Epigenetic control of the genome—lessons from genomic imprinting. Genes (Basel) 5:635

Akk A, Springer LE, Yang L, Hamilton-Burdess S, Lambris JD, Yan H et al (2019) Complement activation on neutrophils initiates endothelial adhesion and extravasation. Mol Immunol 114:629

Alipoor SD, Mortaz E, Jamaati H, Tabarsi P, Bayram H, Varahram M et al (2021) COVID-19: molecular and cellular response. Front Cell Infect Microbiol 11:563085

Almubaid Z, Al-Mubaid H (2021) Analysis and comparison of genetic variants and mutations of the novel coronavirus SARS-CoV-2. Gene Reports [Internet] 23(January):101064. https://doi.org/10.1016/j.genrep.2021.101064

Al-Qaaneh AM, Alshammari T, Aldahhan R, Aldossary H, Alkhalifah ZA, Borgio JF (2021) Genome composition and genetic characterization of SARS-CoV-2. Saudi J Biol Sci 28:1978

Angeletti S, Benvenuto D, Bianchi M, Giovanetti M, Pascarella S, Ciccozzi M (2020) COVID-2019: the role of the nsp2 and nsp3 in its pathogenesis. J Med Virol 92:584

Angelopoulou A, Alexandris N, Konstantinou E, Mesiakaris K, Zanidis C, Farsalinos K et al (2020) Imiquimod—a toll like receptor 7 agonist—is an ideal option for management of COVID 19. Environ Res 188:109858

Anka AU, Tahir MI, Abubakar SD, Alsabbagh M, Zian Z, Hamedifar H et al (2021) Coronavirus disease 2019 (COVID-19): an overview of the immunopathology, serological diagnosis and management. Scand J Immunol 93:e12998

Araya S, Tsegay YG, Atlaw A, Aragaw M, Tadlo G, Tsegaye N et al (2022) Organ function biomarker abnormalities, associated factors and disease outcome among hospitalized patients with COVID-19. Biomark Med 16(6):417–426

Asakura H, Ogawa H (2021) COVID-19-associated coagulopathy and disseminated intravascular coagulation. Int J Hematol 113:45

Ashour HM, Elkhatib WF, Rahman MM, Elshabrawy HA (2020) Insights into the recent 2019 novel coronavirus (Sars-coV-2) in light of past human coronavirus outbreaks. Pathogens 9:186

Barrett S, Goh J, Coughlan B, Ryan E, Stewart S, Cockram A et al (2001) The natural course of hepatitis C virus infection after 22 years in a unique homogenous cohort: spontaneous viral clearance and chronic HCV infection. Gut 49:423

Batlle D, Wysocki J, Satchell K (2020) Soluble angiotensin-converting enzyme 2: a potential approach for coronavirus infection therapy? Clin Sci 134:543

Beacon TH, Su RC, Lakowski TM, Delcuve GP, Davie JR (2020) SARS-CoV-2 multifaceted interaction with the human host. Part II: innate immunity response, immunopathology, and epigenetics. IUBMB Life 72:2331

Bénard A, Jacobsen A, Brunner M, Krautz C, Klösch B, Swierzy I et al (2021) Interleukin-3 is a predictive marker for severity and outcome during SARS-CoV-2 infections. Nat Commun [Internet] 12(1):1–8. https://doi.org/10.1038/s41467-021-21310-4

Bernardes JP, Mishra N, Tran F, Bahmer T, Best L, Blase JI et al (2020) Longitudinal multi-omics analyses identify responses of megakaryocytes, erythroid cells, and plasmablasts as hallmarks of severe COVID-19. Immunity 53:1296

Bertolini A, van de Peppel IP, Bodewes FAJA, Moshage H, Fantin A, Farinati F et al (2020) Abnormal liver function tests in patients with COVID-19: relevance and potential pathogenesis. Hepatology 72:1864

Biswas I, Khan GA (2020) Coagulation disorders in covid-19: role of toll-like receptors. J Inflamm Res 13:823

Bogdanović Z, Marinović-Terzić I, Kuret S, Jerončić A, Bradarić N, Forempoher G et al (2016) The impact of IL-6 and IL-28B gene polymorphisms on treatment outcome of chronic hepatitis C infection among intravenous drug users in Croatia. PeerJ 4:e2576

Brest P, Refae S, Mograbi B, Hofman P, Milano G (2020) Host polymorphisms may impact SARS-CoV-2 infectivity. Trends Genet 36:813

Camuzi D, De Amorim ÍSS, Pinto LFR, Trivilin LO, Mencalha AL, Lima SCS (2019) Regulation is in the air: the relationship between hypoxia and epigenetics in cancer. Cells 8(4):1–22

Cao Y, Li L, Feng Z, Wan S, Huang P, Sun X et al (2020) Comparative genetic analysis of the novel coronavirus (2019-nCoV/SARS-CoV-2) receptor ACE2 in different populations. Cell Discov 6:11

Casanova JL, Su HC, Abel L, Aiuti A, Almuhsen S, Arias AA et al (2020) A global effort to define the human genetics of protective immunity to SARS-CoV-2 infection. Cell 181:1194

Cavalli G, Larcher A, Tomelleri A, Campochiaro C, Della-Torre E, De Luca G et al (2021) Interleukin-1 and interleukin-6 inhibition compared with standard management in patients with COVID-19 and hyperinflammation: a cohort study. Lancet Rheumatol 3:e253

Chang D, Lin M, Wei L, Xie L, Zhu G, Dela Cruz CS et al (2020) Epidemiologic and clinical characteristics of novel coronavirus infections involving 13 patients outside Wuhan, China. JAMA 323:1092

Channappanavar R, Perlman S (2017) Pathogenic human coronavirus infections: causes and consequences of cytokine storm and immunopathology. Semin Immunopathol 39:529

Channappanavar R, Fehr AR, Vijay R, Mack M, Zhao J, Meyerholz DK et al (2016) Dysregulated type I interferon and inflammatory monocyte-macrophage responses cause lethal pneumonia in SARS-CoV-infected mice. Cell Host Microbe 19:181

Chen YY, Zhang P, Zhou XM, Liu D, Zhong JC, Zhang CJ, Jin LJYH (2018) Relationship between genetic variants of ACE2 gene and circulating levels of ACE2 and its metabolites. J Clin Pharm Ther 43(2):189–195

Chen G, Wu D, Guo W, Cao Y, Huang D, Wang H et al (2020a) Clinical and immunological features of severe and moderate coronavirus disease 2019. J Clin Invest 130:2620

Chen Y, Liu Q, Guo D (2020b) Emerging coronaviruses: genome structure, replication, and pathogenesis. J Med Virol 92:2249

Cherian S, Potdar V, Jadhav S, Yadav P, Gupta N, Das M et al (2021) Sars-cov-2 spike mutations, l452r, t478k, e484q and p681r, in the second wave of covid-19 in Maharashtra, India. Microorganisms 9:1542

Choudhary S, Sharma K, Silakari O (2021) The interplay between inflammatory pathways and COVID-19: a critical review on pathogenesis and therapeutic options. Microb Pathog 150:104673

Chua RL, Lukassen S, Trump S, Hennig BP, Wendisch D, Pott F et al (2020) COVID-19 severity correlates with airway epithelium–immune cell interactions identified by single-cell analysis. Nat Biotechnol 38:970

Ciaglia E, Vecchione C, Puca AA (2020) COVID-19 infection and circulating ACE2 levels: protective role in women and children. Front Pediatr 8:206

Coker R, Rushton J, Mounier-Jack S, Karimuribo E, Lutumba P, Kambarage D et al (2011) Towards a conceptual framework to support one-health research for policy on emerging zoonoses. Lancet Infect Dis 11:326

Corman VM, Muth D, Niemeyer D, Drosten C (2018) Hosts and sources of endemic human coronaviruses. Adv Virus Res 100:163

Cugno M, Meroni PL, Gualtierotti R, Griffini S, Grovetti E, Torri A et al (2021) Complement activation and endothelial perturbation parallel COVID-19 severity and activity. J Autoimmun 116:102560

da Silva Motta Junior J, Miggiolaro AFRDS, Nagashima S, de Paula CBV, Baena CP, Scharfstein J et al (2020) Mast cells in alveolar septa of COVID-19 patients: a pathogenic pathway that may link interstitial edema to immunothrombosis. Front Immunol 11:574862

Darbani B (2020) The expression and polymorphism of entry machinery for covid-19 in human: juxtaposing population groups, gender, and different tissues. Int J Environ Res Public Health 17:3433

De Carvalho H, Richard MC, Chouihed T, Goffinet N, Le Bastard Q, Freund Y et al (2021) Electrolyte imbalance in COVID-19 patients admitted to the emergency department: a

case–control study. Intern Emerg Med [Internet] 16(7):1945–1950. https://doi.org/10.1007/s11739-021-02632-z
De Sanctis JB, García AH, Moreno D, Hajduch M (2021) Coronavirus infection: an immunologists' perspective. Scand J Immunol 93:e13043
Del Ser T, Fernández-Blázquez MA, Valentí M, Zea-Sevilla MA, Frades B, Alfayate E et al (2021) Residence, clinical features, and genetic risk factors associated with symptoms of COVID-19 in a cohort of older people in Madrid. Gerontology 67:281
Delanghe JR, Speeckaert MM, De Buyzere ML (2020) The host's angiotensin-converting enzyme polymorphism may explain epidemiological findings in COVID-19 infections. Clin Chim Acta 505:192
Dendrou CA, Petersen J, Rossjohn J, Fugger L (2018) HLA variation and disease. Nat Rev Immunol 18:325
Dobrindt K, Hoagland DA, Seah C, Kassim B, O'Shea CP, Murphy A et al (2021) Common genetic variation in humans impacts in vitro susceptibility to SARS-CoV-2 infection. Stem Cell Rep 16:505
Duncan CJA, Mohamad SMB, Young DF, Skelton AJ, Leahy TR, Munday DC et al (2015) Human IFNAR2 deficiency: lessons for antiviral immunity. Sci Transl Med 7:307ra154
Fang S, Wang H, Lu L, Jia Y, Xia Z (2020) Decreased complement C3 levels are associated with poor prognosis in patients with COVID-19: a retrospective cohort study. Int Immunopharmacol 89:107070
Feuillet V, Canard B, Trautmann A (2021) Combining antivirals and Immunomodulators to fight COVID-19. Trends Immunol 42:31
Gaebler C, Wang Z, Lorenzi JCC, Muecksch F, Finkin S, Tokuyama M et al (2021) Evolution of antibody immunity to SARS-CoV-2. Nature 591:639
Gemmati D, Bramanti B, Serino ML, Secchiero P, Zauli G, Tisato V (2020) COVID-19 and individual genetic susceptibility/receptivity: role of ACE1/ACE2 genes, immunity, inflammation and coagulation. Might the double x-chromosome in females be protective against SARS-COV-2 compared to the single x-chromosome in males? Int J Mol Sci 21:3474
Ghahramani S, Tabrizi R, Lankarani KB, Kashani SMA, Rezaei S, Zeidi N et al (2020) Laboratory features of severe vs. non-severe COVID-19 patients in Asian populations: a systematic review and meta-analysis. Eur J Med Res 25:30
Goel R, Bloch EM, Pirenne F, Al-Riyami AZ, Crowe E, Dau L et al (2021) ABO blood group and COVID-19: a review on behalf of the ISBT COVID-19 working group. Vox Sang 116:849
Gralinski LE, Menachery VD (2020) Return of the coronavirus: 2019-nCoV. Viruses 12:135
Grover A, Oberoi M (2021) A systematic review and meta-analysis to evaluate the clinical outcomes in COVID-19 patients on angiotensin-converting enzyme inhibitors or angiotensin receptor blockers. Eur Heart J Cardiovasc Pharmacother 7(2):148–157
Gu X, Huang L, Cui D, Wang Y, Wang Y, Xu J et al (2022) Association of acute kidney injury with 1-year outcome of kidney function in hospital survivors with COVID-19: a cohort study. eBioMedicine 76:1–12
Guan W, Ni Z, Hu Y, Liang W, Ou C, He J et al (2020) Clinical characteristics of coronavirus disease 2019 in China. N Engl J Med 382:1708
Halaji M, Farahani A, Ranjbar R, Heiat M, Dehkordi FS (2020) Emerging coronaviruses: first SARS, second MERS and third SARS-COV-2: epidemiological updates of COVID-19. Infez Med 28:6–17
Hashizume M, Gonzalez G, Ono C, Takashima A, Iwasaki M (2021) Population-specific ace2 single-nucleotide polymorphisms have limited impact on sars-cov-2 infectivity in vitro. Viruses 13:67
Hilser JR, Han Y, Biswas S, Gukasyan J, Cai Z, Zhu R et al (2021) Association of serum HDL-cholesterol and apolipoprotein A1 levels with risk of severe SARS-CoV-2 infection. J Lipid Res 62:100061

Hoffmann M, Kleine-Weber H, Schroeder S, Krüger N, Herrler T, Erichsen S et al (2020) SARS-CoV-2 cell entry depends on ACE2 and TMPRSS2 and is blocked by a clinically proven protease inhibitor. Cell 181:271

Hoseinyazdi M, Esmaeilian S, Jahankhah R, Teimouri A, Sherbaf FG, Rafiee F et al (2021) Clinical, laboratory, and chest CT features of severe versus non-severe pediatric patients with COVID-19 infection among different age groups. BMC Infect Dis 21:560

Hu B, Guo H, Zhou P, Shi ZL (2021) Characteristics of SARS-CoV-2 and COVID-19. Nat Rev Microbiol 19:141

Huang C, Wang Y, Li X, Ren L, Zhao J, Hu Y et al (2020) Clinical features of patients infected with 2019 novel coronavirus in Wuhan, China. Lancet 395:497

Hulswit RJG, de Haan CAM, Bosch BJ (2016) Coronavirus spike protein and tropism changes. Adv Virus Res 96:29–57

Ibrahim-Kosta M, Bailly P, Silvy M, Saut N, Suchon P, Morange PE et al (2020) ABO blood group, glycosyltransferase activity and risk of venous thromboembolism. Thromb Res 193:31

Jafarzadeh A, Jafarzadeh S, Nozari P, Mokhtari P, Nemati M (2021) Lymphopenia an important immunological abnormality in patients with COVID-19: possible mechanisms. Scand J Immunol 93:e12967

Jaimes JA, André NM, Chappie JS, Millet JK, Whittaker GR (2020) Phylogenetic analysis and structural modeling of SARS-CoV-2 spike protein reveals an evolutionary distinct and proteolytically sensitive activation loop. J Mol Biol 432:3309

Jeltsch A, Broche J, Bashtrykov P (2018) Molecular processes connecting DNA methylation patterns with DNA methyltransferases and histone modifications in mammalian genomes. Genes (Basel) 9:566

Jiang Y, Wu Q, Song P, You C (2022) The variation of SARS-CoV-2 and advanced research on current vaccines. Front Med 8(January):1–19

Johansson MA, Quandelacy TM, Kada S, Prasad PV, Steele M, Brooks JT et al (2021) SARS-CoV-2 transmission from people without COVID-19 symptoms. JAMA Netw Open 4:e2035057

Johnson BA, Xie X, Bailey AL, Kalveram B, Lokugamage KG, Muruato A et al (2021) Loss of furin cleavage site attenuates SARS-CoV-2 pathogenesis. Nature 591:293

Kaltoum ABO (2021) Mutations and polymorphisms in genes involved in the infections by covid 19: a review. Gene Rep 23:101062

Kasparian K, Graykowski D, Cudaback E (2020) Commentary: APOE e4 genotype predicts severe COVID-19 in the UK biobank community cohort. Front Immunol 11:1939

Kawai T, Akira S (2010) The role of pattern-recognition receptors in innate immunity: update on toll-like receptors. Nat Immunol 11:373

Khanmohammadi S, Rezaei N (2021) Role of toll-like receptors in the pathogenesis of COVID-19. J Med Virol 93:2735

Kragstrup TW, Singh HS, Grundberg I, Nielsen ALL, Rivellese F, Mehta A et al (2021) Plasma ACE2 predicts outcome of COVID-19 in hospitalized patients. PLoS One 16:e0252799

Kumar SU, Priya NM, Nithya SR, Kannan P, Jain N, Kumar DT et al (2021) A review of novel coronavirus disease (COVID-19): based on genomic structure, phylogeny, current shreds of evidence, candidate vaccines, and drug repurposing. 3 Biotech 11(4):198

La Porta E, Baiardi P, Fassina L, Faragli A, Perna S, Tovagliari F et al (2022) The role of kidney dysfunction in COVID-19 and the influence of age. Sci Rep [Internet] 12(1):1–9. https://doi.org/10.1038/s41598-022-12652-0

Latini A, Agolini E, Novelli A, Borgiani P, Giannini R, Gravina P et al (2020) COVID-19 and genetic variants of protein involved in the SARS-CoV-2 entry into the host cells. Genes (Basel) 11:1010

Lauer SA, Grantz KH, Bi Q, Jones FK, Zheng Q, Meredith HR et al (2020) The incubation period of coronavirus disease 2019 (CoVID-19) from publicly reported confirmed cases: estimation and application. Ann Intern Med 172:577

Lei X, Dong X, Ma R, Wang W, Xiao X, Tian Z et al (2020) Activation and evasion of type I interferon responses by SARS-CoV-2. Nat Commun 11:3810

Letko M, Marzi A, Munster V (2020) Functional assessment of cell entry and receptor usage for SARS-CoV-2 and other lineage B betacoronaviruses. Nat Microbiol 5:562

Leulseged TW, Hassen IS, Ayele BT, Tsegay YG, Abebe DS, Edo MG et al (2021) Laboratory biomarkers of covid-19 disease severity and outcome: findings from a developing country. PLoS One 16:e0246087

Li MY, Li L, Zhang Y, Wang XS (2020a) Expression of the SARS-CoV-2 cell receptor gene ACE2 in a wide variety of human tissues. Infect Dis Poverty 9:45

Li Q, Cao Z, Rahman P (2020b) Genetic variability of human angiotensin-converting enzyme 2 (hACE2) among various ethnic populations. Mol Genet Genomic Med 8(8):1–6

Liao M, Liu Y, Yuan J, Wen Y, Xu G, Zhao J et al (2020) Single-cell landscape of bronchoalveolar immune cells in patients with COVID-19. Nat Med 26:842

Liao X, Li D, Ma Z, Zhang L, Zheng B, Li Z et al (2022) 12-month post-discharge liver function test abnormalities among patients with COVID-19: a single-center prospective cohort study. Front Cell Infect Microbiol 12(April):1–8

Lin P, Chen W, Huang H, Lin Y, Cai M, Lin D et al (2021) Delayed discharge is associated with higher complement C3 levels and a longer nucleic acid-negative conversion time in patients with COVID-19. Sci Rep 11:1233

Liu Y, Rocklöv J (2021) The reproductive number of the Delta variant of SARS-CoV-2 is far higher compared to the ancestral SARS-CoV-2 virus. J Travel Med 28:taab124

Lu R, Zhao X, Li J, Niu P, Yang B, Wu H et al (2020) Genomic characterisation and epidemiology of 2019 novel coronavirus: implications for virus origins and receptor binding. Lancet 395:565

Maiers M, Gragert L, Klitz W (2007) High-resolution HLA alleles and haplotypes in the United States population. Hum Immunol 68:779

Masselli E, Vaccarezza M, Carubbi C, Pozzi G, Presta V, Mirandola P et al (2020) NK cells: a double edge sword against SARS-CoV-2. Adv Biol Regul 77:100737

Maxwell AJ, Ding J, You Y, Dong Z, Chehade H, Alvero A et al (2021) Identification of key signaling pathways induced by SARS-CoV2 that underlie thrombosis and vascular injury in COVID-19 patients. J Leukoc Biol 109:35

Merle NS, Church SE, Fremeaux-Bacchi V, Roumenina LT (2015) Complement system part I—molecular mechanisms of activation and regulation. Front Immunol 6:262

Moreno-Eutimio MA, López-Macías C, Pastelin-Palacios R (2020) Bioinformatic analysis and identification of single-stranded RNA sequences recognized by TLR7/8 in the SARS-CoV-2, SARS-CoV, and MERS-CoV genomes. Microbes Infect 22:226

Nattermann J, Vogel M, Berg T, Danta M, Axel B, Mayr C et al (2007) Effect of the interleukin-6 C174G gene polymorphism on treatment of acute and chronic hepatitis C in human immunodeficiency virus coinfected patients. Hepatology 46:1016

Nguyen A, David JK, Maden SK, Wood MA, Weeder BR, Nellore A et al (2020) Human leukocyte antigen susceptibility map for severe acute respiratory syndrome coronavirus 2. J Virol 94:e00510–e00520

Nwanaji-Enwerem JC, Colicino E (2020) DNA methylation–based biomarkers of environmental exposures for human population studies. Curr Environ Health Rep 7(2):121

Ottestad W, Seim M, Mæhlen JO (2020) COVID-19 with silent hypoxemia. Tidsskr Nor Laegeforen

Ovsyannikova IG, Haralambieva IH, Crooke SN, Poland GA, Kennedy RB (2020) The role of host genetics in the immune response to SARS-CoV-2 and COVID-19 susceptibility and severity. Immunol Rev 296:205

Papa G, Mallery DL, Albecka A, Welch LG, Cattin-Ortolá J, Luptak J et al (2021) Furin cleavage of SARS-CoV-2 spike promotes but is not essential for infection and cell-cell fusion. PLoS Pathog 17:e1009246

Parameshwar K, Pamu S, Sandeep K, Suresh C (2020) A review novel coronavirus. Asian J Pharm Clin Res 13:12–17

Paraskevis D, Kostaki EG, Magiorkinis G, Panayiotakopoulos G, Sourvinos G, Tsiodras S (2020) Full-genome evolutionary analysis of the novel corona virus (2019-nCoV) rejects the hypothesis of emergence as a result of a recent recombination event. Infect Genet Evol 79:104212

Pascarella S, Ciccozzi M, Zella D, Bianchi M, Benedetti F, Benvenuto D et al (2021) SARS-CoV-2 B.1.617 Indian variants: are electrostatic potential changes responsible for a higher transmission rate? J Med Virol 93:6551
Peacock TP, Goldhill DH, Zhou J, Baillon L, Frise R, Swann OC et al (2021) The furin cleavage site in the SARS-CoV-2 spike protein is required for transmission in ferrets. Nat Microbiol 6:899
Perico L, Benigni A, Casiraghi F, Ng LFP, Renia L, Remuzzi G (2021) Immunity, endothelial injury and complement-induced coagulopathy in COVID-19. Nat Rev Nephrol 17:46
Porfidia A, Valeriani E, Pola R, Porreca E, Rutjes AWS, Di Nisio M (2020) Venous thromboembolism in patients with COVID-19: systematic review and meta-analysis. Thromb Res 196:67
Qin C, Zhou L, Hu Z, Zhang S, Yang S, Tao Y et al (2020) Dysregulation of immune response in patients with COVID-19 in Wuhan, China. SSRN Electron J
Quarleri J, Delpino MV (2021) Type I and III IFN-mediated antiviral actions counteracted by SARS-CoV-2 proteins and host inherited factors. Cytokine Growth Factor Rev 58:55
Rahman MM, Hasan M, Ahmed A (2021) Potential detrimental role of soluble ACE2 in severe COVID-19 comorbid patients. Rev Med Virol 31:1
Ratre YK, Kahar N, Bhaskar LVKS, Bhattacharya A, Verma HK (2021) Molecular mechanism, diagnosis, and potential treatment for novel coronavirus (COVID-19): a current literature review and perspective. 3 Biotech 11(2):94
Rebendenne A, Chaves Valadão AL, Tauziet M, Maarifi G, Bonaventure B, McKellar J et al (2021) SARS-CoV-2 triggers an MDA-5-dependent interferon response which is unable to control replication in lung epithelial cells. J Virol 95:e02415–e02420
Reid MA, Dai Z, Locasale JW (2017) The impact of cellular metabolism on chromatin dynamics and epigenetics. Nat Cell Biol 19:1298
Ren S-Y, Wang W-B, Gao R-D, Zhou A-M (2022) Omicron variant (B.1.1.529) of SARS-CoV-2: mutation, infectivity, transmission, and vaccine resistance. World J Clin Cases 10:1
Reusch N, De Domenico E, Bonaguro L, Schulte-Schrepping J, Baßler K, Schultze JL et al (2021) Neutrophils in COVID-19. Front Immunol 12(March):1–9
Salgado-Albarrán M, Navarro-Delgado EI, Del Moral-Morales A, Alcaraz N, Baumbach J, González-Barrios R et al (2021) Comparative transcriptome analysis reveals key epigenetic targets in SARS-CoV-2 infection. npj Syst Biol Appl 7:21
Sapkal G, Shete-Aich A, Jain R, Yadav P, Sarkale P, Lakra R et al (2020) Development of indigenous IgG ELISA for the detection of anti-SARS-CoV-2 IgG. Indian J Med Res 151(5):444–449
Schultze JL, Aschenbrenner AC (2021) COVID-19 and the human innate immune system. Cell 184:1671
Scialo F, Daniele A, Amato F, Pastore L, Matera MG, Cazzola M et al (2020) ACE2: the major cell entry receptor for SARS-CoV-2. Lung 198:867
Sebastian-Valverde M, Pasinetti GM (2020) The NLRP3 inflammasome as a critical actor in the inflammaging process. Cells 9:1552
Sette A, Crotty S (2021) Adaptive immunity to SARS-CoV-2 and COVID-19. Cell 184:861
Shen L, Wang C, Zhao J, Tang X, Shen Y, Lu M et al (2020) Delayed specific IgM antibody responses observed among COVID-19 patients with severe progression. Emerg Microbes Infect 9:1096
Smith JD (2000) Apolipoprotein E4: an allele associated with many diseases. Ann Med 32:118
Solmaz İ, Araç S (2021) ABO blood groups in COVID-19 patients; cross-sectional study. Int J Clin Pract 75:e13927
Sriram K, Insel PA (2020) A hypothesis for pathobiology and treatment of COVID-19: the centrality of ACE1/ACE2 imbalance. Br J Pharmacol 177:4825
Stawiski EW, Diwanji D, Suryamohan K, Gupta R, Fellouse FA, Sathirapongsasuti JF et al (2020) Human ACE2 receptor polymorphisms predict SARS-CoV-2 susceptibility. bioRxiv
Sun B, Feng Y, Mo X, Zheng P, Wang Q, Li P et al (2020) Kinetics of SARS-CoV-2 specific IgM and IgG responses in COVID-19 patients. Emerg Microbes Infect 9:940

Tang T, Jaimes JA, Bidon MK, Straus MR, Daniel S, Whittaker GR (2021) Proteolytic activation of SARS-CoV-2 spike at the S1/S2 boundary: potential role of proteases beyond Furin. ACS Infect Dis 7:264
Thomas G (2002) Furin at the cutting edge: from protein traffic to embryogenesis and disease. Nat Rev Mol Cell Biol 3:753
Tomita Y, Ikeda T, Sato R, Sakagami T (2020) Association between HLA gene polymorphisms and mortality of COVID-19: an in silico analysis. Immun Inflamm Dis 8:684
Tomo S, Kumar KP, Roy D, Sankanagoudar S, Purohit P, Yadav D et al (2021) Complement activation and coagulopathy—an ominous duo in COVID19. Expert Rev Hematol 14:155
Tseng YH, Yang RC, Lu TS (2020) Two hits to the renin-angiotensin system may play a key role in severe COVID-19. Kaohsiung J Med Sci 36:389
Udugama B, Kadhiresan P, Kozlowski HN, Malekjahani A, Osborne M, Li VYC et al (2020) Diagnosing COVID-19: the disease and tools for detection. ACS Nano 14:3822
Van Dyke AL, Cote ML, Wenzlaff AS, Land S, Schwartz AG (2009) Cytokine SNPs: comparison of allele frequencies by race and implications for future studies. Cytokine 46:236
Velavan TP, Kuk S, Linh LTK, Lamsfus Calle C, Lalremruata A, Pallerla SR et al (2021) Longitudinal monitoring of laboratory markers characterizes hospitalized and ambulatory COVID-19 patients. Sci Rep 11:14471
Verdecchia P, Cavallini C, Spanevello A, Angeli F (2020) The pivotal link between ACE2 deficiency and SARS-CoV-2 infection. Eur J Intern Med 76:14
Villarreal IM, Morato M, Martínez-RuizCoello M, Navarro A, Garcia-Chillerón R, Ruiz Á et al (2021) Olfactory and taste disorders in healthcare workers with COVID-19 infection. Eur Arch Otorhinolaryngol 278:2123
Wang D, Hu B, Hu C, Zhu F, Liu X, Zhang J et al (2020a) Clinical characteristics of 138 hospitalized patients with 2019 novel coronavirus-infected pneumonia in Wuhan, China. JAMA 323:1061
Wang W, Zhang W, Zhang J, He J, Zhu F (2020b) Distribution of HLA allele frequencies in 82 Chinese individuals with coronavirus disease-2019 (COVID-19). HLA 96:194
Wang H, Yuan Z, Pavel MA, Hansen S (2020c) The role of high cholesterol in aged related COVID19 lethality. bioRxiv
Wang C, Zhang M, Garcia G, Tian E, Cui Q, Chen X et al (2021) ApoE-isoform-dependent SARS-CoV-2 neurotropism and cellular response. Cell Stem Cell 28:331
Wen W, Su W, Tang H, Le W, Zhang X, Zheng Y et al (2020) Immune cell profiling of COVID-19 patients in the recovery stage by single-cell sequencing. Cell Discov 6:31
Wrapp D, Wang N, Corbett KS, Goldsmith JA, Hsieh CL, Abiona O et al (2020) Cryo-EM structure of the 2019-nCoV spike in the prefusion conformation. Science 367(6483):1260–1263
Wrobel AG, Benton DJ, Xu P, Roustan C, Martin SR, Rosenthal PB et al (2020) SARS-CoV-2 and bat RaTG13 spike glycoprotein structures inform on virus evolution and furin-cleavage effects. Nat Struct Mol Biol 27(8):763
Wu F, Zhao S, Yu B, Chen YM, Wang W, Song ZG et al (2020) A new coronavirus associated with human respiratory disease in China. Nature 579:265
Xu Z, Shi L, Wang Y, Zhang J, Huang L, Zhang C et al (2020a) Pathological findings of COVID-19 associated with acute respiratory distress syndrome. Lancet Respir Med 8:420
Xu G, Qi F, Li H, Yang Q, Wang H, Wang X et al (2020b) The differential immune responses to COVID-19 in peripheral and lung revealed by single-cell RNA sequencing. Cell Discov 6:73
Yamamoto N, Ariumi Y, Nishida N, Yamamoto R, Bauer G, Gojobori T et al (2020) SARS-CoV-2 infections and COVID-19 mortalities strongly correlate with ACE1 I/D genotype. Gene 758:144944
Yang B, Chang X, Huang J, Pan W, Si Z, Zhang C et al (2021) The role of IL-6/lymphocyte ratio in the peripheral blood of severe patients with COVID-19. Int Immunopharmacol 97:107569
Yin YW, Sheng YJ, Wang M, Ma YQ, Ding HM (2021) Interaction of serum proteins with SARS-CoV-2 RBD. Nanoscale 13:12865

Yuen CK, Lam JY, Wong WM, Mak LF, Wang X, Chu H et al (2020) SARS-CoV-2 nsp13, nsp14, nsp15 and orf6 function as potent interferon antagonists. Emerg Microbes Infect 9:1418

Zhang L, Shen FM, Chen F, Lin Z (2020a) Origin and evolution of the 2019 novel coronavirus. Clin Infect Dis 71:882

Zhang Q, Liu Z, Moncada-Velez M, Chen J, Ogishi M, Bigio B et al (2020b) Inborn errors of type I IFN immunity in patients with life-threatening COVID-19. Science 370(6515):eabd4570

Zhang Y, Zhang Y, Qin L, Zhao Y, Zhang P, Zhang P et al (2020c) Interferon-induced transmembrane protein 3 genetic variant rs12252-C associated with disease severity in coronavirus disease 2019. J Infect Dis 222:34

Zheng H, Cao JJ (2020) ACE gene polymorphism and severe lung injury in patients with COVID-19. Am J Pathol 190(10):2013

Zhou P, Lou YX, Wang XG, Hu B, Zhang L, Zhang W et al (2020) A pneumonia outbreak associated with a new coronavirus of probable bat origin. Nature 588:E6

# Immuno-Metabolic Interplay in Health and Disease

**Rishabh Agrahari, Swati Mohapatra, K. S. Kumar, Debapriya Mohanty, Rahul Kumar, and Ramchander Merugu**

**Abstract** The immune system of the mammalian body interacts with a foreign particle in a diversified way and such interplay helps the immune system in proper functioning and development, in order to protect our body. Majorly microbiota, foreign particles, and other intermediate toxic chemicals boost the power of an innate and adaptive immune system, this orchestrates serves a healthy symbiosis between them. Here in this chapter, we have sketched out the major metabolic function, an important part of the immune system and its reaction in different disease conditions. This chapter emphasized the recent aspect interplay of the immune system in different disease conditions.

**Keywords** Inflammation · Immunometabolism · Metabolic reprogramming · Quiescence · Signaling · Transcriptional cascade

R. Agrahari
Amity Institute of Microbial Technology (AIMT), Amity University, Noida, Uttar Pradesh, India

S. Mohapatra (✉)
Department of Infection Biology, School of Medicine, Wonkwang University, Iksan, Republic of Korea

School of Science, GSFC University, Vadodara, India

K. S. Kumar
School of Science, GSFC University, Vadodara, India

D. Mohanty
Department of Microbiology, CBSH, OUAT, Bhubaneswar, Odisha, India

R. Kumar
Department of Environmental Sciences, H.N.B. Garhwal University, Srinagar, Uttarakhand, India

R. Merugu
Department of Biochemistry, Mahatma Gandhi University, Nalgonda, Telangana, India



A. Das Mohapatra, P. S. Sahu (eds.), *Advances in Immunology and Immuno-techniques*, Interdisciplinary Biotechnological Advances,
https://doi.org/10.1007/978-981-97-5508-0_11

## 1 Introduction

In a crucial experimentation by Urey and Miller, the generation of amino acids and RNA nucleobases is demonstrated using metabolites precursors and conditions resembling that of the early earth's atmosphere (Ferus et al. 2017). Metabolites are biochemical intermediates or end product of metabolic pathways. Metabolites represent the basic foundation of life and are considered a driving force for evolution. The term "metabolism" is used to describe biochemical reactions; these reactions have two categories, anabolism which builds up biomass and catabolism which breaks down molecules for energy production or provides precursor molecules for anabolic processes (Lercher et al. 2020). When it comes to immunity, cytokines are the key components; metabolites are found to have a key role in communication signals apart from bioenergetic and biosynthetic roles. Cell communication is rather an evolutionarily conserved mechanism from prokaryotes (e.g., quorum sensing between bacteria) to eukaryotes (Gallio et al. 2002). Under certain environmental conditions, any extracellular cues could result in the shifting of the metabolic homeostasis of cells. These shifts are recognized by the central metabolic regulator like mTOR, which coalesce existing environmental information and other stimuli to direct the metabolic adaptation and cells proliferation (Saxton and Sabatini 2017). This ability to metabolic adaptation to environmental cues has a very significant role in immunity. These environmental-driven adaptations can cost the organisms up to 30% of their basal metabolic rate (Muehlenbein et al. 2010). It proves the significance of metabolic adaptation in a multicellular organism to regulate the mechanism of tissue repair, pathogenesis, and other vital functions (Geiger et al. 2016).

The immune system comprises a diverse population of cells that are dormant in a stable state but retort vigorously to infection, inflammation, and other perturbation. The responses of immune cells are regulated through a broad range of activating or inhibitory receptors that are activated by intrinsic signals of the immune system or perhaps pathogen-stimulated signals. These responses by immune cells cause the activation of several gene expressions and help in obtaining a new function, such as elevated stimulation of cytokines, lipid mediators, tissue remodeling enzymes, diapedesis, and proliferation (Pearce and Pearce 2013). After acknowledging the key role of nutrition at the time of immune cell transition from quiescent to activated state to support the functional shift, scientists are more interested in how these metabolic pathways are regulated during the immune response.

The immune cells in either of the states require energy, which obligates ATP production. ATP production is fueled by glucose consumption via two integrated pathways which are glycolysis and TCA pathways. In glycolysis, the glucose molecule produces ATP and pyruvate as the end product. The pyruvate produced by the glycolysis that is used in the TCA pathway helps in the OXPHOS (oxidative phosphorylation) process by generating the reduced NADH and $FADH_2$, which donates an electron to the Electron Transport Chain and helps in ATP production in mitochondria (OXPHOS process). The two pathways are integrated by the conversion of pyruvate (end product of glycolysis) into acetyl-CoA (precursor of TCA) which enters in TCA cycle. However, cells are also capable of utilizing other sources for energy production such as glutamine (glutaminolysis) or fatty acid (β-oxidation) to replenish the TCA cycle which results in OXPHOS. In the condition of hypoxia cells can solely utilize glucose for ATP production via glycolysis and at the end pyruvate is directed toward lactate rather than acetyl-CoA. Sometimes cells preferentially go for glycolysis even when the oxygen is plenty; this process is termed as Warburg metabolism or commonly called aerobic glycolysis. Besides oxygen availability, the relative abundance of the substrates (glucose, glutamine, and fatty acids) has significance in the preference of the substrate to be consumed for ATP production (Thompson 2011). Under specific circumstances, the metabolic profiling of a cell is critically linked with pathway-specific genes or proteins involved in post-transcriptional moderation and controlled expression. Besides, activation of TLRs on myeloid cells and T-cell receptors along with cytokines has high significance for the coordination in the immune system (Pearce and Pearce 2013).

## 2 Overview of Different Metabolic Pathway and Their Roles in Immunity

Metabolic pathways activity is regulated by the cell's intrinsic and extrinsic signals which synchronize metabolic machinery that coordinate the production of crucial products to fulfill the sustainable necessity of cells. Though, the metabolic alteration together with immune effector functions brings changes that are more visible in cytokines production. Metabolic pathways are usually controlled by the availability of oxygen. However, specific immune molecules like interleukin-4 (IL-4), or PRRs, can diversely encourage the explicit metabolic pathways in the cell. Immune cells cope with the energy requirement by promoting different metabolic pathways which meet all the energy requirements, and their numerous intermediate metabolites help in biosynthesis which in effect allows cellular growth and proliferation (O'Neill et al. 2016). Although it seems that metabolic pathways have diverse end products, they are closely linked in a cascade reaction where the end product from one pathway serves as a precursor into the alternate pathway. For example, the precursor of fatty acid synthesis is the intermediate of glycolysis and the TCA cycle fulfills the needs for cell proliferation by the synthesis of the cell membrane and other lipid-based structures; availability of these intermediate metabolites acts as a limiting factor for fatty acid synthesis (FAS) (Qian et al. 2018). Here we are going to discuss key metabolic pathways which have a very critical role in the cell's growth and survival. Pathways are interconnected and are synchronized by cellular signaling to meet cell needs.

The glycolytic process starts with glucose cell uptake from the extracellular surrounding of the cell. This pathway eventually produces pyruvate along with several other products. When it comes to ATP production, glycolysis is not an efficient pathway liberating only two ATP per glucose molecule. However, the glycolytic pathway may not be efficient in ATP production but provides a key benefit to the cell by generating NADH (by reducing NAD+) which is an essential enzyme cofactor. Additionally, the glycolytic intermediate can be diverted into biosynthetic growth pathways resulting in anabolic growth. In order to maintain glycolysis continuity; the NADH is recycled in the cell by reducing pyruvate into lactate and producing $NAD^+$ in the process. Several biosynthetic pathways directly or indirectly depend upon the intermediates of glycolysis. For example, the synthesis of nucleotides (pentose phosphate from glucose-6-phosphate as a precursor), amino acid (serine biosynthesis from 3-phosphoglycerate as a precursor), fatty acid, etc. (O'Neill et al. 2016) (Fig. 1).

Glycolysis has a very significant role in immune processes. Affirmative evidence is found in activated immune cells like macrophages and T cells which show a ravenous appetite for glucose (Newsholme et al. 1986). Likewise, several studies have shown how crucial glycolysis is for immune cell function. For example, immune cells like macrophages (LPS-activated), Dendritic cells (DC), Natural killer (NK) cells, T cells, and B cells require elevated glycolysis on their activation (Donnelly et al. 2014; Doughty et al. 2006; Krawczyk et al. 2010; Michalek et al. 2011; Rodríguez-Prados et al. 2010). Following activation, all effector T-cell subsets, including T helper 17 (TH17) cells, TH1 and TH2 cells, and activated effector

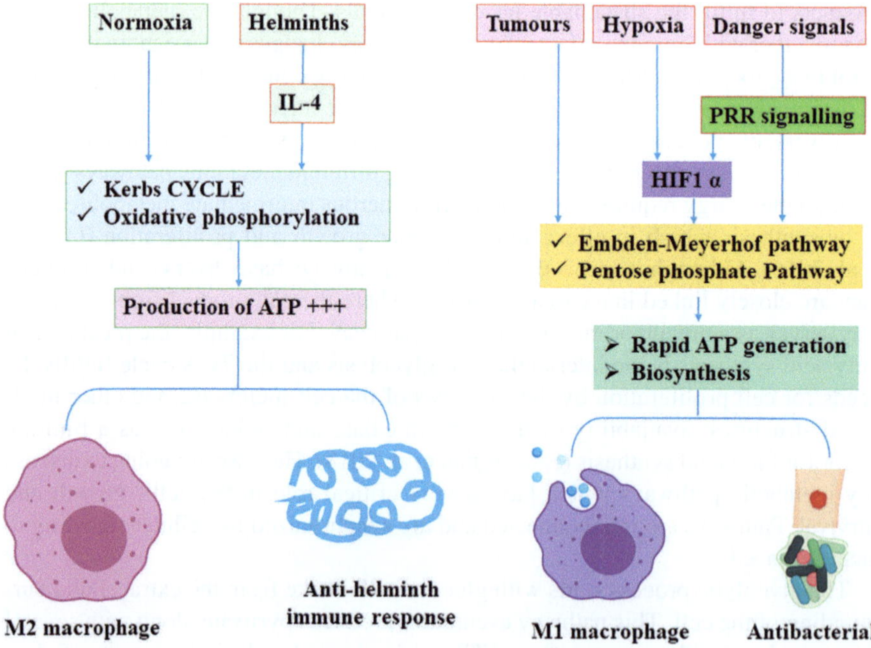


**Fig. 1** Metabolic reprograming by the immune system

CD8+ T cells, display an increase in glycolysis (Gubser et al. 2013; Michalek et al. 2011; Shi et al. 2011).

We have discovered several molecular clues to the signaling pathways that cause immune cells to activate glycolysis. For example, hypoxia-inducible factor 1α (HIF1α) gets activated by LPS; HIF1α is a transcription factor needed for the introduction of numerous glycolysis-related enzymes (Tannahill et al. 2013). The cells which do not possess HIF1α still get induced by LPS and result in enhanced glycolysis. It is perhaps involved in the activation of nuclear factor-$_{\kappa}$B (NF-$_{\kappa}$B) as the ubiquitous isoform of phosphofructokinase-2 (uPFK-2)—essential for the proper functioning of glycolysis (Rodríguez-Prados et al. 2010). Similarly, in HIF1α-independent dendritic cells, LPS activates TANK-binding kinase 1 (TBK1) and/or inhibitor of NF-$_{\kappa}$B kinase ε (IKKε) and hexokinase 2 for the upsurge in glycolysis (Huynh et al. 2015). The cell which is in its quiescent or non-proliferating state prefers the TCA cycle for energy requirement. However, some quiescent stem cells have been observed to primarily use glycolysis while other quiescent cells go for the TCA cycle and OXPHOS for energy requirements and cell endurance. The TCA cycle is a nexus for numerous nutrients input. α-ketoglutarate is an intermediate metabolite of the TCA cycle which can be obtained by the direct conversion of glutamate and can be used as critical fuel for the TCA cycle. In most cell types, this mechanism ensures basic survival. The intermediates of the TCA cycle can be diverted by the growth signals to promote the synthesis of amino acids and lipids similar to what transpires in glycolysis when growth signals are stimulated. However, the process requires huge nutritional diversions (O'Neill et al. 2016).

However, the TCA cycle in some of the immune cells has a different function rather than ATP production. In M2 macrophages, a distinct macrophage subtype is observed with the intact TCA cycle coupled with OXPHOS, thus helping in requisite UDP-GlcNAc intermediate production for glycosylation of M2-associated receptor (Jha et al. 2015). However, a broken TCA cycle (after citrate and succinate) is well documented in M1-type macrophages (Jha et al. 2015; Tannahill et al. 2013). Citrate which is accumulated in mitochondria is transported via a citrate transporter; citrate is bestowed for fatty acid production which in turn accomplishes membrane biogenesis. A similar pulverized TCA cycle is reported in the stimulated dendritic cells, where it performs a very crucial role in antigen presentation by substantial membrane fabrication (Everts et al. 2014). Nitric oxide and prostaglandins can be synthesized by utilizing the excess citrate, and these are crucial effector molecules for macrophages (Infantino et al. 2011). Macrophages exhibit metabolic rewiring as a response to the LPS activation by using cis-aconitate (TCA cycle intermediate) to synthesize itaconic acid which is a very effective bacteriocidal against *Salmonella enterica and Mycobacterium tuberculosis*. This example validates how metabolic rewiring can produce a potential result in the production of antimicrobial metabolites (Michelucci et al. 2013).

In M1 macrophages, the TCA cycle break leads straightly to the accumulation of succinate due to its direct impression on macrophage cytokine production. Prolyl hydroxylase inhibition by succinate is one of the mechanisms involved, which results in HIF1α offset and continued synthesis of IL 1β (Tannahill et al. 2013). This pathway can operate regardless of the oxygen gradient (normoxia or hypoxia), thus provides scope for HIF1α activation under an aerobic environment. Collectively, these studies depict the immune function of M1 macrophages by the diversion of the intermediate metabolites of the TCA cycle. These occurrences are believed to be one of the reasons for nitric oxide production in macrophages, which eventually hinders the electron transport chain (Clementi et al. 1998).

The pentose phosphate pathway has various crucial roles in cell growth and proliferation; it helps in nucleotide production and is a precursor for the amino acid. From the perspective of immune cell function, the pentose phosphate pathway offers several key functionalities. It has two important outcomes: nucleotide and NADPH. Both are deputed with unique purposes in the immune cell; nucleotide helps in the cell's proliferation and growth; however, NADPH performs several functions. Immune cells like neutrophils and macrophages respond to an infection by the rapid production of ROS (reactive oxygen species); this functionality gain is possible due to the presence of NADPH which is required by NADPH oxidase. As a defense mechanism of cells, NADPH is required for the production of glutathione and several other antioxidants which minimizes tissue damage if ROS leakage happens. In dendritic cells, endoplasmic reticulum synthesis is regulated by NADPH and lipid availability and has a critical role in the activation of DC and cytokine secretion (Everts et al. 2014). Sedoheptulose kinase (SHK; a carbohydrate kinase-like protein) is a crucial regulatory enzyme that constrains the pentose phosphate pathway flux, and it is highly expressed in M2 macrophages (Haschemi et al. 2012).

Several products like acetyl-CoA, NADH, and $FADH_2$ can be achieved by fatty acid oxidation in case it is also a source of energy. Fatty acid activation is the foremost step for fatty acid oxidation. "Thiokinase or acyl-CoA synthetases" is the enzyme that activates the fatty acid by converting it into acyl-CoA at the expense of ATP—coenzyme-A and $mg^{2+}$ are essential for the purpose. The actual β-oxidation starts when the acyl-CoA enters the mitochondria; the oxidation of acyl-CoA subsequently results in the acetyl-CoA production and eventually generates NADH and $FADH_2$ via TCA cycle and promote OXPHOS—one molecule of palmitate molecule can synthesize up to 129 ATP molecules through β-oxidation (Bhagavan 2002; Schulz 2013). However, fatty acid oxidation is regulated through a distinct mechanism of CPT1A and CPT1B genes. At the time of lipid synthesis, the carnitine shuttle is impaired by a lipid synthesis intermediate, and malonyl-CoA and acyl-CoA are diverted for the anabolic pathway (Schreurs et al. 2010; van Weeghel et al. 2018).

The fatty acid synthesis pathway has a very crucial role in cell growth and proliferation. The fatty acid synthesis is regulated via mTOR signaling; mTOR signals regulate several enzymes which eventually promote de novo lipid synthesis. The mTOR signaling regulates proteins such as SREBP-1 (sterol regulatory element-binding protein-1) which in turn induces FASN (fatty acid synthase), ACC (acetyl-CoA carboxylase), and SCD-1 (stearoyl-CoA desaturase 1). Several intermediate metabolites from different metabolic pathways contribute to fatty acid synthesis—remarkably from the HMP shunt, TCA cycle, and EMP pathway (Laplante and Sabatini 2009). In this case, the detected increase in fatty acid biosynthesis was critical for macrophage differentiation and inflammatory activity (Ecker et al. 2010). Further studies suggest that the NLRP3 inflammasome pathway gets activated through FASN regulation by mitochondrial uncoupling protein 2 (UCP2), resulting in a harmful inflammatory response during polymicrobial sepsis (Moon et al. 2015).

## 3 Roles of Metabolic Pathways in Quiescent and Activated Immune Cells

The immune system constantly detects and reacts to ecological risks during the organism's existence, a process carried on a significant amount of bioenergetic expense. For instance, innate immune cells respond to any stimuli through sets of messenger molecules (inflammatory mediators, chemokines, and cytokines) often linked to pathogenic or ecological stress. For a momentous response, immune cells require numerous extracellular nutrition sources, as they lack nutritional stores. Indeed, as discovered over a century ago, glucose is essential for the amplification of an effector innate response (Levene and Meyer 1912). Similarly, extracellular glutamine is required for mitogen-driven adaptive immune cell proliferation (Ardawi and Newsholme 1982; Newsholme et al. 1986). This ravenous nutritional uptake helps the immune cells in two ways: First, it provides a source for ATP synthesis to meet the bioenergetic requirements for the proper functioning of activated immune cells. Second, it plays an anaplerotic function in activated immune cells and their proliferation. Therefore, it helps in immune cell maturation by reprogramming their cellular metabolism.

### *3.1 Granulocytes: Cell Dynamics in Their Activation and Quiescence*

Myeloid cells are thought to use the EMP pathway as their primary ATP source (Kominsky et al. 2010). Neutrophils are an excellent example of this axiom due to their main reliance on glycolysis for ATP, considering the research dating back to the 1950s (Sbarra and Karnovsky 1959; Valentine 1960). Besides this, neutrophils are characterized by fewer mitochondria; hence, little oxygen consumption although the oxygen, as well as glucose need, rises on subsequent activation of TLR or phagocytosis (Borregaard and Herlin 1982; van Raam et al. 2006). However, this increase in oxygen and glucose uptake by neutrophils is not an indication of mitochondrial ATP synthesis, rather it is required for aerobic glycolysis and increased pentose phosphate pathway for NADPH generation; in turn, NADPH oxidase promotes the production of neutrophil microbicidal product $H_2O_2$ by consuming NADPH and oxygen (Dale et al. 2008).

A cell with non-functional mitochondria poses a threat to the cell itself; non-functional mitochondria can lose their membrane potential which might cause the cytoplasmic release of the cytochrome C, which can end up in cell apoptosis (Galluzzi et al. 2012). Though mitochondria in neutrophils are designated with a different function rather than ATP synthesis—mitochondria are indulged in the electron acceptance from glycolysis via complex III of ETC by utilizing the glycerol-3-phosphate shuttle and hence help in the retention of membrane potential (van Raam et al. 2008). Therefore, neutrophils avoid apoptosis by indulging mitochondria in redox balancing and promoting flux through the glycolytic pathway. This high dependency on aerobic glycolysis provides a chance to remain feasible throughout NETosis—a discharging process of extracellular trap chromatins and embedded microbicidal materials (Remijsen et al. 2011). NETosis is an NADPH oxidase-dependent process that indicates neutrophil's microbicidal functionality dependency on glycolysis (Kirchner et al. 2012).

There is supporting evidence and various albeit scant speculating metabolic similarities among eosinophils, neutrophils, and basophils (Sumbayev et al. 2009; Venge et al. 2003). Granulocytes are critically distinguished and are mostly incompetent of proliferating in the fringe, which is a common feature. It is intriguing to hypothesize that this facet of their cellular mechanisms is inextricably linked to their mitochondrial dysfunction.

## *3.2 Dendritic Cells (DCs): Cells Dynamic at Active and Quiescence State*

The DCs are multifarious that can be branched into a wide-ranging and mounting number of subsets (Satpathy et al. 2012). DCs are the key regulatory point in inflammation initiation crucial for the launch of the T-cell response, hence conjointly function as a checkpoint amidst innate and adaptive immunity (Banchereau et al. 2000). The response generated by DCs is regulated by numerous distinct receptors—these receptors recognize PAMPS and alarmins—it promotes DCs response against infection or any other kind of environmental changes (Bianchi 2007). During inflammation, DCs may emerge from monocytes, just as macrophages do, and there are metabolic similarities between specific stimulus response of macrophages and monocyte-derived DCs (O'Neill and Hardie 2013).

DCs are often derived from bone marrow culture in granulocyte macrophage-colony stimulating factor (GM-CSF) as a model for monocyte-derived and TNF- and inducible nitric oxide synthase (iNOS) generating (TiP) DCs. At the quiescent stage, cells undergo combined glycolysis and OXPHOS—thus producing a slight quantity of lactate (Krawczyk et al. 2010). BMDCs (resting GM-CSF-induced bone marrow-derived DCs) aren't similar to stimulated DCs in many aspects such as being comparatively sessile, producing a low secretion, and couldn't activate or interact with T cells to a greater extent. BMDCs are also reported with the usage of

FAO to fuel OXPHOS. Whereas BMDCs also utilize glucose, but it is still unambiguous that glucose is primarily used for OXPHOS or to fulfill any other anabolic pathways (Pearce and Everts 2015). However, at the active stage, DCs show a notable metabolic transformation due to which they shift to aerobic glycolysis, for example, stimulation with TLR agonists (Krawczyk et al. 2010). The activated immune cells are solitarily dependent upon aerobic glycolysis for energy requirement, in effect the mitochondrial oxygen consumption ceases. The respiration failure of mitochondria in these cells coincides with the TLR-driven expression of iNOS, responsible for the production of nitric oxide (NO) from arginine at a tremendous scale (Everts et al. 2012). DC subsets that produce NO are only the DCs that exhibit the obligation on glycolytic metabolism; the produced NO is responsible for the disintegration of OXPHOS—which includes complex-I, II, and IV of Electron Transport chain due to NO resulting in the nitrosylation of proteins having iron and sulfur (Beltran et al. 2000; Cleeter et al. 1994; Clementi et al. 1998).

## 3.3 *Macrophages: Metabolic Response Upon Activation*

Macrophages have a substantial importance in the immune system. These cells reside all over the body in almost all tissues. Macrophages are embryonically derived cells, sown into tissues in utero, and maintained by in situ proliferation (Geissmann et al. 2010; Schulz et al. 2012). The macrophages are flexible with the course of action; they can be stimulated locally and/or through pathogenic stimuli and provide a well-established response by determining appropriate activation states. Macrophage is recognized with two types of activation: (1) CAM (classically activated macrophages or M1), Interferon-γ (IFN-γ) in consolidation with TLR agonists promotes its activation. (2) AAM (Alternatively activated macrophages or M2), cytokines IL-4 and IL-3 play roles in its activation (Murray and Wynn 2011). From the perspective of defense (host) the role of the M1 macrophage is rather inflammatory, involved in the secretion of IL-12; which triggers NK and T cells for IFN-γ production, and TNF-α which stimulates NO (cytotoxic) and various immune cells. Hence, it plays a substantial part in the elimination of several microbial infections. In contrast, M2 macrophages are equipped with the ability to produce an array of molecules which plays a key role in the regulation of immunity, modulation of inflammation, and fostering of tissue repair. M2 macrophages are most notably seen in an immune response against helminth parasites (Anthony et al. 2007). M2 macrophages, it turns out, play another important role in metabolic homeostasis, especially in adipose and liver tissues (Biswas and Mantovani 2012; Odegaard and Chawla 2011). M1 and M2 when activated have different metabolic states if compared to quiescent macrophages (Rodríguez-Prados et al. 2010). The utilization of arginine by these two cells in different metabolic pathways manifests the defining characterization between these cells; M1 macrophages utilize arginine as a precursor for iNOS, while M2 macrophages utilize arginine as a precursor for Arginase 1 (Murray and Wynn 2011) (Fig. 2).

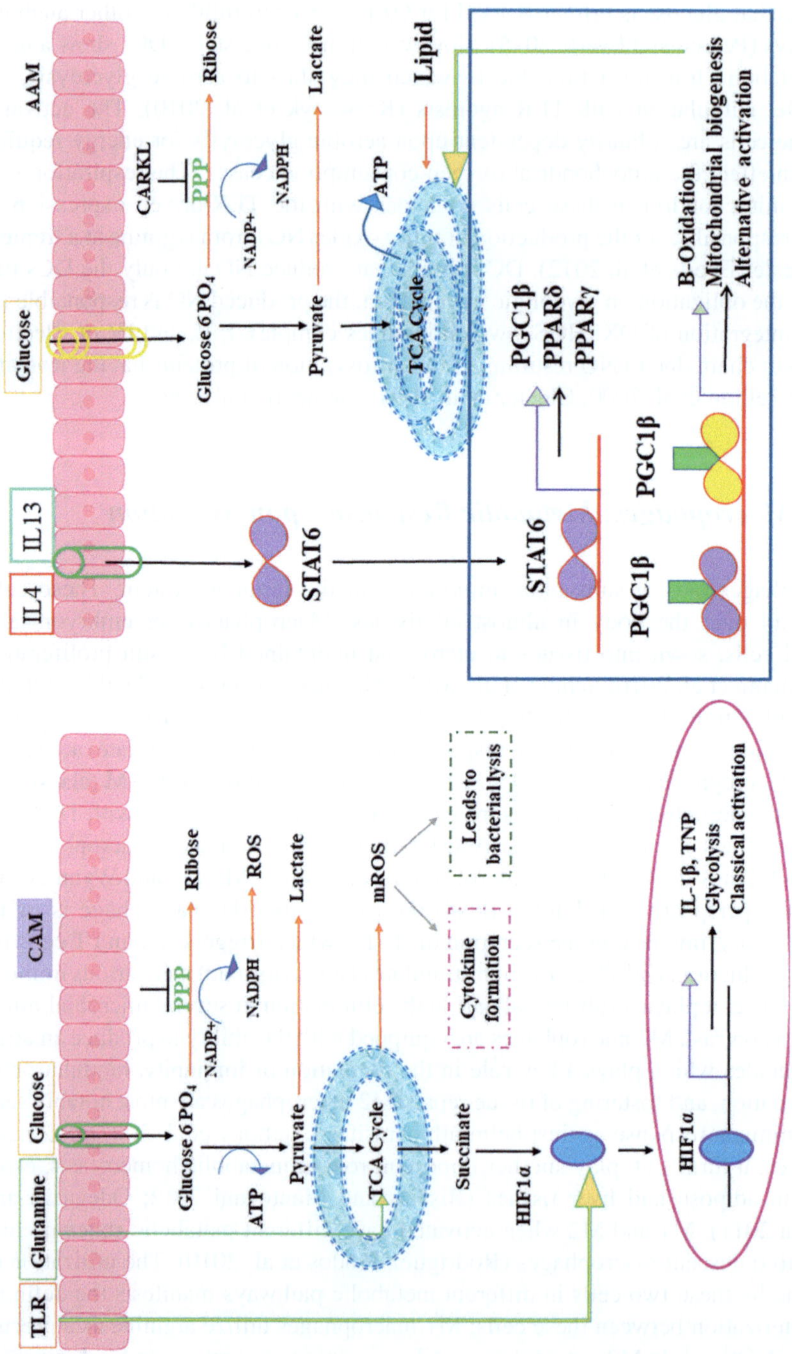


**Fig. 2** Illustration of discrete metabolic programs opted by CAMs and AAMs for their functioning

M1 macrophages metabolically resemble bone marrow-derived dendritic cells—having aerobic glycolysis. The phosphofructokinase 2 (PFK2) PFKFB3, which produces fructose-2,6-bisphosphate by utilizing fructose-6-phosphate to promote glycolytic flux, regulates the progression to glycolytic metabolism in M1 macrophages (Rodríguez-Prados et al. 2010). Usually, a non-functional or broken mitochondrial respiration produces ROS (reactive oxygen species) which is also important in host defense. Typically, NADPH oxidase is expected to be accountable for the ROS production process, but it was recently discovered that in TLRs activated macrophages enroll the mitochondria to perform the function of phagolysosomes for the production of ROS which is necessary for eliminating phagocytosed pathogen; this response is generated relying upon the TRAF6 (TNF receptor-associated factor 6) (West et al. 2011). When infected with specific bacteria or viruses or exposed to threat signals such as extracellular ATP or monosodium urate, or adjuvant such as alum; macrophages respond by organized raise in ROS (produced by mitochondria), IL-1β, and release of caspase-1 (Sorbara and Girardin 2011; Tschopp and Schroder 2010). The inflammatory activation of macrophages has been linked to two mitochondria-derived signals: First, the NLRP3 inflammasome detects mitochondria-derived ROS, which activates caspase 1 and causes the exudation of mature IL-1β (Zhou et al. 2011). The voltage-dependent anion channels of mitochondria were also required for IL-1β release from macrophages, which gives a non-dependent link between mitochondria and inflammasome. In line with this theory, autophagy inhibition—the process of elimination of rogue mitochondria (ROS producing) get impaired—resulted in NLRP3 inflammasome instigation and proIL-1β cleavage (Nakahira et al. 2011; Zhou et al. 2011). Second, succinate, a TCA cycle intermediate, is another mitochondrial signal involved in the transmission of the effector phenotype of CAMs. In a study, it was found that LPS-stimulated macrophages result in augmented IL-1β mRNA expression relying on glycolysis and HIF-1a (Tannahill et al. 2013). It was all clear after metabolomic analysis of LPS-activated macrophages which proved the anticipated Warburg effect; in addition, it also provided the reason behind the accumulation of TCA cycle intermediates, especially succinate. The role of mitochondrial succinate in HIF-1α activation and its stabilization is well known but to get a deeper insight into the synthesis of these succinates the author has planned the study to identify its precursor (by using the metabolic tracer technique). Remarkably, in macrophages stimulated by LPS, the primary source of carbon that assimilates into succinate was anticipated to be pyruvate obtained by glycolysis, but it was found to be α-ketoglutarate which is obtained from the glutaminolysis process (Tannahill et al. 2013).

Alternatively activated macrophages (M2), contrasting to M1, M2 choose oxidative metabolism, fuel preference in M1 and M2 is evocative of the fast (Type II) and slow (Type I) twitch muscle fibers. For illustration, Type-II muscle fibers have high-intensity contraction (short and rapid burst) which is powered by anaerobic glycolysis resembling M1 macrophages. Likewise, M2 macrophages resemble Type I muscle fibers that rely on their mitochondrial fatty acid oxidation to endure low contraction but for a longer period (Picard et al. 2012). Interestingly, M1 macrophages and Type-II muscle fibers are controlled by HIF-1α (Mason et al. 2004), whereas FAO and mitochondrial oxidative metabolism in M1 macrophage and Type I muscle are controlled by PPARs and PGCs (Lin et al. 2002; Narkar et al. 2008; Wang et al. 2004).

When macrophages are stimulated with Type 2 cytokines like IL-4 or IL-13, a transcriptional cascade is triggered that regulates and maintains their cellular oxidative metabolism program. Although prior research had shown that STAT6 is needed for alternative macrophage activation by IL-4 and IL-13, the transcription factor's function in metabolic programming was unknown. The switch mechanism of M1 macrophages was initiated using fatty acid through β-oxidation metabolic processes that helps to run the mitochondrial metabolic processes which help in STAT6 generation bidirectionally; initially, the strand of genomic part accelerates the generating energy from the fats by the processes of oxidation, OXPHOS, and mitochondrial biogenesis which were transcriptionally activated by STAT6 (Vats et al. 2006). Second, STAT6 induced the expression of transcriptional regulators and co-activator proteins, such as PPARγ, PPARδ, and PGC1β, which synergized with STAT6 to carry on metabolic switch to oxidative pathway (Kang et al. 2008; Odegaard et al. 2007, 2008; Vats et al. 2006). Furthermore, the targeted regulators were discovered to work in tandem with STAT6 on the promoters of alternative activation genes including Arg1 and Clec10A, enhancing and preserving the cell's effector phenotype. It is not yet understood how this matches with the facts that (1) AAMs mainly use catabolic pathways like fatty acid oxidation and (2) IL-4 induces carbohydrate kinase-like protein (CARKL) expression, which uses PPP to prevent the flux—constraining NADPH output and decreasing glycolytic flux. Whereas, overexpression of CARKL shifts macrophage metabolism toward oxidative metabolism and vitiates inflammatory activation (Haschemi et al. 2012). This is an intriguing subject for forthcoming research.

## 3.4 T Cells: Cells Dynamic at Active and Quiescence State

T cells perform a critical role in controlling infection and cancer, facilitating long-term protective immunity, considering this it is quite logical why these cells are targeted for therapeutic use. The functioning of T cells and other immune cells is stimulated by metabolic pathways; however, immunity can be influenced by metabolic interventions but it is very less occasional. T cells are different from other innate immune cells in several aspects. For example, T cells display rapid proliferation on their activation. These characteristics of activated T cells resemble tumor cells, hence are often compared metabolically. They both engage in aerobic glycolysis while proliferating, whereas other immune cells utilize aerobic glycolysis on their activation but not when they proliferate (Fox et al. 2005; Jones and Thompson 2007; Vander Heiden et al. 2009). Adaptive immunity is capable of generating long-lived memory cells that are antigen-specific and protects against re-infection or re-emergence of the tumor. These characteristics are limited to cells of adaptive immunity. The memory T cells are metabolically unlike their counterpart—T--effector cell, memory T cell does not indulge in the Warburg metabolism rather it depends upon mitochondrial FAO for survival (van der Windt et al. 2012).

On antigen recognition by a T cell (naive) through the proper co-stimulatory process, the T cell undergoes several processes important for gaining specially designed effector function including cell growth and proliferation. This kind of process requires metabolic reprogramming because the cell has to go through anabolic processes to create the biomass for the daughter cell which requires lots of energy. The cell shifts to anabolic metabolism from catabolic metabolism (nutrients are utilized for cell maintenance and homeostasis). Similarly, T-cell transformation to the T memory cell is marked by the metabolic shift toward catabolism which is required to maintain quiescence and longevity and may obstruct their terminal differentiation. It has been thoroughly examined how molecules such as MTOR, P13K, Akt, Myc, and HIF relate immune signals and metabolic signals to the activation, growth, function, and maintenance of T cells (Hirata and Narumiya 2012; Seth et al. 2006; Waickman and Powell 2012; Wang and Green 2012). Various T helper cell subsets opt for different metabolic pathways while differentiation, which is an interesting area of research (Chi 2012; Gerriets and Rathmell 2012; Zemirli and Arnoult 2012). For instance, Treg cells mainly utilize oxidative phosphorylation and mitochondrial FAO for their sustainability (Michalek et al. 2011) although glycolysis is required for generating T helper-17 cells (Shi et al. 2011). However, the reason behind these metabolic phenotypic alterations while T-cell differentiation is still unclear.

### 3.5 B Cells: Metabolic Response Upon Activation

B cells, which are the only cells that generate antibodies, are important for developing and upholding humoral immunity. There are several distinctions between B and T cells in terms of stimulation and multiplication. Although B cells and T cells share some basic metabolic characteristics in their activated phase like the initiation of glycolysis hence elevating the glucose uptake, still very little is known about how metabolism guides the outcome of B cells (Doughty et al. 2006; Dufort et al. 2007). For example, B cells get activated through the B cells receptor which in turn stimulates PI3K/mTOR/Akt pathways to promote aerobic glycolysis—a metabolic shift required for their proliferation (Zhang et al. 2005). B cells come to have several unique features like affinity maturation, plasma cell formation, and antibody class switching, but their metabolic basis is still indescribable. Although, the occurrence of immune response can be followed by B cells differentiating into plasma cells and spending years in bone marrow—through continuous biogenesis of antibodies—fulfilling the role of balancing the physiologically relevant concentration of antibodies in plasma. However, there is no precise explanation of how B cells can have combined effects of longevity and high productivity; thus, it has the potential of providing insight into regular humoral responses and vaccine development (Pearce and Pearce 2013). Therefore, it is a challenging subject for further work.

## 4 Immuno-Metabolic Response During Infection

The rhythmic action of the immune system is very mesmerizing. For the proper functioning of immune cells, an excellent metabolic balance is vital for protection from a range of pathogens while preventing immune-mediated pathologies. The immune cells exhibit different metabolic activity at their different states; activated and quiescence states. Activated immune cells rely on a different metabolic pattern compared to resting immune cells; these metabolic fluctuation helps in the immune cell functions and other cellular activities (O'Neill et al. 2016). The study of immunometabolism has offered insight into the cellular functioning of immune cells and their dependency on nutritional availability and its utilization amid the immune response (Klein Geltink et al. 2018; Teijlingen Bakker and Pearce 2020).

The study of immunometabolism focuses on the transformation in the metabolism of activated immune cells (Fukuzumi et al. 1996; Oren et al. 1963). However, the activation of immune cells increases glucose consumption to meet the requirement of their effector functions (Krawczyk et al. 2010; Michalek et al. 2011). The metabolic shift in an activated immune cell is not a kind of on/off mechanism but somewhat of a qualitative engagement and systematization of distinct metabolic pathways that result in a specific effector outcome, for example, the polarization of T helper cells or macrophage effector response.

Both innate and adaptive immunity undergo DNA methylation which requires contributor metabolites like s-adenosylmethionine (SAM) and α-ketoglutarate (α-KG) for one-carbon metabolism. Studies have provided strong evidence indicating a strong connection between metabolic pathways and gene expression (epigenetic regulation) (Lio and Huang 2020). α-KG is an intermediate metabolite of the TCA cycle; it also plays a crucial role in the epigenetic regulation of inflammation-associated genes, the dioxygenase enzymes use α-KG as a cofactor (Tahiliani et al. 2009). The ten-eleven translocation (TET) methylcytosine oxidases are among these dioxygenases, and they are needed for proper lymphocyte production and stimulation, demonstrating the significance of metabolic regulation for gene expression.

Iron upholds major roles in several functions in organisms cell; it is essential for DNA synthesis and functioning of mitochondria as it provides a cofactor for essential enzymes involved in it. Geros and his colleagues have discussed the role of iron in metabolic regulators in host and pathogen; the same goes for TET methylcytosine where iron plays a major role (Sousa Gerós et al. 2020; Tahiliani et al. 2009). Host immune eliminates pathogens by upholding the iron availability; they do so by capturing iron and affect the pathogen functionality, and these all occur in the host (human) guts which do affect the normal intestinal microbiota of the host (human). Geros et al. have described the process of metabolic regulation among different ecological relationships (commensalism, mutualism, amensalism, or parasitism) that influence health and dysbiosis (Sousa Gerós et al. 2020). Considering the impact of iron on hosts and pathogens, it provides a scope for metabolic interventions for therapeutic purposes.

Nowadays, research has been focused on the metabolic processes occurring in hosts and pathogen and the development of a novel mechanism to control the immune cells metabolically. They define a probable linkage between the hexosamine biosynthesis pathway (HBP) and O-GlcNAcylation—post-translational glycosylation form. The O-GlcNAcylation can refashion the protein's functional properties through modification—with transcription factors and epigenetic regulators being two of the most common targets (Wang et al. 2020; Zhu and Hart 2021).

## 5 Immuno-Metabolic Response During Cancer

The current studies in the field of immunometabolism have led to the modernization of cancer therapy. Host defense against cancer involves both kinds of immunity—innate and adaptive immunity—through various mechanisms. The immune cells have the capability of sensing a range of surrounding signals and triggering a function in response to it. Cellular metabolic pathways govern the immune response; immune cells have specific responses toward specific alterations in tissue metabolism, such as nutrients exhaustion, elevated oxygen utilization, and generation of intermediate metabolites (Chen et al. 2020; Huang et al. 2020; Terry et al. 2020). Likewise, the metabolites from the tumor cells can also impact the differentiation of immune cells and effector functions (Karayama et al. 2021; Yan et al. 2018). Nevertheless, recent work has shown that immune cells compete in the nutrient environment with cancer cells and other proliferating cells (Chang et al. 2015). Hence, it provides a scope for effective immunotherapy through metabolic interventions. Many experimental findings indicated that alteration in the metabolism of cancerous cells is due to the involvement of various immune and inflammatory cells (Ramapriyan et al. 2019). Subsequently, several researchers have noticed that several activities of immune cells are the outcome of abnormal metabolites or metabolites from a cancerous cell (Cronin et al. 2018; Moon et al. 2018; de Mendonça Oliveira et al. 2018). Findings from recent studies have shown the interrelation between metabolic functions and immune cells which was never anticipated before. Hence, this novel field is termed immunometabolism (Shyer et al. 2020).

### *5.1 Metabolism of Tumor Cells*

Cancerous cells are capable of metabolism reprogramming which is described as a hallmark for cancer; the cancer cells adjust the process of energy metabolism and exploit it for rapid cell growth (Callao and Montoya 1961). Cancer is a heterogeneous disease; tumor heterogeneity delivers it with a complex metabolic pattern. Cancer is a heterogeneous disease characterized by complex metabolic patterns. To support their rapid growth and proliferation, cancer cells utilize various metabolic pathways, including the pentose phosphate and serine metabolic pathways for biosynthesis, while meeting their energy demands through glycolysis, which provides rapid ATP production (Boroughs and DeBerardinis 2015; Hurley et al. 2021). For example, the tumor cells utilize glycolysis pathways regardless of the oxygen availability—known as aerobic glycolysis—the end product from glycolysis that is pyruvate is converted into lactic acid instead of acetyl-CoA and didn't go for mitochondrial respiration (Callao and Montoya 1961; Rodríguez et al. 2021). Tumor cells not only utilize glucose, depending upon the external nutritional availability the tumor cells utilize several macromolecules to produce ATP through a range of metabolic pathways. Moreover, it utilizes all the available biomolecules such as glutamine, arginine, serine, fatty acid, and lipids for their cell growth and proliferation (Altman et al. 2016). For example, under nutrient deficiency stress conditions (e.g., glucose or glutamine), tumor cells activate oncogenic c-Myc to promote tumor cell endurance by adjusting the regulation of metabolic enzymes such as PHGDH, PSAT1, and PSPH in the serine synthesis process (Sun et al. 2015). Moreover, to maintain their growth and proliferation under a hypoxic or nutrient-deficient environment, the tumor cells are capable of utilizing the fatty acid (Mashimo et al. 2014; Schug et al. 2015, 2016). The tumor cells yield intermediate metabolites via the breakdown of ketone bodies that can go for the TCA cycle and provide ATP for the cell energy requirements (Huang et al. 2016). Therefore, the tumor metabolic modes are complex and flexible which function according to the nutritional availability to support the tumor cell survival.

## 5.2 *The Struggle for Nutrient Uptake Among Tumor Cells and Immune Cells*

Cancer cells possess the capability of metabolic transition, but it is not limited to cancer cells, several other cells like activated T cells, Tregs, and neutrophils, are also capable of metabolic transition (Ghesquière et al. 2014). One of the most required biomolecules is glucose as it provides intermediate metabolites for several metabolic pathways; it is one of the most utilized by tumor cells. Moreover, glucose is also required by several immune cells for their function, differentiation, and immune cell activation (Kolb et al. 2020; Palmer et al. 2015). A varying degree and types of immune cells accompany the tumor microenvironment (TME), similar to cancer cells, the tumor-infiltrating lymphocytes (TIL) require nutrition for their differentiation and proliferation which they absorb from the TME (Togo et al. 2020). Studies have shown that tumor-infiltrating T cells get suppressed by tumor cells through the competitive uptake of glucose regardless of the presence of an adequate amount of tumor antigen for T cells (Qiu et al. 2019). Certainly, the studies have demonstrated that the tumor cells escape from TIL T cells by competing uptake of glucose and force T cells to exhaustion and metabolic alteration due to glucose unavailability (Sukumar et al. 2015). For example, renal cell carcinoma (RCC) tissue is characterized by enhanced glucose metabolism but lower $CD8^+$ effector T cells infiltration concluded through microarray analysis of RCC tissues which showed an inverse correlation between GLUT1 expression and T-cell infiltration (Singer et al. 2011).

If the reported literature is to be believed, glycolytic metabolites are not ideal due to their negative consequence on immune function (Harmon et al. 2020). Similar to the competitive uptake of glucose by tumor cells causes damage to T cells, the immune cell functioning is still dependent upon the stimulation of respective transporters on the cell surface, including the competitive uptake of amino acids, fatty acids, and other metabolites (Chang et al. 2015). Furthermore, TME is characterized by increased levels of ROS, high lactate concentration, low pH, and hypoxia which contributes to cancer development and immune escape (Harmon et al. 2020). Therefore, to enhance the tumor immunogenicity and minimize deleterious effects caused by tumors; metabolic pathways in TME should be the target for treatment, and inhibition of nutrient competition among these cells will provide significant success in its therapy (Kareva 2019).

## 6 Immuno-Metabolic Response During Autoimmunity

Immune cell differentiation and functions are closely controlled during immune responses to guarantee defense from pathogens while also preventing the destruction of its own tissues. It is now known that specific changes in cellular metabolism shore up immune cell functions among both innate and adaptive immune cells, while various cell components are applied to distinguish and signify variations in a broad variety of elements present in nutrients and metabolites (Bantug et al. 2018; Jung et al. 2019; Wei et al. 2017). In this context, mTOR and AMPK are superb among immune cells when compared to their nutrient-energy sensing and signaling pathways (Herzig and Shaw 2018; Powell et al. 2012). These complexes play an important role in linking immune functions and nutrient accessibility by regulating metabolic switches between various intracellular metabolic pathways like glycolysis, FAO, OXPHOS, etc. (E. L. Pearce and Pearce 2013) (Fig. 3).

In addition, the number of individuals suffering from autoimmunity majorly comes from the prosperous society which indicates that certain parameters like stress-eater, excessive diet, and lardaceous can mislead to any deregulation of the biochemical/metabolic pathway; hence, it leads to lowering down tolerance in hereditary predisposition client (Alwarawrah et al. 2018; Procaccini et al. 2011). Leptin hormone capable of correlating the condition of metabolic pathway and immunological status, mainly synthesized by adipose tissues, signifies as a crucial moderator at the interface among energy metabolism, inflammation, and immune response (Francisco et al. 2018; Procaccini et al. 2015). In this study, we are going

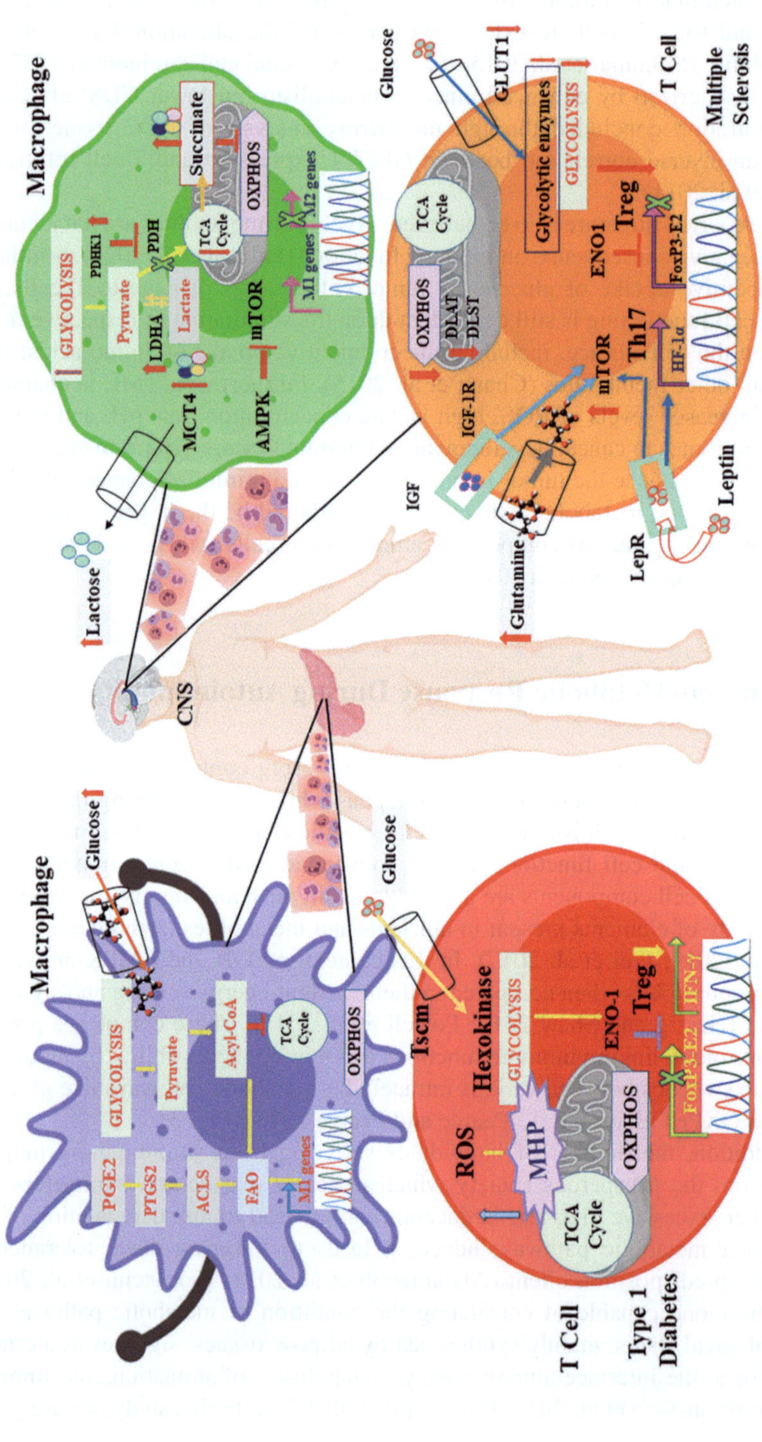


**Fig. 3** Alterations in intracellular and systemic metabolism in autoimmune disorders

to discuss how dysregulation of innate and adaptive immunity is promoted by systemic and intracellular metabolic status, resulting in pathogenic events associated with MS and T1D.

## *6.1 Immunometabolic Alterations in Autoimmune Diseases*

Metabolic programs are now well-established as crucial regulators for the differentiation and functioning of immune cells. In many studies, metabolic programs have been linked to dysregulated immune responses, which contribute to autoimmunity and its progression. Autoimmune disease occurs because of the alteration in the cell's metabolic pathways, metabolites, and metabolic products regardless of whether the disease type is systemic or organ specific. For instance, the hyperactivated autoreactive immune cells share a common malfunction of glucose metabolism—impaired mitochondrial respiration together with oxidative stress results in the production of ROS, which can be seen in both SLE and T1D (Stathopoulou et al. 2019). Moreover, numerous adipokines are the key regulators of food intake and T helper (Th)1 response, like leptin, these adipokines are frequently linked to autoimmune pathogenesis, including RA, SLE, MS, and T1D (Francisco et al. 2018; Procaccini et al. 2015).

## *6.2 Immunometabolism of Multiple Sclerosis*

MS is a type of autoimmune disease that causes demyelinating and neurodegenerative disorder—when autoreactive immune cells infiltrate the central nervous system (CNS) (Filippi et al. 2018). Multiple sclerosis (MS) pathogenesis is linked to and influenced by metabolic status, as concluded from experimental studies (de Jesús Guerrero-García et al. 2016; Huppke et al. 2019; Munger et al. 2009). Research indicates that individuals with obesity or higher body weight at birth are more vulnerable to developing MS (Munger et al. 2009). However, many findings have indicated the detrimental effects of leptin in the occurrence and development of MS. For instance, induction of experimental autoimmune encephalomyelitis (EAE) was tried on mice lacking leptin but mice didn't develop the EAE (Carrieri et al. 2015; De Rosa 2006; Gerriets et al. 2016; Matarese et al. 2001); however, after administering the exogenous leptin into the mice promotes proliferation in T cells, and the response of Th1 cells is also noted (Matarese et al. 2001). Moreover, EAE progression can be cured by leptin neutralization which suppresses the biosynthesis of autoreactive T cells (De Rosa 2006). In the EAE model, it was found that leptin initiates the glycolytic metabolism in T cells, the differentiation of effector T cells through HIF1$\alpha$, a key regulator of the Th17 lineage (Gerriets et al. 2016).

Nutritional status and leptin gradients are interrelated, and experimental findings have indicated the importance of the leptin-mTOR axis in bridging the metabolism and immune cells in pathological conditions. Leptin has been found responsible for the activation of the AKT-mTOR pathway in the EAE model; hence, it endorses the $CD4^+$ T cells proliferation and survival (Galgani et al. 2010). mTOR has several functions other than nutrient sensors; mTOR has been found to be linked with various micro-environmental inputs, including nutrients and growth factors. mTOR functions as more than just a nutrient sensor; it integrates various micro-environmental inputs, including nutrients and growth factors. ASCT2 (alanine-serine-cysteine transporter 2) facilitates glutamine uptake in CD4+ naïve T cells, which is essential for TCR-mediated activation of mTORC1. Inhibiting ASCT2 in mice reduces mTOR activation, resulting in EAE resistance due to a suppressed inflammatory response (Nakaya et al. 2014). In humans, including individuals with the relapsing-remitting (RR)-MS, a link between mTOR and Tregs proliferation was found; it has been observed that mTOR hyperactivation minimizes the Tregs proliferation potential (Carbone et al. 2014). Whereas on inhibiting the mTOR pathway the capability of Tregs proliferation is restored and the EAE symptoms diminish (Procaccini et al. 2010).

## *6.3 Immunometabolism of T1D*

When the autoreactive lymphoid cells attack and destroy the insulin-producing β-cells, it results in the T1D disorder (Chillarón et al. 2014). The main cause for the T1D disorder is overnutrition and obesity which disrupt the normal functioning of immune cells, as per evidence autoimmunity is associated with disrupted metabolism (Corbin et al. 2018). In this context, the main reason for the immune cell intrusion into the islets of the pancreas is caused by leptin. In a study, a non-obese diabetic (OBD) mouse was administered with leptin; leptin was found to be the reason for the increased IFN-γ mRNA expression in T cells. Hence, leptin's role in T1D disorder was verified (Matarese et al. 2002). Remarkably, a mutation occurred in the leptin receptor of an NOD mouse (named NODLepR-db5j) which diminish the occurrence of antigen-specific $CD8^+$ Teff cells, thus preventing T1D progression (Lee et al. 2006). Genetically susceptible individuals are diagnosed with T1D when the major proportion of the β-cells of the pancreas are destroyed, which renders immune cells with prolonged exposure to a high concentration of glucose (Cnop et al. 2005). Insulin is crucial in regulating glucose levels in the blood and influences the metabolism of lipids and proteins. T1D individuals suffer from hyperglycemia which can alter the intracellular metabolic pathways of innate and adaptive immune cells and may result in immune hyperactivation (Amersfoort and Kuiper 2017). In context, the NOD mouse with elevated levels of glucose has been reported with M1 macrophage polarization and prostaglandin (PG)E2 (a proinflammatory molecule) synthesis. Macrophages under hyperglycemia can upregulate ACSL1 (acyl-CoA synthetase long chain 1) which is involved in FAO—several inflammatory molecules are produced like PGE2 through PTGS2 (prostaglandin-endoperoxide synthase 2) enzyme (Kanter et al. 2012).

## 7 Immune-Metabolic Response in Obesity

Autoimmune is visible in high rate in case of overweight individuals. Autoimmune conditions are more prevalent in overweight individuals. Insulin resistance occurs when cells, such as adipocytes, hepatocytes, and myocytes, do not respond properly to insulin, impairing glucose uptake. A prevalent view is that β-cell dysfunction occurs in a subset of obese, insulin-resistant individuals, leading to reduced insulin production, impaired blood glucose control, and the eventual development of type 2 diabetes (T2D) (Muoio and Newgard 2008; Prentki 2006). Since β-cell dysfunction can be identified well before the onset of T2D, it likely develops before or concurrently with insulin resistance (Kahn 2003; Mehran et al. 2012). The mechanisms that cause obesity to be a risk factor for developing insulin resistance and are the subject of intense research, with mounting evidence that inflammation plays a significant role. In particular, over-infiltration of activated immune cells and overproduction of proinflammatory cytokines closely link excess adipose tissue (AT) development to chronic inflammation. Proinflammatory cytokines, such as TNF-α, mechanistically can induce serine phosphorylation and disable insulin receptor substrate 1 and thus, in several cell types, including adipocytes and hepatocytes, insulin receptors. Proinflammatory cytokines like TNF-α can cause serine phosphorylation and inactivation of insulin receptor substrate 1, thereby blocking insulin receptor signaling in a variety of cell types, including adipocytes and hepatocytes (Hirosumi et al. 2002; Yuan 2001). Visceral adiposity is associated with metabolic abnormalities, potentially because visceral adipose tissue (VAT) is more prone to losing immune regulation, leading to increased inflammation (Bentham Science Publisher 2006; Lebovitz and Banerji 2005). In obese individuals, visceral adipose tissue (AT) shows significantly higher mast cell infiltration and a relative decrease in regulatory T cells (Tregs) compared to subcutaneous AT, contributing to a more inflammatory profile (Altintas et al. 2011; Feuerer et al. 2009; Winer et al. 2009).

### *7.1 Regulation of AT Inflammation by Immune Cell*

Proinflammatory cytokines are easily observed in obese AT. However, recent studies indicate that the source of these cytokines is actually both innate and adaptive immune cells in obese AT. One of the distinguishing characteristics of AT inflammation in obesity is a significant rise due to deposition of macrophages around adipocytes in which look like "crown-like structures" (Murano et al. 2008; Weisberg et al. 2003). The macrophages in inflamed AT are mostly inflammation-causing M1 macrophages—which produce significant quantities of proinflammatory cytokines as TNF-α. Whereas M2 macrophages are found in lean AT (Lumeng et al. 2007). Besides macrophages, the function of other innate immune cells is increasingly evident. For instance, CD11c high F4/80 low dendritic cells also infiltrate AT in obese mice—promotes macrophage infiltration (Stefanovic-Racic et al. 2012) and encourages Th17 cells (proinflammatory) differentiation (Bertola et al. 2012). Similarly, AT inflammation of obesity was also shown to be stimulated by mast cells whose numbers increase in obese AT (Liu et al. 2009). Even 3 days after initiation of a high-fat diet (HFD) in mice, neutrophils infiltrate the AT transiently, and their elastase output contributes to inflammation of AT and may directly cause insulin resistance (Elgazar-Carmon et al. 2008; Talukdar et al. 2012). Indeed, serum from obese human subjects is spotted with elevated activity of neutrophil elastase; however, mice lacking neutrophil elastase are found protected from HFD-induced obesity and insulin resistance (Mansuy-Aubert et al. 2013).

Adaptive immune cells have also been linked to AT inflammation. The vascular AT of obese mice could be accumulated by both T cells and B cells which can promote their phenotypic alteration toward proinflammatory subtypes (DeFuria et al. 2013; Feuerer et al. 2009; Nishimura et al. 2009; Winer et al. 2011; Winer et al. 2009). Both CD4+ and CD8+ T cells in obese AT, for example, express markers of effector memory cells and produce high levels of IFN-γ (Winer et al. 2009; Yang et al. 2010), implying that an overactive Th1 cell response can contribute to AT inflammation. Infiltrating CD8+ T cells often generate high chemokine levels such as CCL5 and CXCL1, contributing to macrophages being further recruited into AT (Nishimura et al. 2009; Yang et al. 2010). Interestingly, because of the inadequate TCR diversity of obese AT-resident T cells (Yang et al. 2010), particular antigens may induce the response from these T cells. However, researchers couldn't find solid evidence of any antigen involved though there is some evidence suggesting the involvement of ingested intestinal antigens (Wang et al. 2010).

## *7.2 Immune Regulation Is Lost in Obese AT*

For instance, in obese AT the transformation of M2 macrophages (produces anti-inflammatory IL-10 in lean AT) into M1 macrophages occurs (Fujisaka et al. 2009; Lumeng et al. 2007; Morris et al. 2011). Usually, IL-4 maintains the major phenotype-M2 macrophages, while cytokines sources are believed to be Th2 cells and eosinophils. Eosinophils are crucial in maintaining M2 macrophages in healthy AT concluded on the basis that IL-4 and IL-13 are produced by eosinophils cells—the eosinophils have a major role in IL-4 production which account for up to the 90% of the total IL-4 being produced in AT (Wu et al. 2011). However, the loss of Th2 cytokines derived by cells probably affects the state of AT macrophage polarization; the transfer of T cells to the immunodeficient mouse may also cause colitis, and consequences on weight may confuse interpretations of the AT effects with a microbiome and immune homeostasis (Cipolletta et al. 2011). The invariant NKT (iNKT) cell is another subset of regulatory immune cells that are usually located in AT. iNKT cells are more abundant in lean AT than in peripheral lymphoid organs, but their levels are significantly lower in obese AT (Lynch et al. 2012; Schipper et al. 2012). In addition, the mouse missing iNKT cells exhibits increased resistance to insulin and increased weight on HFD, correlated to increased proinflammatory macrophage infiltration into the AT (Lynch et al. 2012).

In addition to high proportionate M2 macrophages, VAT in lean mice has a noticeably higher ratio of $CD4^{+}Foxp3^{+}$Tregs within $CD4^{+}$ T cells as compared to peripheral lymphoid organs; however (Feuerer et al. 2009), in obese mice this ratio is noticeably lower. AT-resident Tregs could usually regulate the function of several, if not all, of the immune cells that lead to AT inflammation due to their wide range of suppressive activities (Josefowicz et al. 2012). Remarkably, Tregs that reside in AT are different from Tregs in peripheral lymphoid organs; Tregs in AT exhibit increased IL-10 expression (Cipolletta et al. 2012; Feuerer et al. 2009). IL-10 performs classical anti-inflammatory functions and is found to play a role in the suppression of TNF-α-induced insulin resistance in 3T3-L1 cells (differentiated into adipocytes) (Lumeng et al. 2007). Overall, regulatory failure in immune-metabolic pathway specifically leads to the evolution in intensification of AT inflammation during obesity, where the combined number and/or function deficits of multiple regulatory cell types lead to a failure to maintain M2 macrophages. Although it is unclear what causes this immunoregulatory mechanism to malfunction, inappropriate proinflammatory cell stimulation may be one of the reasons.

### 7.3 *Factors Contributing to the Development of AT Inflammation*

Probable cause of immune cells turning into proinflammatory subtypes in the obese AT could be the integrated threat signals of endogenous and exogenous nature. By considering endogenous signals, TLR4 and TLR2 are triggered in macrophages and adipocytes as well directly through the presence of saturated fatty acids—marks the production of proinflammatory cytokines (Lee et al. 2004; Shi et al. 2006). Inflammasome NLRP3 can also be activated by saturated acids, specifically, palmitates, which cause IL-1β to mature and release by macrophages (Wen et al. 2011). Besides dietary resources, during lipolysis, fatty acids are also released from adipocytes, which occurs at a higher pace in obese AT (Duncan et al. 2007). Macrophages are more susceptible and probably the first cell among the immune cells to endogenous danger signals as it surrounds the adipocytes; the possible threats could be a fatty acid and/or an ATP from dying adipocytes (Murano et al. 2008; Weisberg et al. 2003). Notably, TNF-a secreted by activated macrophages can promote lipolysis (Zhang et al. 2002), resulting in a positive feedback loop that aggravates AT inflammation. Fatty acids, interestingly, inhibit IFN-γ-producing Th1 cell differentiation (Michalek et al. 2011). Along with the endogenous danger signals, PAMPs such as LPS are found in the plasma of obese mice at a chronically elevated level (Cani et al. 2007). However, LPS surge in human plasma can be induced in a short span of 2 h by meals containing high-fat, high-carbohydrates (Ghanim et al. 2009). Whereas low levels of LPS fail to induce endotoxin tolerance in macrophages (Maitra et al.

2012); however, chronic inflammation can occur due to a high-fat, high-carbohydrate diet. Furthermore, an HFD induces a significant change in the composition of gut microbiota (Parks et al. 2013; Turnbaugh et al. 2008), which may have a significant impact on immune cell function (Maynard et al. 2012).

Compared to lean AT, the obese AT is identified to have numerous upregulated chemokines which correspond to the previously described rise in endogenous and foreign danger signals. CCL2, 3, 5, 7, 8, 11, and 20 expressions, along with CXCL14 (Duffaut et al. 2009; Huber et al. 2008; Nara et al. 2007), are elevated in obese AT, which creates a high chemotactic gradient in innate and adaptive immune cells. Much research on chemokines present in AT has centered on CCL2 (MCP1) and the systemic removal of CCL2 prevents the aggregation of macrophages in AT and improves insulin resistance (Kanda 2006). These data indicate that chemokines formed by the immune or adipocyte cells residing in AT are capable of producing chemokines and have a significant part in generating inflammation in the adipose tissue (Jiao et al. 2009).

## 8 Conclusion

Immunometabolism is an important aspect connected to the health of the body and mind. The production of energy, synthesis of macromolecules, intracellular signaling, cell-cell communication, post-translational modifications, cell cycle, and cell survival, all these processes are well organized and driven by the activity of metabolic pathways. To comprehend a disease, it is of primary importance to understand the specific metabolic programs of the immune cells. In this chapter, we have highlighted this interrelationship, which will make it easy to decipher the metabolic reprogramming associated with cancer cells and autoimmune conditions. These understandings will not only enhance our knowledge on basic biology and metabolic dysfunction, but also will help in identifying novel targets for anti-inflammatory and anticancer immunotherapies. Immunometabolism is a rapidly maturing research field, poised to bring new hopes to the existing health system. Appreciating the diversity of metabolic programs associated with immune system will open new avenues for immune modulation and therapies, with increased specificity and reduced toxicity.

## Glossary

**Inflammation** An essential part of our body's healing process mediated through a complex interaction between cells and many secreted soluble mediators. It occurs when different blood cells travel to the place of an injury or foreign body like bacteria.

**Immunometabolism** The interface between immunology and metabolism. Mainly metabolism in immune cells.

**Metabolic reprogramming** The process that cells alter their metabolism to support the increased energy request due to continuous growth, rapid proliferation, and other characteristics typical of immune cells and neoplastic cells.

**Quiescence** A state of quietness or inactivity.

**Signaling** Transmission of information or instructions from cell exterior to cell interior for a desired response by the cell. The dynamic interaction of cells of the immune system with other cells, antigens, and secreted factors determines the nature of an immune response.

**Transcriptional cascade** The cascade or pathway in gene expression.

## References


Altintas MM, Azad A, Nayer B, Contreras G, Zaias J, Faul C et al (2011) Mast cells, macrophages, and crown-like structures distinguish subcutaneous from visceral fat in mice. J Lipid Res 52(3):480–488. https://doi.org/10.1194/jlr.M011338

Altman BJ, Stine ZE, Dang CV (2016) Erratum: from Krebs to clinic: glutamine metabolism to cancer therapy. Nat Rev Cancer 16(11):749. https://doi.org/10.1038/nrc.2016.114

Alwarawrah Y, Kiernan K, MacIver NJ (2018) Changes in nutritional status impact immune cell metabolism and function. Front Immunol 9:1055. https://doi.org/10.3389/fimmu.2018.01055

Amersfoort J, Kuiper J (2017) T cell metabolism in metabolic disease-associated autoimmunity. Immunobiology 222(10):925–936. https://doi.org/10.1016/j.imbio.2017.03.001

Anthony RM, Rutitzky LI, Urban JF, Stadecker MJ, Gause WC (2007) Protective immune mechanisms in helminth infection. Nat Rev Immunol 7(12):975–987. https://doi.org/10.1038/nri2199

Ardawi MSM, Newsholme EA (1982) Maximum activities of some enzymes of glycolysis, the tricarboxylic acid cycle and ketone-body and glutamine utilization pathways in lymphocytes of the rat. Biochem J 208(3):743–748. https://doi.org/10.1042/bj2080743

Banchereau J, Briere F, Caux C, Davoust J, Lebecque S, Liu Y-J et al (2000) Immunobiology of dendritic cells. Annu Rev Immunol 18(1):767–811. https://doi.org/10.1146/annurev.immunol.18.1.767

Bantug GR, Galluzzi L, Kroemer G, Hess C (2018) The spectrum of T cell metabolism in health and disease. Nat Rev Immunol 18(1):19–34. https://doi.org/10.1038/nri.2017.99

Beltran B, Mathur A, Duchen MR, Erusalimsky JD, Moncada S (2000) The effect of nitric oxide on cell respiration: A key to understanding its role in cell survival or death. Proc Natl Acad Sci U S A 97(26):14602–14607. https://doi.org/10.1073/pnas.97.26.14602

Bentham Science Publisher (2006) Metabolic obesity: the paradox between visceral and subcutaneous fat. Curr Diabetes Rev 2(4):367–373. https://doi.org/10.2174/157339810602040367

Bertola A, Ciucci T, Rousseau D, Bourlier V, Duffaut C, Bonnafous S et al (2012) Identification of adipose tissue dendritic cells correlated with obesity-associated insulin-resistance and inducing Th17 responses in mice and patients. Diabetes 61(9):2238–2247. https://doi.org/10.2337/db11-1274

Bhagavan NV (2002) Lipids I: fatty acids and eicosanoids. In: Medical biochemistry. Elsevier, pp 365–399. https://doi.org/10.1016/B978-012095440-7/50020-2

Bianchi ME (2007) DAMPs, PAMPs and alarmins: all we need to know about danger. J Leukoc Biol 81(1):1–5. https://doi.org/10.1189/jlb.0306164

Biswas SK, Mantovani A (2012) Orchestration of metabolism by macrophages. Cell Metab 15(4):432–437. https://doi.org/10.1016/j.cmet.2011.11.013

Boroughs LK, DeBerardinis RJ (2015) Metabolic pathways promoting cancer cell survival and growth. Nat Cell Biol 17(4):351–359. https://doi.org/10.1038/ncb3124

Borregaard N, Herlin T (1982) Energy metabolism of human neutrophils during phagocytosis. J Clin Invest 70(3):550–557. https://doi.org/10.1172/JCI110647

Callao V, Montoya E (1961) Toxohormone-like factor from microorganisms with impaired respiration. Science 134(3495):2041–2042. https://doi.org/10.1126/science.134.3495.2041

Cani PD, Amar J, Iglesias MA, Poggi M, Knauf C, Bastelica D et al (2007) Metabolic Endotoxemia initiates obesity and insulin resistance. Diabetes 56(7):1761–1772. https://doi.org/10.2337/db06-1491

Carbone F, De Rosa V, Carrieri PB, Montella S, Bruzzese D, Porcellini A et al (2014) Regulatory T cell proliferative potential is impaired in human autoimmune disease. Nat Med 20(1):69–74. https://doi.org/10.1038/nm.3411

Carrieri PB, Carbone F, Perna F, Bruzzese D, La Rocca C, Galgani M et al (2015) Longitudinal assessment of immuno-metabolic parameters in multiple sclerosis patients during treatment with glatiramer acetate. Metabolism 64(9):1112–1121. https://doi.org/10.1016/j.metabol.2015.05.001

Chang C-H, Qiu J, O'Sullivan D, Buck MD, Noguchi T, Curtis JD et al (2015) Metabolic competition in the tumor microenvironment is a driver of cancer progression. Cell 162(6):1229–1241. https://doi.org/10.1016/j.cell.2015.08.016

Chen B, Gao A, Tu B, Wang Y, Yu X, Wang Y et al (2020) Metabolic modulation via mTOR pathway and anti-angiogenesis remodels tumor microenvironment using PD-L1-targeting codelivery. Biomaterials 255:120187. https://doi.org/10.1016/j.biomaterials.2020.120187

Chi H (2012) Regulation and function of mTOR signalling in T cell fate decisions. Nat Rev Immunol 12(5):325–338. https://doi.org/10.1038/nri3198

Chillarón JJ, Flores Le-Roux JA, Benaiges D, Pedro-Botet J (2014) Type 1 diabetes, metabolic syndrome and cardiovascular risk. Metabolism 63(2):181–187. https://doi.org/10.1016/j.metabol.2013.10.002

Cipolletta D, Kolodin D, Benoist C, Mathis D (2011) Tissular Tregs: a unique population of adipose-tissue-resident Foxp3+CD4+ T cells that impacts organismal metabolism. Semin Immunol 23(6):431–437. https://doi.org/10.1016/j.smim.2011.06.002

Cipolletta D, Feuerer M, Li A, Kamei N, Lee J, Shoelson SE et al (2012) PPAR-γ is a major driver of the accumulation and phenotype of adipose tissue Treg cells. Nature 486(7404):549–553. https://doi.org/10.1038/nature11132

Cleeter MWJ, Cooper JM, Darley-Usmar VM, Moncada S, Schapira AHV (1994) Reversible inhibition of cytochrome c oxidase, the terminal enzyme of the mitochondrial respiratory chain, by nitric oxide. FEBS Lett 345(1):50–54. https://doi.org/10.1016/0014-5793(94)00424-2

Clementi E, Brown GC, Feelisch M, Moncada S (1998) Persistent inhibition of cell respiration by nitric oxide: crucial role of S-nitrosylation of mitochondrial complex I and protective action of glutathione. Proc Natl Acad Sci U S A 95(13):7631–7636. https://doi.org/10.1073/pnas.95.13.7631

Cnop M, Welsh N, Jonas J-C, Jorns A, Lenzen S, Eizirik DL (2005) Mechanisms of pancreatic -cell death in type 1 and type 2 diabetes: many differences, few similarities. Diabetes 54(Supplement 2):S97–S107. https://doi.org/10.2337/diabetes.54.suppl_2.S97

Corbin KD, Driscoll KA, Pratley RE, Smith SR, Maahs DM, Mayer-Davis EJ (2018) Obesity in type 1 diabetes: pathophysiology, clinical impact, and mechanisms. Endocr Rev 39(5):629–663. https://doi.org/10.1210/er.2017-00191

Cronin SJF, Seehus C, Weidinger A, Talbot S, Reissig S, Seifert M et al (2018) The metabolite BH4 controls T cell proliferation in autoimmunity and cancer. Nature 563(7732):564–568. https://doi.org/10.1038/s41586-018-0701-2

Dale DC, Boxer L, Liles WC (2008) The phagocytes: neutrophils and monocytes. Blood 112(4):935–945. https://doi.org/10.1182/blood-2007-12-077917

de Jesús Guerrero-García J, Carrera-Quintanar L, López-Roa RI, Márquez-Aguirre AL, Rojas-Mayorquín AE, Ortuño-Sahagún D (2016) Multiple sclerosis and obesity: possible roles of adipokines. Mediat Inflamm 2016:4036232. https://doi.org/10.1155/2016/4036232

de Mendonça Oliveira L, Teixeira FME, Sato MN (2018) Impact of retinoic acid on immune cells and inflammatory diseases. Mediat Inflamm 2018:3067126. https://doi.org/10.1155/2018/3067126

De Rosa V (2006) Leptin neutralization interferes with pathogenic T cell autoreactivity in autoimmune encephalomyelitis. J Clin Invest 116(2):447–455. https://doi.org/10.1172/JCI26523

DeFuria J, Belkina AC, Jagannathan-Bogdan M, Snyder-Cappione J, Carr JD, Nersesova YR et al (2013) B cells promote inflammation in obesity and type 2 diabetes through regulation of T-cell function and an inflammatory cytokine profile. Proc Natl Acad Sci U S A 110(13):5133–5138. https://doi.org/10.1073/pnas.1215840110

Donnelly RP, Loftus RM, Keating SE, Liou KT, Biron CA, Gardiner CM, Finlay DK (2014) mTORC1-dependent metabolic reprogramming is a prerequisite for NK cell effector function. J Immunol 193(9):4477–4484. https://doi.org/10.4049/jimmunol.1401558

Doughty CA, Bleiman BF, Wagner DJ, Dufort FJ, Mataraza JM, Roberts MF, Chiles TC (2006) Antigen receptor–mediated changes in glucose metabolism in B lymphocytes: role of phosphatidylinositol 3-kinase signaling in the glycolytic control of growth. Blood 107(11):4458–4465. https://doi.org/10.1182/blood-2005-12-4788

Duffaut C, Zakaroff-Girard A, Bourlier V, Decaunes P, Maumus M, Chiotasso P et al (2009) Interplay between human adipocytes and T lymphocytes in obesity. Arterioscler Thromb Vasc Biol 29(10):1608–1614. https://doi.org/10.1161/ATVBAHA.109.192583

Dufort FJ, Bleiman BF, Gumina MR, Blair D, Wagner DJ, Roberts MF et al (2007) Cutting edge: IL-4-mediated protection of primary B lymphocytes from apoptosis via Stat6-dependent regulation of glycolytic metabolism. J Immunol 179(8):4953–4957. https://doi.org/10.4049/jimmunol.179.8.4953

Duncan RE, Ahmadian M, Jaworski K, Sarkadi-Nagy E, Sul HS (2007) Regulation of lipolysis in adipocytes. Annu Rev Nutr 27(1):79–101. https://doi.org/10.1146/annurev.nutr.27.061406.093734

Ecker J, Liebisch G, Englmaier M, Grandl M, Robenek H, Schmitz G (2010) Induction of fatty acid synthesis is a key requirement for phagocytic differentiation of human monocytes. Proc Natl Acad Sci U S A 107(17):7817–7822. https://doi.org/10.1073/pnas.0912059107

Elgazar-Carmon V, Rudich A, Hadad N, Levy R (2008) Neutrophils transiently infiltrate intra-abdominal fat early in the course of high-fat feeding. J Lipid Res 49(9):1894–1903. https://doi.org/10.1194/jlr.M800132-JLR200

Everts B, Amiel E, van der Windt GJW, Freitas TC, Chott R, Yarasheski KE et al (2012) Commitment to glycolysis sustains survival of NO-producing inflammatory dendritic cells. Blood 120(7):1422–1431. https://doi.org/10.1182/blood-2012-03-419747

Everts B, Amiel E, Huang SC-C, Smith AM, Chang C-H, Lam WY et al (2014) TLR-driven early glycolytic reprogramming via the kinases TBK1-IKKε supports the anabolic demands of dendritic cell activation. Nat Immunol 15(4):323–332. https://doi.org/10.1038/ni.2833

Ferus M, Pietrucci F, Saitta AM, Knížek A, Kubelík P, Ivanek O et al (2017) Formation of nucleobases in a Miller–Urey reducing atmosphere. Proc Natl Acad Sci U S A 114(17):4306–4311. https://doi.org/10.1073/pnas.1700010114

Feuerer M, Herrero L, Cipolletta D, Naaz A, Wong J, Nayer A et al (2009) Lean, but not obese, fat is enriched for a unique population of regulatory T cells that affect metabolic parameters. Nat Med 15(8):930–939. https://doi.org/10.1038/nm.2002

Filippi M, Bar-Or A, Piehl F, Preziosa P, Solari A, Vukusic S, Rocca MA (2018) Multiple sclerosis. Nat Rev Dis Primers 4(1):43. https://doi.org/10.1038/s41572-018-0041-4

Fox CJ, Hammerman PS, Thompson CB (2005) Fuel feeds function: energy metabolism and the T-cell response. Nat Rev Immunol 5(11):844–852. https://doi.org/10.1038/nri1710

Francisco V, Pino J, Campos-Cabaleiro V, Ruiz-Fernández C, Mera A, Gonzalez-Gay MA et al (2018) Obesity, fat mass and immune system: role for leptin. Front Physiol 9:640. https://doi.org/10.3389/fphys.2018.00640

Fujisaka S, Usui I, Bukhari A, Ikutani M, Oya T, Kanatani Y et al (2009) Regulatory mechanisms for adipose tissue M1 and M2 macrophages in diet-induced obese mice. Diabetes 58(11):2574–2582. https://doi.org/10.2337/db08-1475

Fukuzumi M, Shinomiya H, Shimizu Y, Ohishi K, Utsumi S (1996) Endotoxin-induced enhancement of glucose influx into murine peritoneal macrophages via GLUT1. Infect Immun 64(1):108–112. https://doi.org/10.1128/IAI.64.1.108-112.1996

Galgani M, Procaccini C, De Rosa V, Carbone F, Chieffi P, La Cava A, Matarese G (2010) Leptin modulates the survival of autoreactive CD4 + T cells through the nutrient/energy-sensing mammalian target of rapamycin signaling pathway. J Immunol 185(12):7474–7479. https://doi.org/10.4049/jimmunol.1001674

Gallio M, Sturgill G, Rather P, Kylsten P (2002) A conserved mechanism for extracellular signaling in eukaryotes and prokaryotes. Proc Natl Acad Sci U S A 99(19):12208–12213. https://doi.org/10.1073/pnas.192138799

Galluzzi L, Kepp O, Kroemer G (2012) Mitochondria: master regulators of danger signalling. Nat Rev Mol Cell Biol 13(12):780–788. https://doi.org/10.1038/nrm3479

Geiger R, Rieckmann JC, Wolf T, Basso C, Feng Y, Fuhrer T et al (2016) L-arginine modulates T cell metabolism and enhances survival and anti-tumor activity. Cell 167(3):829–842.e13. https://doi.org/10.1016/j.cell.2016.09.031

Geissmann F, Manz MG, Jung S, Sieweke MH, Merad M, Ley K (2010) Development of monocytes, macrophages, and dendritic cells. Science 327(5966):656–661. https://doi.org/10.1126/science.1178331

Gerriets VA, Rathmell JC (2012) Metabolic pathways in T cell fate and function. Trends Immunol 33(4):168–173. https://doi.org/10.1016/j.it.2012.01.010

Gerriets VA, Danzaki K, Kishton RJ, Eisner W, Nichols AG, Saucillo DC et al (2016) Leptin directly promotes T-cell glycolytic metabolism to drive effector T-cell differentiation in a mouse model of autoimmunity. Eur J Immunol 46(8):1970–1983. https://doi.org/10.1002/eji.201545861

Ghanim H, Abuaysheh S, Sia CL, Korzeniewski K, Chaudhuri A, Fernandez-Real JM, Dandona P (2009) Increase in plasma endotoxin concentrations and the expression of toll-like receptors and suppressor of cytokine signaling-3 in mononuclear cells after a high-fat, high-carbohydrate meal: implications for insulin resistance. Diabetes Care 32(12):2281–2287. https://doi.org/10.2337/dc09-0979

Ghesquière B, Wong BW, Kuchnio A, Carmeliet P (2014) Metabolism of stromal and immune cells in health and disease. Nature 511(7508):167–176. https://doi.org/10.1038/nature13312

Gubser PM, Bantug GR, Razik L, Fischer M, Dimeloe S, Hoenger G et al (2013) Rapid effector function of memory CD8+ T cells requires an immediate-early glycolytic switch. Nat Immunol 14(10):1064–1072. https://doi.org/10.1038/ni.2687

Harmon C, O'Farrelly C, Robinson MW (2020) The immune consequences of lactate in the tumor microenvironment, pp 113–124. https://doi.org/10.1007/978-3-030-43093-1_7

Haschemi A, Kosma P, Gille L, Evans CR, Burant CF, Starkl P et al (2012) The sedoheptulose kinase CARKL directs macrophage polarization through control of glucose metabolism. Cell Metab 15(6):813–826. https://doi.org/10.1016/j.cmet.2012.04.023

Herzig S, Shaw RJ (2018) AMPK: guardian of metabolism and mitochondrial homeostasis. Nat Rev Mol Cell Biol 19(2):121–135. https://doi.org/10.1038/nrm.2017.95

Hirata T, Narumiya S (2012) Prostanoids as regulators of innate and adaptive immunity, pp 143–174. https://doi.org/10.1016/B978-0-12-394300-2.00005-3

Hirosumi J, Tuncman G, Chang L, Görgün CZ, Uysal KT, Maeda K et al (2002) A central role for JNK in obesity and insulin resistance. Nature 420(6913):333–336. https://doi.org/10.1038/nature01137

Huang D, Li T, Wang L, Zhang L, Yan R, Li K et al (2016) Hepatocellular carcinoma redirects to ketolysis for progression under nutrition deprivation stress. Cell Res 26(10):1112–1130. https://doi.org/10.1038/cr.2016.109

Huang B, Song B, Xu C (2020) Cholesterol metabolism in cancer: mechanisms and therapeutic opportunities. Nat Metab 2(2):132–141. https://doi.org/10.1038/s42255-020-0174-0

Huber J, Kiefer FW, Zeyda M, Ludvik B, Silberhumer GR, Prager G et al (2008) CC chemokine and CC chemokine receptor profiles in visceral and subcutaneous adipose tissue are altered in human obesity. J Clin Endocrinol Metab 93(8):3215–3221. https://doi.org/10.1210/jc.2007-2630

Huppke B, Ellenberger D, Hummel H, Stark W, Röbl M, Gärtner J, Huppke P (2019) Association of obesity with multiple sclerosis risk and response to first-line disease modifying drugs in children. JAMA Neurol 76(10):1157. https://doi.org/10.1001/jamaneurol.2019.1997

Hurley HJ, Dewald H, Rothkopf ZS, Singh S, Jenkins F, Deb P et al (2021) Frontline science: AMPK regulates metabolic reprogramming necessary for interferon production in human plasmacytoid dendritic cells. J Leukoc Biol 109(2):299–308. https://doi.org/10.1002/JLB.3HI0220-130

Huynh A, DuPage M, Priyadharshini B, Sage PT, Quiros J, Borges CM et al (2015) Control of PI(3) kinase in Treg cells maintains homeostasis and lineage stability. Nat Immunol 16(2):188–196. https://doi.org/10.1038/ni.3077

Infantino V, Convertini P, Cucci L, Panaro MA, Di Noia MA, Calvello R et al (2011) The mitochondrial citrate carrier: a new player in inflammation. Biochem J 438(3):433–436. https://doi.org/10.1042/BJ20111275

Jha AK, Huang SC-C, Sergushichev A, Lampropoulou V, Ivanova Y, Loginicheva E et al (2015) Network integration of parallel metabolic and transcriptional data reveals metabolic modules that regulate macrophage polarization. Immunity 42(3):419–430. https://doi.org/10.1016/j.immuni.2015.02.005

Jiao P, Chen Q, Shah S, Du J, Tao B, Tzameli I et al (2009) Obesity-related upregulation of monocyte chemotactic factors in adipocytes: involvement of nuclear factor- B and c-Jun NH2-terminal kinase pathways. Diabetes 58(1):104–115. https://doi.org/10.2337/db07-1344

Jones RG, Thompson CB (2007) Revving the engine: signal transduction fuels T cell activation. Immunity 27(2):173–178. https://doi.org/10.1016/j.immuni.2007.07.008

Josefowicz SZ, Lu L-F, Rudensky AY (2012) Regulatory T cells: mechanisms of differentiation and function. Annu Rev Immunol 30(1):531–564. https://doi.org/10.1146/annurev.immunol.25.022106.141623

Jung J, Zeng H, Horng T (2019) Metabolism as a guiding force for immunity. Nat Cell Biol 21(1):85–93. https://doi.org/10.1038/s41556-018-0217-x

Kahn SE (2003) The relative contributions of insulin resistance and beta-cell dysfunction to the pathophysiology of type 2 diabetes. Diabetologia 46(1):3–19. https://doi.org/10.1007/s00125-002-1009-0

Kanda H (2006) MCP-1 contributes to macrophage infiltration into adipose tissue, insulin resistance, and hepatic steatosis in obesity. J Clin Invest 116(6):1494–1505. https://doi.org/10.1172/JCI26498

Kang K, Reilly SM, Karabacak V, Gangl MR, Fitzgerald K, Hatano B, Lee C-H (2008) Adipocyte-derived Th2 cytokines and myeloid PPAR$δ$ regulate macrophage polarization and insulin sensitivity. Cell Metab 7(6):485–495. https://doi.org/10.1016/j.cmet.2008.04.002

Kanter JE, Kramer F, Barnhart S, Averill MM, Vivekanandan-Giri A, Vickery T et al (2012) Diabetes promotes an inflammatory macrophage phenotype and atherosclerosis through acyl-CoA synthetase 1. Proc Natl Acad Sci U S A 109(12):E715–E724. https://doi.org/10.1073/pnas.1111600109

Karayama M, Masuda J, Mori K, Yasui H, Hozumi H, Suzuki Y et al (2021) Comprehensive assessment of multiple tryptophan metabolites as potential biomarkers for immune checkpoint inhibitors in patients with non-small cell lung cancer. Clin Transl Oncol 23(2):418–423. https://doi.org/10.1007/s12094-020-02421-8

Kareva I (2019) Metabolism and gut microbiota in cancer immunoediting, CD8/Treg ratios, immune cell homeostasis, and cancer (immuno)therapy: concise review. Stem Cells 37(10):1273–1280. https://doi.org/10.1002/stem.3051

Kirchner T, Möller S, Klinger M, Solbach W, Laskay T, Behnen M (2012) The impact of various reactive oxygen species on the formation of neutrophil extracellular traps. Mediat Inflamm 2012:849136. https://doi.org/10.1155/2012/849136
Klein Geltink RI, Kyle RL, Pearce EL (2018) Unraveling the complex interplay between T cell metabolism and function. Annu Rev Immunol 36(1):461–488. https://doi.org/10.1146/annurev-immunol-042617-053019
Kolb D, Kolishetti N, Surnar B, Sarkar S, Guin S, Shah AS, Dhar S (2020) Metabolic modulation of the tumor microenvironment leads to multiple checkpoint inhibition and immune cell infiltration. ACS Nano 14(9):11055–11066. https://doi.org/10.1021/acsnano.9b10037
Kominsky DJ, Campbell EL, Colgan SP (2010) Metabolic shifts in immunity and inflammation. J Immunol 184(8):4062–4068. https://doi.org/10.4049/jimmunol.0903002
Krawczyk CM, Holowka T, Sun J, Blagih J, Amiel E, DeBerardinis RJ et al (2010) Toll-like receptor–induced changes in glycolytic metabolism regulate dendritic cell activation. Blood 115(23):4742–4749. https://doi.org/10.1182/blood-2009-10-249540
Laplante M, Sabatini DM (2009) An emerging role of mTOR in lipid biosynthesis. Curr Biol 19(22):R1046–R1052. https://doi.org/10.1016/j.cub.2009.09.058
Lebovitz HE, Banerji MA (2005) Point: visceral adiposity is causally related to insulin resistance. Diabetes Care 28(9):2322–2325. https://doi.org/10.2337/diacare.28.9.2322
Lee JY, Zhao L, Youn HS, Weatherill AR, Tapping R, Feng L et al (2004) Saturated fatty acid activates but polyunsaturated fatty acid inhibits toll-like receptor 2 dimerized with toll-like receptor 6 or 1. J Biol Chem 279(17):16971–16979. https://doi.org/10.1074/jbc.M312990200
Lee C-H, Chen Y-G, Chen J, Reifsnyder PC, Serreze DV, Clare-Salzler M et al (2006) Novel leptin receptor mutation in NOD/LtJ mice suppresses type 1 diabetes progression: II. Immunologic analysis. Diabetes 55(1):171–178. https://doi.org/10.2337/diabetes.55.01.06.db05-1129
Lercher A, Baazim H, Bergthaler A (2020) Systemic Immunometabolism: challenges and opportunities. Immunity 53(3):496–509. https://doi.org/10.1016/j.immuni.2020.08.012
Levene PA, Meyer GM (1912) The action of leucocytes on glucose. J Biol Chem 11(4):361–370. https://doi.org/10.1016/S0021-9258(18)88742-7
Lin J, Wu H, Tarr PT, Zhang C-Y, Wu Z, Boss O et al (2002) Transcriptional co-activator PGC-1α drives the formation of slow-twitch muscle fibres. Nature 418(6899):797–801. https://doi.org/10.1038/nature00904
Lio CJ, Huang SC (2020) Circles of life: linking metabolic and epigenetic cycles to immunity. Immunology 161(3):165–174. https://doi.org/10.1111/imm.13207
Liu J, Divoux A, Sun J, Zhang J, Clément K, Glickman JN et al (2009) Genetic deficiency and pharmacological stabilization of mast cells reduce diet-induced obesity and diabetes in mice. Nat Med 15(8):940–945. https://doi.org/10.1038/nm.1994
Lumeng CN, Bodzin JL, Saltiel AR (2007) Obesity induces a phenotypic switch in adipose tissue macrophage polarization. J Clin Invest 117(1):175–184. https://doi.org/10.1172/JCI29881
Lynch L, Nowak M, Varghese B, Clark J, Hogan AE, Toxavidis V et al (2012) Adipose tissue invariant NKT cells protect against diet-induced obesity and metabolic disorder through regulatory cytokine production. Immunity 37(3):574–587. https://doi.org/10.1016/j.immuni.2012.06.016
Maitra U, Deng H, Glaros T, Baker B, Capelluto DGS, Li Z, Li L (2012) Molecular mechanisms responsible for the selective and low-grade induction of proinflammatory mediators in murine macrophages by lipopolysaccharide. J Immunol 189(2):1014–1023. https://doi.org/10.4049/jimmunol.1200857
Mansuy-Aubert V, Zhou QL, Xie X, Gong Z, Huang J-Y, Khan AR et al (2013) Imbalance between neutrophil elastase and its inhibitor α1-antitrypsin in obesity alters insulin sensitivity, inflammation, and energy expenditure. Cell Metab 17(4):534–548. https://doi.org/10.1016/j.cmet.2013.03.005
Mashimo T, Pichumani K, Vemireddy V, Hatanpaa KJ, Singh DK, Sirasanagandla S et al (2014) Acetate is a bioenergetic substrate for human glioblastoma and brain metastases. Cell 159(7):1603–1614. https://doi.org/10.1016/j.cell.2014.11.025

Mason SD, Howlett RA, Kim MJ, Olfert IM, Hogan MC, McNulty W et al (2004) Loss of skeletal muscle HIF-1α results in altered exercise endurance. PLoS Biol 2(10):e288. https://doi.org/10.1371/journal.pbio.0020288
Matarese G, Di Giacomo A, Sanna V, Lord GM, Howard JK, Di Tuoro A et al (2001) Requirement for leptin in the induction and progression of autoimmune encephalomyelitis. J Immunol 166(10):5909–5916. https://doi.org/10.4049/jimmunol.166.10.5909
Matarese G, Sanna V, Lechler RI, Sarvetnick N, Fontana S, Zappacosta S, La Cava A (2002) Leptin accelerates autoimmune diabetes in female NOD mice. Diabetes 51(5):1356–1361. https://doi.org/10.2337/diabetes.51.5.1356
Maynard CL, Elson CO, Hatton RD, Weaver CT (2012) Reciprocal interactions of the intestinal microbiota and immune system. Nature 489(7415):231–241. https://doi.org/10.1038/nature11551
Mehran AE, Templeman NM, Brigidi GS, Lim GE, Chu K-Y, Hu X et al (2012) Hyperinsulinemia drives diet-induced obesity independently of brain insulin production. Cell Metab 16(6):723–737. https://doi.org/10.1016/j.cmet.2012.10.019
Michalek RD, Gerriets VA, Jacobs SR, Macintyre AN, MacIver NJ, Mason EF et al (2011) Cutting edge: distinct glycolytic and lipid oxidative metabolic programs are essential for effector and regulatory CD4 + T cell subsets. J Immunol 186(6):3299–3303. https://doi.org/10.4049/jimmunol.1003613
Michelucci A, Cordes T, Ghelfi J, Pailot A, Reiling N, Goldmann O et al (2013) Immune-responsive gene 1 protein links metabolism to immunity by catalyzing itaconic acid production. Proc Natl Acad Sci U S A 110(19):7820–7825. https://doi.org/10.1073/pnas.1218599110
Moon J-S, Lee S, Park M-A, Siempos II, Haslip M, Lee PJ et al (2015) UCP2-induced fatty acid synthase promotes NLRP3 inflammasome activation during sepsis. J Clin Invest 125(2):665–680. https://doi.org/10.1172/JCI78253
Moon J-Y, Zolnik CP, Wang Z, Qiu Y, Usyk M, Wang T et al (2018) Gut microbiota and plasma metabolites associated with diabetes in women with, or at high risk for, HIV infection. EBioMedicine 37:392–400. https://doi.org/10.1016/j.ebiom.2018.10.037
Morris DL, Singer K, Lumeng CN (2011) Adipose tissue macrophages: phenotypic plasticity and diversity in lean and obese states. Curr Opin Clin Nutr Metab Care 14(4):341–346. https://doi.org/10.1097/MCO.0b013e328347970b
Muehlenbein MP, Hirschtick JL, Bonner JZ, Swartz AM (2010) Toward quantifying the usage costs of human immunity: altered metabolic rates and hormone levels during acute immune activation in men. Am J Hum Biol 22(4):546–556. https://doi.org/10.1002/ajhb.21045
Munger KL, Chitnis T, Ascherio A (2009) Body size and risk of MS in two cohorts of US women. Neurology 73(19):1543–1550. https://doi.org/10.1212/WNL.0b013e3181c0d6e0
Muoio DM, Newgard CB (2008) Molecular and metabolic mechanisms of insulin resistance and β-cell failure in type 2 diabetes. Nat Rev Mol Cell Biol 9(3):193–205. https://doi.org/10.1038/nrm2327
Murano I, Barbatelli G, Parisani V, Latini C, Muzzonigro G, Castellucci M, Cinti S (2008) Dead adipocytes, detected as crown-like structures, are prevalent in visceral fat depots of genetically obese mice. J Lipid Res 49(7):1562–1568. https://doi.org/10.1194/jlr.M800019-JLR200
Murray PJ, Wynn TA (2011) Protective and pathogenic functions of macrophage subsets. Nat Rev Immunol 11(11):723–737. https://doi.org/10.1038/nri3073
Nakahira K, Haspel JA, Rathinam VAK, Lee S-J, Dolinay T, Lam HC et al (2011) Autophagy proteins regulate innate immune responses by inhibiting the release of mitochondrial DNA mediated by the NALP3 inflammasome. Nat Immunol 12(3):222–230. https://doi.org/10.1038/ni.1980
Nakaya M, Xiao Y, Zhou X, Chang J-H, Chang M, Cheng X et al (2014) Inflammatory T cell responses rely on amino acid transporter ASCT2 facilitation of glutamine uptake and mTORC1 kinase activation. Immunity 40(5):692–705. https://doi.org/10.1016/j.immuni.2014.04.007
Nara N, Nakayama Y, Okamoto S, Tamura H, Kiyono M, Muraoka M et al (2007) Disruption of CXC motif chemokine Ligand-14 in mice ameliorates obesity-induced insulin resistance. J Biol Chem 282(42):30794–30803. https://doi.org/10.1074/jbc.M700412200

Narkar VA, Downes M, Yu RT, Embler E, Wang Y-X, Banayo E et al (2008) AMPK and PPARδ agonists are exercise mimetics. Cell 134(3):405–415. https://doi.org/10.1016/j.cell.2008.06.051

Newsholme P, Curi R, Gordon S, Newsholme EA (1986) Metabolism of glucose, glutamine, long-chain fatty acids and ketone bodies by murine macrophages. Biochem J 239(1):121–125. https://doi.org/10.1042/bj2390121

Nishimura S, Manabe I, Nagasaki M, Eto K, Yamashita H, Ohsugi M et al (2009) CD8+ effector T cells contribute to macrophage recruitment and adipose tissue inflammation in obesity. Nat Med 15(8):914–920. https://doi.org/10.1038/nm.1964

O'Neill LAJ, Hardie DG (2013) Metabolism of inflammation limited by AMPK and pseudo-starvation. Nature 493(7432):346–355. https://doi.org/10.1038/nature11862

O'Neill LAJ, Kishton RJ, Rathmell J (2016) A guide to immunometabolism for immunologists. Nat Rev Immunol 16(9):553–565. https://doi.org/10.1038/nri.2016.70

Odegaard JI, Chawla A (2011) Alternative macrophage activation and metabolism. Annu Rev Pathol 6(1):275–297. https://doi.org/10.1146/annurev-pathol-011110-130138

Odegaard JI, Ricardo-Gonzalez RR, Goforth MH, Morel CR, Subramanian V, Mukundan L et al (2007) Macrophage-specific PPARγ controls alternative activation and improves insulin resistance. Nature 447(7148):1116–1120. https://doi.org/10.1038/nature05894

Odegaard JI, Ricardo-Gonzalez RR, Red Eagle A, Vats D, Morel CR, Goforth MH et al (2008) Alternative M2 activation of Kupffer cells by PPARδ ameliorates obesity-induced insulin resistance. Cell Metab 7(6):496–507. https://doi.org/10.1016/j.cmet.2008.04.003

Oren R, Farnham AE, Saito K, Milofsky E, Karnovsky ML (1963) Metabolic patterns in three types of phagocytizing cells. J Cell Biol 17(3):487–501. https://doi.org/10.1083/jcb.17.3.487

Palmer CS, Ostrowski M, Balderson B, Christian N, Crowe SM (2015) Glucose metabolism regulates T cell activation, differentiation, and functions. Front Immunol 6:1. https://doi.org/10.3389/fimmu.2015.00001

Parks BW, Nam E, Org E, Kostem E, Norheim F, Hui ST et al (2013) Genetic control of obesity and gut microbiota composition in response to high-fat, high-sucrose diet in mice. Cell Metab 17(1):141–152. https://doi.org/10.1016/j.cmet.2012.12.007

Pearce EJ, Everts B (2015) Dendritic cell metabolism. Nat Rev Immunol 15(1):18–29. https://doi.org/10.1038/nri3771

Pearce EL, Pearce EJ (2013) Metabolic pathways in immune cell activation and quiescence. Immunity 38(4):633–643. https://doi.org/10.1016/j.immuni.2013.04.005

Picard M, Hepple RT, Burelle Y (2012) Mitochondrial functional specialization in glycolytic and oxidative muscle fibers: tailoring the organelle for optimal function. Am J Phys Cell Phys 302(4):C629–C641. https://doi.org/10.1152/ajpcell.00368.2011

Powell JD, Pollizzi KN, Heikamp EB, Horton MR (2012) Regulation of immune responses by mTOR. Annu Rev Immunol 30(1):39–68. https://doi.org/10.1146/annurev-immunol-020711-075024

Prentki M (2006) Islet cell failure in type 2 diabetes. J Clin Invest 116(7):1802–1812. https://doi.org/10.1172/JCI29103

Procaccini C, De Rosa V, Galgani M, Abanni L, Calì G, Porcellini A et al (2010) An oscillatory switch in mTOR kinase activity sets regulatory T cell responsiveness. Immunity 33(6):929–941. https://doi.org/10.1016/j.immuni.2010.11.024

Procaccini C, Carbone F, Galgani M, La Rocca C, De Rosa V, Cassano S, Matarese G (2011) Obesity and susceptibility to autoimmune diseases. Expert Rev Clin Immunol 7(3):287–294. https://doi.org/10.1586/eci.11.18

Procaccini C, Pucino V, Mantzoros CS, Matarese G (2015) Leptin in autoimmune diseases. Metabolism 64(1):92–104. https://doi.org/10.1016/j.metabol.2014.10.014

Qian X, Yang Z, Mao E, Chen E (2018) Regulation of fatty acid synthesis in immune cells. Scand J Immunol 88(5):e12713. https://doi.org/10.1111/sji.12713

Qiu J, Villa M, Sanin DE, Buck MD, O'Sullivan D, Ching R et al (2019) Acetate promotes T cell effector function during glucose restriction. Cell Rep 27(7):2063–2074.e5. https://doi.org/10.1016/j.celrep.2019.04.022

Ramapriyan R, Caetano MS, Barsoumian HB, Mafra ACP, Zambalde EP, Menon H et al (2019) Altered cancer metabolism in mechanisms of immunotherapy resistance. Pharmacol Therap 195:162–171. https://doi.org/10.1016/j.pharmthera.2018.11.004
Remijsen Q, Kuijpers TW, Wirawan E, Lippens S, Vandenabeele P, Vanden Berghe T (2011) Dying for a cause: NETosis, mechanisms behind an antimicrobial cell death modality. Cell Death Diff 18(4):581–588. https://doi.org/10.1038/cdd.2011.1
Rodríguez C, Puente-Moncada N, Reiter RJ, Sánchez-Sánchez AM, Herrera F, Rodríguez-Blanco J et al (2021) Regulation of cancer cell glucose metabolism is determinant for cancer cell fate after melatonin administration. J Cell Physiol 236(1):27–40. https://doi.org/10.1002/jcp.29886
Rodríguez-Prados J-C, Través PG, Cuenca J, Rico D, Aragonés J, Martín-Sanz P et al (2010) Substrate fate in activated macrophages: a comparison between innate, classic, and alternative activation. J Immunol 185(1):605–614. https://doi.org/10.4049/jimmunol.0901698
Satpathy AT, Wu X, Albring JC, Murphy KM (2012) Re(de)fining the dendritic cell lineage. Nat Immunol 13(12):1145–1154. https://doi.org/10.1038/ni.2467
Saxton RA, Sabatini DM (2017) mTOR signaling in growth, metabolism, and disease. Cell 168(6):960–976. https://doi.org/10.1016/j.cell.2017.02.004
Sbarra AJ, Karnovsky ML (1959) The biochemical basis of phagocytosis. J Biol Chem 234(6):1355–1362. https://doi.org/10.1016/S0021-9258(18)70011-2
Schipper HS, Rakhshandehroo M, van de Graaf SFJ, Venken K, Koppen A, Stienstra R et al (2012) Natural killer T cells in adipose tissue prevent insulin resistance. J Clin Invest 122(9):3343–3354. https://doi.org/10.1172/JCI62739
Schreurs M, Kuipers F, van der Leij FR (2010) Regulatory enzymes of mitochondrial β-oxidation as targets for treatment of the metabolic syndrome. Obes Rev 11(5):380–388. https://doi.org/10.1111/j.1467-789X.2009.00642.x
Schug ZT, Peck B, Jones DT, Zhang Q, Grosskurth S, Alam IS et al (2015) Acetyl-CoA Synthetase 2 promotes acetate utilization and maintains cancer cell growth under metabolic stress. Cancer Cell 27(1):57–71. https://doi.org/10.1016/j.ccell.2014.12.002
Schug ZT, Vande Voorde J, Gottlieb E (2016) The metabolic fate of acetate in cancer. Nat Rev Cancer 16(11):708–717. https://doi.org/10.1038/nrc.2016.87
Schulz H (2013) Fatty acid oxidation. In: Encyclopedia of biological chemistry. Elsevier, pp 281–284. https://doi.org/10.1016/B978-0-12-378630-2.00071-2
Schulz C, Perdiguero EG, Chorro L, Szabo-Rogers H, Cagnard N, Kierdorf K et al (2012) A lineage of myeloid cells independent of Myb and hematopoietic stem cells. Science 336(6077):86–90. https://doi.org/10.1126/science.1219179
Seth RB, Sun L, Chen ZJ (2006) Antiviral innate immunity pathways. Cell Res 16(2):141–147. https://doi.org/10.1038/sj.cr.7310019
Shi H, Kokoeva MV, Inouye K, Tzameli I, Yin H, Flier JS (2006) TLR4 links innate immunity and fatty acid–induced insulin resistance. J Clin Invest 116(11):3015–3025. https://doi.org/10.1172/JCI28898
Shi LZ, Wang R, Huang G, Vogel P, Neale G, Green DR, Chi H (2011) HIF1α–dependent glycolytic pathway orchestrates a metabolic checkpoint for the differentiation of TH17 and Treg cells. J Exp Med 208(7):1367–1376. https://doi.org/10.1084/jem.20110278
Shyer JA, Flavell RA, Bailis W (2020) Metabolic signaling in T cells. Cell Res 30(8):649–659. https://doi.org/10.1038/s41422-020-0379-5
Singer K, Kastenberger M, Gottfried E, Hammerschmied CG, Büttner M, Aigner M et al (2011) Warburg phenotype in renal cell carcinoma: high expression of glucose-transporter 1 (GLUT-1) correlates with low CD8+ T-cell infiltration in the tumor. Int J Cancer 128(9):2085–2095. https://doi.org/10.1002/ijc.25543
Sorbara MT, Girardin SE (2011) Mitochondrial ROS fuel the inflammasome. Cell Res 21(4):558–560. https://doi.org/10.1038/cr.2011.20
Sousa Gerós A, Simmons A, Drakesmith H, Aulicino A, Frost JN (2020) The battle for iron in enteric infections. Immunology 161(3):186–199. https://doi.org/10.1111/imm.13236

Stathopoulou C, Nikoleri D, Bertsias G (2019) Immunometabolism: an overview and therapeutic prospects in autoimmune diseases. Immunotherapy 11(9):813–829. https://doi.org/10.2217/imt-2019-0002

Stefanovic-Racic M, Yang X, Turner MS, Mantell BS, Stolz DB, Sumpter TL et al (2012) Dendritic cells promote macrophage infiltration and comprise a substantial proportion of obesity-associated increases in CD11c+ cells in adipose tissue and liver. Diabetes 61(9):2330–2339. https://doi.org/10.2337/db11-1523

Sukumar M, Roychoudhuri R, Restifo NP (2015) Nutrient competition: a new axis of tumor immunosuppression. Cell 162(6):1206–1208. https://doi.org/10.1016/j.cell.2015.08.064

Sumbayev VV, Nicholas SA, Streatfield CL, Gibbs BF (2009) Involvement of hypoxia-inducible factor-1 HiF(1α) in IgE-mediated primary human basophil responses. Eur J Immunol 39(12):3511–3519. https://doi.org/10.1002/eji.200939370

Sun L, Song L, Wan Q, Wu G, Li X, Wang Y et al (2015) cMyc-mediated activation of serine biosynthesis pathway is critical for cancer progression under nutrient deprivation conditions. Cell Res 25(4):429–444. https://doi.org/10.1038/cr.2015.33

Tahiliani M, Koh KP, Shen Y, Pastor WA, Bandukwala H, Brudno Y et al (2009) Conversion of 5-methylcytosine to 5-hydroxymethylcytosine in mammalian DNA by MLL partner TET1. Science 324(5929):930–935. https://doi.org/10.1126/science.1170116

Talukdar S, Oh DY, Bandyopadhyay G, Li D, Xu J, McNelis J et al (2012) Neutrophils mediate insulin resistance in mice fed a high-fat diet through secreted elastase. Nat Med 18(9):1407–1412. https://doi.org/10.1038/nm.2885

Tannahill GM, Curtis AM, Adamik J, Palsson-McDermott EM, McGettrick AF, Goel G et al (2013) Succinate is an inflammatory signal that induces IL-1β through HIF-1α. Nature 496(7444):238–242. https://doi.org/10.1038/nature11986

Teijlingen Bakker N, Pearce EJ (2020) Cell-intrinsic metabolic regulation of mononuclear phagocyte activation: findings from the tip of the iceberg. Immunol Rev 295(1):54–67. https://doi.org/10.1111/imr.12848

Terry S, Engelsen AST, Buart S, Elsayed WS, Venkatesh GH, Chouaib S (2020) Hypoxia-driven intratumor heterogeneity and immune evasion. Cancer Lett 492:1–10. https://doi.org/10.1016/j.canlet.2020.07.004

Thompson CB (2011) Rethinking the regulation of cellular metabolism. Cold Spring Harb Symp Quant Biol 76:23–29. https://doi.org/10.1101/sqb.2012.76.010496

Togo M, Yokobori T, Shimizu K, Handa T, Kaira K, Sano T et al (2020) Diagnostic value of 18F-FDG-PET to predict the tumour immune status defined by tumoural PD-L1 and CD8+tumour-infiltrating lymphocytes in oral squamous cell carcinoma. Br J Cancer 122(11):1686–1694. https://doi.org/10.1038/s41416-020-0820-z

Tschopp J, Schroder K (2010) NLRP3 inflammasome activation: the convergence of multiple signalling pathways on ROS production? Nat Rev Immunol 10(3):210–215. https://doi.org/10.1038/nri2725

Turnbaugh PJ, Bäckhed F, Fulton L, Gordon JI (2008) Diet-induced obesity is linked to marked but reversible alterations in the mouse distal gut microbiome. Cell Host Microbe 3(4):213–223. https://doi.org/10.1016/j.chom.2008.02.015

Valentine WN (1960) The metabolism of the leukemic leukocyte. Am J Med 28(5):699–710. https://doi.org/10.1016/0002-9343(60)90128-5

van der Windt GJW, Everts B, Chang C-H, Curtis JD, Freitas TC, Amiel E et al (2012) Mitochondrial respiratory capacity is a critical regulator of CD8+ T cell memory development. Immunity 36(1):68–78. https://doi.org/10.1016/j.immuni.2011.12.007

van Raam B, Verhoeven A, Kuijpers T (2006) Mitochondria in neutrophil apoptosis. Int J Hematol 84(3):199–204. https://doi.org/10.1532/IJH97.06131

van Raam BJ, Sluiter W, de Wit E, Roos D, Verhoeven AJ, Kuijpers TW (2008) Mitochondrial membrane potential in human neutrophils is maintained by complex III activity in the absence of supercomplex organisation. PLoS One 3(4):e2013. https://doi.org/10.1371/journal.pone.0002013

van Weeghel M, Abdurrachim D, Nederlof R, Argmann CA, Houtkooper RH, Hagen J et al (2018) Increased cardiac fatty acid oxidation in a mouse model with decreased malonyl-CoA sensitivity of CPT1B. Cardiovasc Res 114(10):1324–1334. https://doi.org/10.1093/cvr/cvy089

Vander Heiden MG, Cantley LC, Thompson CB (2009) Understanding the Warburg effect: the metabolic requirements of cell proliferation. Science 324(5930):1029–1033. https://doi.org/10.1126/science.1160809

Vats D, Mukundan L, Odegaard JI, Zhang L, Smith KL, Morel CR et al (2006) Oxidative metabolism and PGC-1β attenuate macrophage-mediated inflammation. Cell Metab 4(1):13–24. https://doi.org/10.1016/j.cmet.2006.05.011

Venge P, Moberg L, Björnsson E, Bergström M, Långström B, Håkansson L (2003) Mechanisms of basal and cytokine-induced uptake of glucose in normal human eosinophils: relation to apoptosis. Respir Med 97(10):1109–1119. https://doi.org/10.1016/S0954-6111(03)00143-4

Waickman AT, Powell JD (2012) mTOR, metabolism, and the regulation of T-cell differentiation and function. Immunol Rev 249(1):43–58. https://doi.org/10.1111/j.1600-065X.2012.01152.x

Wang R, Green DR (2012) Metabolic checkpoints in activated T cells. Nat Immunol 13(10):907–915. https://doi.org/10.1038/ni.2386

Wang Y-X, Zhang C-L, Yu RT, Cho HK, Nelson MC, Bayuga-Ocampo CR et al (2004) Regulation of muscle fiber type and running endurance by PPARδ. PLoS Biol 2(10):e294. https://doi.org/10.1371/journal.pbio.0020294

Wang Y, Li J, Tang L, Wang Y, Charnigo R, de Villiers W, Eckhardt E (2010) T-lymphocyte responses to intestinally absorbed antigens can contribute to adipose tissue inflammation and glucose intolerance during high fat feeding. PLoS One 5(11):e13951. https://doi.org/10.1371/journal.pone.0013951

Wang X, Lin Y, Liu S, Zhu Y, Lu K, Broering R, Lu M (2020) O -GlcNAcylation modulates HBV replication through regulating cellular autophagy at multiple levels. FASEB J 34(11):14473–14489. https://doi.org/10.1096/fj.202001168RR

Wei J, Liu B, Fan S, Li H, Chen M, Zhang B et al (2017) Differentially expressed immune-related genes in hemocytes of the pearl oyster Pinctada fucata against allograft identified by transcriptome analysis. Fish Shellfish Immunol 62:247–256. https://doi.org/10.1016/j.fsi.2017.01.025

Weisberg SP, McCann D, Desai M, Rosenbaum M, Leibel RL, Ferrante AW (2003) Obesity is associated with macrophage accumulation in adipose tissue. J Clin Invest 112(12):1796–1808. https://doi.org/10.1172/JCI19246

Wen H, Gris D, Lei Y, Jha S, Zhang L, Huang MT-H et al (2011) Fatty acid–induced NLRP3-ASC inflammasome activation interferes with insulin signaling. Nat Immunol 12(5):408–415. https://doi.org/10.1038/ni.2022

West AP, Brodsky IE, Rahner C, Woo DK, Erdjument-Bromage H, Tempst P et al (2011) TLR signalling augments macrophage bactericidal activity through mitochondrial ROS. Nature 472(7344):476–480. https://doi.org/10.1038/nature09973

Winer S, Chan Y, Paltser G, Truong D, Tsui H, Bahrami J et al (2009) Normalization of obesity-associated insulin resistance through immunotherapy. Nat Med 15(8):921–929. https://doi.org/10.1038/nm.2001

Winer DA, Winer S, Shen L, Wadia PP, Yantha J, Paltser G et al (2011) B cells promote insulin resistance through modulation of T cells and production of pathogenic IgG antibodies. Nat Med 17(5):610–617. https://doi.org/10.1038/nm.2353

Wu D, Molofsky AB, Liang H-E, Ricardo-Gonzalez RR, Jouihan HA, Bando JK et al (2011) Eosinophils sustain adipose alternatively activated macrophages associated with glucose homeostasis. Science 332(6026):243–247. https://doi.org/10.1126/science.1201475

Yan Y, Chang L, Tian H, Wang L, Zhang Y, Yang T et al (2018) 1-pyrroline-5-carboxylate released by prostate cancer cell inhibit T cell proliferation and function by targeting SHP1/cytochrome c oxidoreductase/ROS axis. J Immunother Cancer 6(1):148. https://doi.org/10.1186/s40425-018-0466-z

Yang H, Youm Y-H, Vandanmagsar B, Ravussin A, Gimble JM, Greenway F et al (2010) Obesity increases the production of proinflammatory mediators from adipose tissue T cells and com-

promises TCR repertoire diversity: implications for systemic inflammation and insulin resistance. J Immunol 185(3):1836–1845. https://doi.org/10.4049/jimmunol.1000021
Yuan M (2001) Reversal of obesity- and diet-induced insulin resistance with salicylates or targeted disruption of Ikkbeta. Science 293(5535):1673–1677. https://doi.org/10.1126/science.1061620
Zemirli N, Arnoult D (2012) Mitochondrial anti-viral immunity. Int J Biochem Cell Biol 44(9):1473–1476. https://doi.org/10.1016/j.biocel.2012.05.018
Zhang HH, Halbleib M, Ahmad F, Manganiello VC, Greenberg AS (2002) Tumor necrosis factor- stimulates lipolysis in differentiated human adipocytes through activation of extracellular signal-related kinase and elevation of intracellular cAMP. Diabetes 51(10):2929–2935. https://doi.org/10.2337/diabetes.51.10.2929
Zhang Y, Joe G, Hexner E, Zhu J, Emerson SG (2005) Host-reactive CD8+ memory stem cells in graft-versus-host disease. Nat Med 11(12):1299–1305. https://doi.org/10.1038/nm1326
Zhou R, Yazdi AS, Menu P, Tschopp J (2011) A role for mitochondria in NLRP3 inflammasome activation. Nature 469(7329):221–225. https://doi.org/10.1038/nature09663
Zhu Y, Hart GW (2021) Targeting O-GlcNAcylation to develop novel therapeutics. Mol Asp Med 79:100885. https://doi.org/10.1016/j.mam.2020.100885

# Role of Human Milk Immune Modulators on Infant's Health

**Debapriya Mohanty and Snigdha Misra**

**Abstract** Human milk is an excellent source of nutrients along with immune-enhancing components which nourish the infant by promoting their health outcomes. At birth, the newborn's immune system contains minimal anti-inflammatory and anti-infective properties in their gastro-intestinal tract (GIT) with underdeveloped physical and chemical barrier as well as a delayed T-cell activity and decreased immunoglobulin secretion. Exclusive breastfeeding generally enhances newborn's immune system by providing them with several nutrients, immunomodulatory and anti-inflammatory components along with beneficial gut flora which protect them from serious health risk during their early stage of life. Hence, it is strongly recommended for the first 6 months of life. Potential probiotic bacteria and other immune components which are transmitted to newborn's gut through breast feeding also play a crucial part in the GIT. Research has established that the placenta transfers the SARS-CoV-2 antibodies from earlier infected mother to its babies through lactation and can protect the baby against COVID-19 for at least 6 months. This review article mainly focuses on the essential and immunomodulatory and immunostimulatory components found in human milk along with their function in GIT and its protective effects against SARS-CoV2 (COVID-19) in newborns.

**Keywords** Bioactive peptides · Colostrum · Growth factors · Immunomodulator · Lactoferrin · Milk · Probiotics

D. Mohanty (✉)
National Institute of Foot and Mouth Disease, Indian Council of Agricultural Research, Jatni, Odisha, India

S. Misra
Jeffrey Cheah School of Medicine & Health Sciences, Monash University Malaysia, Bandar Sunway, Selangor DE, Malaysia



A. Das Mohapatra, P. S. Sahu (eds.), *Advances in Immunology and Immuno-techniques*, Interdisciplinary Biotechnological Advances,
https://doi.org/10.1007/978-981-97-5508-0_12

# 1 Introduction

Human milk is considered as a natural supply of nutrition for newborns and infants as it provides all the nutrients in the right amounts. It also helps in building and supporting their immune system. It contains antibodies that aid in preventing a number of prevalent pediatric ailments, and it is secure and hygienic. Human colostrum refers to the first milk made following birth (HC), and abundant source of immunoglobulins (Ig), bioactive molecules, growth factors, etc. These molecules are crucial for infants' nutrition, growth, and development as well as passive immunity (Godhia and Patel 2013). It is packed with immunity-boosting factors, particularly oligosaccharides (HMOs), lactose, heterogeneous mixture of proteins (casein and whey), amino acids, serum albumin, lipids, hormones, enzymes, growth factors, nucleotides, bioactive peptides, antibodies, immune-stimulating cells, white blood cells macrophages, anti-inflammatory factors, and several other components which are helpful in fighting against several infections in many different ways, facilitate passive and active immunity, reduce inflammation, maintain gut microbiome, and influence organ growth in newborns.

Interestingly, human milk contains probiotic factors that are transferred to the infant through breast feeding resulting in a healthy gut flora. This in turn may provide a head start to a lifelong protection against gastroenteritis, among infants. They also help to strengthen the immune system. Apart from immunomodulation, probiotics in human milk increases the good bacteria count in a baby's gut and aids in faster recovery from infections. Particularly, human milk's antibacterial activity against a number of bacteria, protozoa, and viruses may lower the prevalence of infection in young children (Morrow et al. 2004; Ogra et al. 2006; Chirico et al. 2008; Palmeira and Carneiro-Sampaio 2016; Carr et al. 2021; Quitadamo et al. 2021).

Evidences have shown that infants who nurse for more than 6 months are less likely to develop gastroenteritis, childhood diarrhea, enterocolitis, vomiting, allergies, pneumonia, asthma, bacteremia, otitis media, necrotizing enterocolitis, childhood leukemia, lymphoma, , type 1 diabetes mellitus, and other chronic diseases (Bachrach et al. 2003; Marild et al. 2004; Chulada et al. 2003; Gdalevich et al. 2001; Sadauskaite-Kuehne et al. 2004; van Odijk et al. 2003; Donovan 2006; Carr et al. 2021; Quitadamo et al. 2021) as compared to those fed with formula milk. This might be due to the presence of cytokines in human milk, which have a major impact on the immune system in newborns and also contain substances having anti-infective, hormonal, enzymatic, and bioactive action.

# 2 Nutrients in Human Milk

Human milk is a complex physiological fluid which is a rich source of essential nutrients such as carbohydrates or human milk oligosaccharides (HMOs), a diverse collection of proteins (casein and whey), amino acids, serum albumin, bioactive peptides, lipids, nucleotides, hormones, enzymes, antipathogenic, anti-infective, anti-inflammatory bioactive factors, and microbiota. The milk produced in preliminary days following delivery is known as colostrum. It is whey-predominant and contains significant amounts of immune-suppressive proteins such as immunoglobulins, lactoferrin, haptocorrin, kappa-casein, lysozyme, cytokines, complement system components, and leukocytes (Table 1) (Lönnerdal 2003; Baldi et al. 2005; Donovan 2006; Hosea Blewett et al. 2008; Palmeira and Carneiro-Sampaio 2016; Carr et al. 2021; Quitadamo et al. 2021). These components provide antimicrobial protection, inhibit infections/inflammation, help the GI tract's healthy microorganisms to colonize and establish itself. Additionally, they promote immunocompetence, create an optimal ratio of intestinal immune cells, and promote intestinal cellular proliferation. They also have a central role in mucosal immune function development in neonates (Table 1) (Goldman 2000; Walker 2002; Lönnerdal 2003; Baldi et al. 2005; Donovan 2006; Ballard and Morrow 2013; Palmeira and Carneiro-Sampaio 2016; Carr et al. 2021; Quitadamo et al. 2021).

## 2.1 Carbohydrates

Lactose and HMOs are the main carbohydrates present in human milk, along with some amounts of glycoconjugates. Oligosaccharides are present in abundance, approximately 22 g/L in the colostrum. HMOs act as prebiotics that support the growth and colonization of probiotic microbiota (lactobacilli and bifidobacteria) by producing short chain fatty acids (SCF) through the process fermentation and also helps in the growth of probiotic bacteria. They primarily do this by lowering gut pH, creating bacteriocins and organic acids that have an anti-adhesive action, and limiting the proliferation and penetration of harmful pathogenic bacteria. These oligosaccharides also suppress NF-kB activation, protect against dysbiosis, and decrease intestinal permeability and inflammatory markers in the gut. Additionally, by increasing the synthesis of toll-like receptors (TLRs), human dendritic cells, T cells as well as immunological signaling pathways (TLR3, TLR5, PAMP), HMOs promote the immune system both locally and systemically (Carr et al. 2021; Coppa et al. 2004; Palmeira and Carneiro-Sampaio 2016; Quitadamo et al. 2021).

## 2.2 Lipids

In colostrum, human milk has a fat concentration of 3–4 g/L. Greater levels of omega-3 are present. NF-kB activity, which triggers a number of pro-inflammatory genes, is decreased by PUFAs, while n-6 PUFAs primarily control immunological and inflammatory responses. Lipids found in human milk have nutritional and development benefits along with its anti-infective properties against viruses and

**Table 1** Summary of the role of nutrients present in human milk

| Major nutrients class in milk | Specific nutrients | Role in the infant's body |
|---|---|---|
| Carbohydrates | Human milk oligosaccharides (HMOs), Lactose, Glycolconjugates | 1. Supports the growth and colonization of probiotic microbiota and prevents the gut dysbiosis<br>2. Increases the expression of toll-like receptors (TLRs), human dendritic cells, regulatory T cells, which helps the immune system mature<br>3. Enhances immune signaling pathways (TLR5, TLR3, PAMP)<br>4. ↓ inflammatory markers, pro-inflammatory cytokine expression<br>5. Suppresses activation of the NF-kB |
| Lipids | Omega-3, PUFAs | 1. Regulates the inflammatory and immune responses<br>2. Decreases the activity of NF-kB<br>3. Has an anti-infective effect against viruses and protozoa in infants |
| Proteins | Heterogeneous mixture of proteins (casein/whey), amino acids, serum albumin, bioactive peptides, enzymes, hormones, haptocorrin, immunoglobulins, lactoferrin, kappa-casein, lysozyme, cytokines, complement components | 1. Provides essential nutrients<br>2. Helps in absorption of nutrients<br>3. Prevents bacterial and viral infections<br>4. Strengthens the immune system |
| Growth factors | Epidermal and neuronal growth factors and insulin-like growth factor (IGF) | 1. Stimulates cell proliferation<br>2. Provides protection against damage from hypoxia and ischemia<br>3. Enhances angiogenesis and tissue repair activity<br>4. Facilitates the GI tract growth and development |

(continued)

**Table 1** (continued)

| Major nutrients class in milk | Specific nutrients | Role in the infant's body |
|---|---|---|
| 1.Other essential components | Nucleic acids, nucleosides, nucleotides | 1. Increases iron absorption<br>2. Develops and repairs the gastrointestinal mucosa<br>3. Enhances the growth of probiotic bacteria<br>4. Enhances IL-2 production and NK cell activity. |
| | Antioxidants | 1. Reduces oxidative stress and eliminates free radical damage |

protozoa in infants (Coppa et al. 2004; Palmeira and Carneiro-Sampaio 2016; Quitadamo et al. 2021).

## 2.3 *Proteins*

Human milk contains a diverse collection of proteins (casein and whey), amino acids, serum albumin, bioactive peptides, enzymes, hormones, growth factors, immunoglobulins, lactoferrin, haptocorrin, kappa-casein, lysozyme, cytokines, complement system components, leukocytes, etc. These proteins provide essential nutrients to the newborns, help in nutrient absorption, prevent from bacterial and viral infections and also strengthen their immune system. β-casein is the primary casein in the family of glycosylated proteins with immunologic action in newborns. In the small intestine, it is thought that β-casein may promote MUC2 genes expression and raise the proportion of paneth cells and goblet cells. Additionally, κ-Casein can function as a receptor analog to prevent disease-causing bacteria from adhering to human stomach mucosa in a lab setting (Plaisancie et al. 2013; Jakaitis and Denning 2014; Palmeira and Carneiro-Sampaio 2016; Donovan 2006).

In addition, human milk contains a number of enzymes that affect how other milk components are digested (Chowanadisai and Lönnerdal 2002; Donovan 2006; Palmeira and Carneiro-Sampaio 2016; Quitadamo et al. 2021). Human milk contains large amounts of lysozymes, which collaborate with lactoferrin to prevent bacterial and viral illnesses. In particular, lactoferrin allows lysozyme to penetrate and breakdown the proteoglycan matrix in Gram-negative bacteria while also disrupting the outer membrane lipopolysaccharides of those bacteria. Besides, it solely cleaves the β,1–4 glycoside linkages of bacterial cells walls to destroy them (Lönnerdal 2003; Jakaitis and Denning 2014; Palmeira and Carneiro-Sampaio 2016).

### 2.4 Growth Factors

Several growth factors are also found in human milk that have a major role in stimulation of cell proliferation, provide protection against damage from hypoxia and ischemia, and enhance angiogenesis and tissue repair activity. These milk growth factors also encourage the gastrointestinal tract's expansion and development (GI) (Ballard and Morrow 2013).

### 2.5 Other Essential Components

In human milk, nucleic acids, nucleosides, nucleotides, and similar substances make up about 15–20% of the non-protein nitrogen. They probably increase iron absorption, develop and repair gastrointestinal mucosa, promote probiotic bacterial growth, especially *Bifidobacterium* and IL-2 production and NK cell activity. Free radicals are eliminated by a number of antioxidant components, limiting the harm brought on by oxidative stress (Palmeira and Carneiro-Sampaio 2016). A summary of the role of nutrients present in human milk is presented in Table 1.

## 3 Immunological Components in Human Milk and Their Mechanistic Behavior

Human milk contains several essential immunological components that aid neonatal gut function and immunity. Exclusive administration of human milk generally enhances innate immunity and other immunological components in premature infants to protect them from serious health risks. Well-studied immune-booster substances found in human milk and their mechanistic activity in providing protection in infants are discussed in this section (Tables 2 and 3) (Jakaitis and Denning 2014; Palmeira and Carneiro-Sampaio 2016; Quitadamo et al. 2021).

### 3.1 Secretory Immunoglobulins

Interestingly, human milk has the antibodies or immunological memory against several infectious organisms, which the mother had a contact prior. These immunoglobulins are transferred to the newborn during lactation to provide them passive immunity. Though all types of immunoglobulins are present in human milk, sIgA is being most predominant followed by sIgM and sIgG. Among all secretory immunoglobulins, IgA (sIgA) is found in abundance, accounting up to 90% of total immunoglobulins in the milk. They remain always active throughout the infant's GI tract

**Table 2** Immunoboosters in human milk (Chirico et al. 2008)

| Type of immunoboosters | Specific immunoboosters |
|---|---|
| Adaptive immunity components | Immunoglobulins 7S IgA, sIgA (11S), IgM, IgG, IgD, IgE |
| Cytokines, chemokines, and receptors | IL-(1β, 2, 4, 5, 6, 8, 12, 13, 16, 18), TNFα, IFNϒ, M-CSF, G-CSF, GROα, GM-CSF, TGFβ1 and 2, monocyte chemotactic protein-1, Toll-like receptor, sCD14, sFas, sFasL |
| Cell types | Neutrophils (25%), macrophages (60%); epithelial cells; lymphocytes (10%); leukocyte |
| Proteins and enzymes | Lipoprotein lipase, lactoferrin, platelet-activating factor-acetyl-hydrolase, lysozyme, antiproteases |
| Immunomodulatory bioactive peptides | β-Casein (residues 54–59), β-Casein (residues 60–62), α-lactalbumin (residues 51–53) |
| Innate immunity components | Interferon, chemotactic factors, complement, α-fetoprotein, properdin factors, anti-adherence components (mucins, oligosaccharides, glycosaminoglycans, lactadherin, glycolipids, and ḱ-casein), anti-staphylococci factors, mannose-binding lectin, antiviral factors (fatty acids and monoglycerides), β-defensin-1 |

and have major impact on developing and strengthening the newborn's immunity. Their importance in the defense of the mucous membranes is well documented in several studies. As a crucial component of the innate immune system, they mainly encapsulate harmful microflora to prevent mucosal barrier adherence, colonization, and penetration to inactivate certain viruses (Brandtzaeg 2010; Palmeira and Carneiro-Sampaio 2016). These antibodies also have anti-inflammatory properties, i.e., these antibodies prevent pathogens without triggering inflammatory reactions (known as immune exclusion) which is a harmful reaction to the infants (Lepage and Van de Perre 2012; Brandtzaeg 2010; Palmeira and Carneiro-Sampaio 2016).

IgM is also another most prevalent immunoglobulin in colostrum and serves as the infant's first line of defense against foreign antigens by protecting the mucosal surfaces. Human milk contains IgG, but it is also linked to long-lasting immunity. It exhibits an opsonizing action in the GI mucosal surfaces of the infants that can trigger the complement system and antibody-dependent cytotoxicity. These antibodies also provide protection against several viral infections (including rotavirus infections) at the mucosal level through neutralization in infants. IgE is also found in human milk which reduces allergic reactions, viral (parvovirus B19, HIV-1) and parasitic infections. Human milk contains IgD, which is produced on mature B cells and binds to certain bacteria to stimulate and activate B cells (Hanson et al. 2003; Palmeira and Carneiro-Sampaio 2016; Carr et al. 2021).

**Table 3** Mechanism of action of immune-stimulating components in human milk

| Immune-stimulating components | | Mode of action |
|---|---|---|
| Secretory immunoglobulin | sIgA | 1. Protects against commensal and pathogenic microorganisms from penetrating the epithelial barrier by coating them<br>2. Anti-inflammatory property |
| | IgM | 1. As the infant's initial line of protection against foreign antigens, safeguards the mucosal surfaces |
| | IgG | 1. Cytotoxicity that is antibody-dependent and can activate the complement system on the mucosal surfaces of infants |
| | | 1. Binds iron in the gut, reducing the amount of time that microorganisms can use it<br>2. Reduces inflammatory cytokine production<br>3. Stimulates intestinal epithelial cell proliferation and differentiation |

(continued)

**Table 3** (continued)

| Immune-stimulating components | | Mode of action |
|---|---|---|
| Cytokine and chemokines | IL-1 and IL-1RA | 1. Decreases pro-inflammatory responses of NF-kB signaling |
| | IL-2 | 1. Increases T-cell activity to trigger an immunological response specific for an antigen<br>2. IL-6 present in human milk preferentially differentiates newborn T cells toward a Th1 |
| | IL-6 | 1. Pro-inflammatory properties |
| | IL-10 | 1. Reduces intestinal inflammation<br>2. Encourage the generation and control of immunoglobulin<br>3. Lessens the severity of newborn infants' necrotizing enterocolitis |
| | TGFβ | 1. Regulates dendritic cells, lymphocytes, macrophages, natural killer cells, and granulocytes and<br>2. Inhibits the inflammation and decrease in the production of pro-inflammatory cytokines<br>3. Lowers the risk of wheezing and atopic dermatitis in childhood<br>4. Initiates local production of IgA to strengthen the innate immunity in infants.<br>5. Stimulates intestinal maturation and defense of the newborn |
| | CXC | 1. Activator neutrophils<br>2. Chemotactic activity for intraepithelial lymphocytes<br>3. Essential in the host's defense against various infections |
| Human milk cell | Epithelial cells, macrophages, neutrophils, lymphocyte | 1. Enhance and strengthen immunity of newborn |
| Lactoferrin | | 1. Binds receptors to interfere with pathogen binding to the epithelial barrier<br>2. Reduces the availability of iron to microorganisms by binding it in the gut<br>3. Reduces inflammatory cytokine production<br>4. Stimulates intestinal epithelial cell proliferation and differentiation |

(continued)

**Table 3** (continued)

| Immune-stimulating components | | Mode of action |
|---|---|---|
| Immunomodulatory bioactive peptides | β-Casein (residues 54–59) | 1. Stimulates phagocytic activity of human macrophages |
| | α-lactalbumin | 1. Stimulates phagocytic activity of human macrophages<br>2. Enhances oxidative burst in human polymorpho nuclear leukocytes |
| Other immunostimulatory proteins | Soluble CD14 (sCD14) | 1. Modulates adaptive and innate immune responses<br>2. Controls homeostasis in the infant's intestine |
| | Osteoprotegerin | 1. Develops newborn's immune system |

### 3.2 *Cytokines and Chemokines*

Several cytokines are present in human milk which plays a key role in immune stimulation or immunomodulation in neonate. They prevent allergies and hypersensitivities, increase thymocyte proliferation, suppress IgE production, induce the immunoglobulin development, differentiation, and synthesis by B cell in infants. A newborn human neonate is unable to respond to foreign infections appropriately and is unable to limit an inflammatory response.

Signaling molecules like the anti-inflammatory cytokine IL-10, which encourages the manufacture and control of immunoglobulins, are abundant in human milk. The pro-inflammatory effects of NF-kB signaling are reduced by IL-1 and IL-1RA, whereas IL-2 increases the activity of T cells to elicit an immune response specific to an antigen. In addition, it causes newborn T cells to preferentially develop into Th1 rather than Th2 cells. Human milk contains the pro-inflammatory compound IL-6. Additionally, these lessen the severity of necrotizing enterocolitis in newborn babies by reducing the inflammatory response (Dvorak et al. 2003; Jakaitis and Denning 2014; Palmeira and Carneiro-Sampaio 2016; Quitadamo et al. 2021). Human milk supplies sufficient levels of immunoregulatory cytokines TGFβ to infants which play in major role in regulation of dendritic cells, lymphocytes, macrophages, natural killer cells, and granulocytes and reduce the production of pro-inflammatory cytokines to minimize inflammation. They are said to promote intestinal maturation and the defense of the newborn by assisting with the transition of B lymphocytes from IgM to IgA (Penttila 2010; Jakaitis and Denning 2014; Palmeira and Carneiro-Sampaio 2016). In their intestinal mucosal system, newborns have a deficient number of sIgA producing cells and a T-helper 2 (Th2)-biased cytokine response. Human milk administration colonizes the commensal microbiota, which in turn promotes the generation of sIgA. Thus, by triggering a suitable immune response in neonata, the immunoregulatory components of human milk influence the development of the commensal microbiota (Forchielli and Walker 2005; Weng and Walker 2006; Brandtzaeg 2003; Field 2005; Field et al. 2001; Donovan 2006). Chemokines CXC are found in human milk in large quantities and act as a activator of neutrophils. They are crucial for the host's defense against numerous pathogens and exhibit chemotactic action for intraepithelial lymphocytes (Palmeira and Carneiro-Sampaio 2016).

## 3.3 Human Milk Cells

Breast milk contains CD4+ T lymphocytes that are in an active state and produce the immunological memory-related surface protein CD45RO. In human milk, more than 70% of the B cells have the capacity to release antibodies. The concentration of leukocytes is higher in colostrum and interestingly, a newborn baby can administer up to $10^{10}$ mother leukocytes every day. Granulocyte-macrophage colony-stimulating factor can be produced by human milk's phagocytes on their own, and they can also spontaneously differentiate into dendritic cells (Ballard and Morrow 2013; Jakaitis and Denning 2014; Ichikawa et al. 2003; Tuaillon et al. 2009; Lepage and Van de Perre 2012; Palmeira and Carneiro-Sampaio 2016).

## 3.4 Lactoferrin

Human milk contains the bioactive protein lactoferrin, which acts as a component of the innate immune system. It has antipathogenic, immunomodulatory, and anti-inflammatory properties. By binding or reducing the availability of free iron, a required nutrient for microbe growth, lactoferrin, an iron-binding glycoprotein, protects newborns from pathogenic bacteria and has a bacteriostatic effect. Lactoferrin administration can lower the occurrence of sepsis brought on by bacterial infections and fungal infections. These also have antiviral effects against rotavirus and adenovirus (Manzoni et al. 2012; Valenti and Antonini 2005; Jakaitis and Denning 2014; Carr et al. 2021).

Furthermore, it has been reported that lactoferrin is degraded to produce certain antimicrobial agent, lactoferricin while exposure to stomach pepsin. These inhibit Gram-negative bacteria by disturbing their cell membrane. They bind to several receptors, including TLRs and CD14, to prevent pathogenic bacteria from adhering to the intestinal epithelium. Lactoferrin also has a major immunomodulatory action. It helps in limiting immune responses by blocking several inflammatory cytokines. The impact of lactoferrin was found to increase crypt cell proliferation and serum immunoglobulin G (IgG). Through the regulation of nuclear factor-B activation in monocytes, lactoferrin causes apoptosis in infected intestinal epithelial cells and lowers the production of inflammatory cytokines. The battle against systemic infections and newborn infant necrotizing enterocolitis is still led by these (Sherman 2013; Buccigrossi et al. 2007; Haversen et al. 2002; Jakaitis and Denning 2014; Palmeira and Carneiro-Sampaio 2016; Carr et al. 2021).

### 3.5 Immunomodulatory Bioactive Peptides

Infants' immune systems are strengthened by the immunomodulatory peptides from human milk. Human casein is typically hydrolyzed to produce immunomodulatory peptides that enhance the phagocytic activity of murine and human macrophages, which kill germs (Wada and Lönnerdal 2020).

### 3.6 Other Immunostimulatory Proteins

Human milk contains haptocorrin, which binds to vitamin B12 and prevents an enteropathogenic *E. coli* strain from growing there. In order to maintain intestinal homeostasis in infants, soluble CD14 (sCD14) limits the modification of adaptive and innate immune responses during the time of pathogenic bacterial colonization. In human milk, osteoprotegerin has the ability to attach with TRAIL and trigger caspase-dependent apoptosis, particularly in Th1 cells. This property is important for controlling the balance of Th1/Th2 in the development of immune system in newborns (Hallgren et al. 2008; Adkins and Lönnerdal 2003; Vidal et al. 2004; Palmeira and Carneiro-Sampaio 2016)

## 4 Microbiota in Human Milk

Colostrum is an abundant source of potentially beneficial bacteria that are given to babies through breast milk ($10^4$–$10^6$ bacteria per day based on an average daily intake of 800 mL of milk) and are crucial for the maturation and development of their immune systems (Moossavi et al. 2018). Presence of more than 200 bacterial taxa such as *Lactobacillus* (e.g., *L. gasseri*, *L. salivarius*, *L. rhamnosus*, *L. plantarum*, and *L. fermentum*), *Bifidobacterium* (*B. breve* and *B. longum*), *Enterococcus* (*E. faecium, E. faecalis*), *Staphylococcus* (mainly *S. epidermidis* and also *S. epidermidis, S. hominis, S. capitis, S. aureus*), *Streptococcus* (mainly *S. mitis* spp., *S. salivarius* spp. and also *S. parasanguinis*), *Corynebacterium, Lactococcus, Leuconostoc, Pediococcus, Propionibacterium, Bacteroides, Rothia, Weissella, Eubacterium, Kocuri, Faecalibacterium, Roseburia, Ruminococcus,* and *Veillonella* have been reported from human milk.

The genera *Streptococcus*, *Propionibacterium,* and *Staphylococcus* are the essential microbes found in human milk (Table 4) (Arici et al. 2004; Sakwinska et al. 2016; Palmeira and Carneiro-Sampaio 2016; Amezcua Lopez et al. 2019; Ruiz et al. 2019; Liu et al. 2020; Ogra et al. 2020). For adults, the amount of *Lactobacillus* species in the total gut bacteria is less than 1%; however, in newborns, this proportion is fairly significant (approximately between $10^5$ and $10^8$) (Francisco et al. 2010).

**Table 4** Microbiota present in human milk

| | |
|---|---|
| Microbiota present in human milk | 1. *Lactobacillus* (*L. plantarum, L. rhamnosus, L. salivarius, L. fermentum,* and *L. gasseri*)<br>2. *Bifidobacterium* (*B. longum* and *B. breve*)<br>3. *Enterococcus* (*E. faecium, E. faecalis*)<br>4. *Staphylococcus* (*S. hominis, S. aureus, S. epidermidis, S. capitis, S. epidermidis*)<br>5. *Streptococcus* (*S. mitis* spp., *S. salivarius* spp., *S. parasanguinis, S. peores*)<br>6. *Corynebacterium*<br>7. *Lactococcus*<br>8. *Leuconostoc*<br>9. *Pediococcus*<br>10. *Propionibacterium*<br>11. *Bacteroides*<br>12. *Rothia*<br>13. *Weissella*<br>14. *Eubacterium*<br>15. *Kocuri*<br>16. *Faecalibacterium*<br>17. *Roseburia*<br>18. *Ruminococcus*<br>19. *Veillonella* |

Probiotic microbiota in human milk enhances the immune system in newborns by producing cytokines, stimulates macrophages, increases secretory IgA concentration, regulates T cells proliferation, Th17 response, IgE response, and B cell maturation. The exclusive colonization of the probiotic microbiota in the intestine promotes the development of humoral immunity and protects the newborn babies from severe allergic reactions. Probiotics also prevent atopic dermatitis in infants (Lara-Villoslada et al. 2007; Diaz-Ropero et al. 2006; Francisco et al. 2010; Moossavi et al. 2018; Meenakshi 2020).

Evidences have suggested that probiotic *L. fermentum* and *L. salivarius* present in human milk have immunomodulatory and anti-inflammatory effects whereas *L. gasseri* CECT5714 exert both immunomodulatory and anti-allergic effect among neonates. Both have been reported to significantly induce the production of IL-2 and IL-12, several other cytokines and chemokines and the inflammatory mediator TNF-*a*. These two probiotics are also thought to be powerful NK cell activators, moderate NK cell activators, and moderate CD4+, CD8+, and regulatory T-cell activators. They also lessen the lipopolysaccharide-induced inflammatory response of murine bone marrow-derived macrophages. Figure 1 illustrates schematically the immunomodulatory probiotic bacteria that are found in human milk (Lara-Villoslada et al. 2007; Diaz-Ropero et al. 2006; Francisco et al. 2010; Moossavi et al. 2018; Meenakshi 2020).

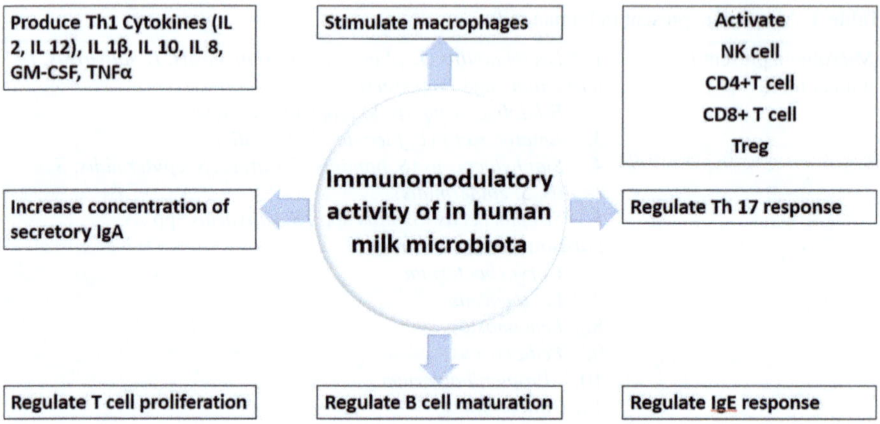


**Fig. 1** A schematic representation of mode of action of immunomodulatory probiotic bacteria present in human milk

## 5 Human Milk and Infant's Gut Health

The gastrointestinal system's epithelial layers and mucus secretions offer a sophisticated mechanical barrier and an innate defensive mechanism against commensal bacteria and several disease-causing substances. In particular, innate and adaptive immunity are the two main groups of the intestinal immune system.

At the time of birth, gastrointestinal immune system of a child is found to be developmentally immature including immature phagocytic response and lower level of antibodies and serum IgA. The gut's adaptive immune system is initially immature and may react inappropriately to environmental and other external factors that increase risk for serious infection (Nanthakumar et al. 2000; Walker 2002; Donovan 2006; Jakaitis and Denning 2014).

More than 400 kinds of commensal bacteria are colonized in the gut of infants, and an exclusively human milk diet offers nutrients that encourage their growth. These components of human milk also support the coexistence of colonizing bacteria with the lamina propria and basal epithelial cells. This microbial colonization in a newborn's stomach contributes to IgA secretion, barrier function, apoptosis, and immunological homeostasis, among other homeostatic processes (Walker 2013; Mirpuri et al. 2010; Jakaitis and Denning 2014). Exclusive breast feeding to a newborn strengthens its innate immunity.

Antimicrobial peptides which are members of the cathelicidin family are secreted into the mucus layer of the intestine to maintain the largely aseptic environment of the intestinal crypt. Additionally, antimicrobial peptides are released into the lumen to aid mucosal defense against a variety of pathogenic flora, such as several Gram-negative and Gram-positive bacteria (McElroy and Weitkamp 2011; Lievin-Le Moal and Servin 2006; Jakaitis and Denning 2014).

Similar to this, a single layer of intestinal epithelial cells (IECs), which are highly polarized, forms a physical barrier beneath the mucus layer in a healthy immune system and is grounded by junctional proteins. These are also in charge of sampling intra-luminal substances that starts transcellular signaling and gene transcription. This triggers a defense response by releasing cytokines and chemokines, which then draw leukocytes. The lamina propria carries several innate immune cells such as macrophages and dendritic cells which induce T-cell differentiation, uptake and transport of antigen and stimulate immunoglobulin production (IgA) (Jakaitis and Denning 2014). Some epidemiological studies have reported that human milk is a rich source of antibodies against several pathogenic bacteria (Palmeira and Carneiro-Sampaio 2016).

Furthermore, *Lactobacillus crispatus* and *Lactobacillus gasseri* present in human colostrum stick to intestinal cells to colonize the good bacteria in the guts of preterm neonates. Among these, *Lactobacillus gasseri*, *Bifidobacterium breve*, and *Streptococcus salivarius* prevent harmful germs from adhering to newborns' intestinal epithelial cells.

Salmonella infection is prevented by *Lactobacillus* spp. and *Bifidobacterium* spp. which also shield the intestine from epithelial injury. Bacteroides and *Bifidobacterium* support the infant's life by enhancing its immune system during its early stages (Carr et al. 2021).

Non-digestible sugars, or oligosaccharides (HMOs), found in human colostrum enhance the development of healthy and protective bacteria in the colon. Due to their indigestibility, oligosaccharides travel into the colon, where they ferment to produce SCF, which in turn foster the growth of probiotic bacteria, maintain a balanced ecology in the gut, and boost host defenses in neonate's intestine (Jakaitis and Denning 2014).

HMOs act as prebiotics that support the growth and colonization of probiotic microbiota (lactobacilli and bifidobacteria) and reduce the growth and invasion of pathogenic bacteria by decreasing intestinal pH, producing bacteriocins and organic acids, serving as anti-adhesive effect and provide gut barrier function. It protects the infants against infections caused by *Vibrio cholerae*, *Escherichia coli* (AIEC), *Entamoeba histolytica* and *Campylobacter jejuni* (Quitadamo et al. 2021). α-lactalbumin inhibits the activities of *E. coli, Klebsiella pneumoniae*, *C. albicans*, streptococcus, and staphylococcus (Pellegrini et al. 1999).

The antibacterial properties of human milk have been shown to defend infants' health against a variety of bacteria, viruses, and protozoa as well as to possibly lower the frequency of gastrointestinal and non-enteric diseases (Morrow et al. 2004; Ogra et al. 2006; Chirico et al. 2008). Nutrients and bioactive components present in human milk establish a symbiotic microflora in the GI tract where the intestinal colonization begins. These components also stimulate cellular growth, digestive maturation of babies, and develop gut-associated lymphoid tissues. All these processes facilitate the successful postnatal adaptation, strengthen and activate mucosal immune system of the newborns preventing them from gastrointestinal infections (Salminen and Isolauri 2006; Weng and Walker 2006; Donovan 2006).

## 6 Does Human Milk Help to Protect Infants from Covid 19?

Coronavirus (COVID-19) outbreak is a great global concern nowadays. A recent study reported, a pregnant woman who was tested positive for COVID-19 earlier was found to have IgA and IgG in her colostrum. Although there have been conflicting reports on the transmission of SARS-Cov-2 antibodies, data indicates that a newborn had high IgM antibodies against SARS-CoV-2 from her mother who had previously contracted the same virus. Dong et al. reported that there is no detection of COVID-19 virus in COVID-positive mother's breast milk after giving birth. Intriguingly, IgG antibodies specific to SARS-CoV-2 were discovered in colostrum only one and a half months after birth, showing placental transfer of COVID antibody. The infant was also determined to be negative for SARS-CoV-2 RNA at neonate's birth. Typically, a newborn retains the mother's IgG antibody for more than 6 months after birth (Dong et al. 2020). More than 800 COVID-19 recovered mothers who had given milk samples every 3–4 weeks participated in a different trial. It was discovered that 95% of colostrum samples contained IgA antibodies specific to SARS-CoV-2 following their longitudinal examination of the immunological response to SARS-CoV-2 in colostrum (Fox et al. 2020a; Spatz et al. 2021).

Similar findings have been made by van Keulen et al. and Pace et al. whose studies production of sIgA against SARS-CoV-2 and the capacity of these sIgA to neutralize a several strains of SARS-CoV-2 (Pace et al. 2020; van Keulen et al. 2020). Another study used colostrum samples from COVID-19-positive mothers. Of these, 76% had IgA specific to SARS-CoV-2, 80% had IgG specific to SARS-CoV-2, and 62% had the ability to neutralize SARS-CoV-2 infectivity, demonstrating the anti-infective and immunomodulatory value of human milk against SARS-CoV-2 (Pace et al. 2021).

Generally, the spike proteins of corona virus are composed by two specific subunits such as S1 and S2 to determine viral specificity and infectivity. Particularly, the receptor-binding domain (RBD) of the S1 subunit participates in cell surface recognition whereas fusion is mostly accomplished by the S2 subunit. Recently, in a study conducted by Fox et al., milk was collected from 15 women who were earlier infected with SARS-CoV-2 and interestingly, antibodies (IgA) specific to SARS-CoV-2 were detected in their colostrums sample. These antibodies were discovered to be reactive to the RBD of the SARS-CoV-2 spike protein in 12 out of 15 cases. Compared to the milk taken from the same volunteers before the pandemic, all 15 samples of milk had increased levels of IgA (Fox et al. 2020a, b; Demers-Mathieu et al. 2021; Pace et al. 2021).

Human milk contains several antimicrobial proteins, including lactoferrin, peptides, HMOs, and fatty acids, which may function as antiviral components and protect against these diseases (Fan et al. 2020; Chang et al. 2020; Conzelmann et al. 2019; Spatz et al. 2021; Vassilopoulou et al. 2021; Lang et al. 2011)

## 7 Conclusion

Human milk has been shown to have positive benefits on growth, development, and anti-infective defense beginning in infancy and continuing into later life according to scientific research. Human milk can be considered as the most powerful functional food, with varied benefits. Nutritional, immune-stimulatory, and anti-inflammatory components present in human milk are not only to nourish the newborn but also confer protection against several infections in the gut, strengthening the gastric immune system. Therefore, it is directly and indirectly involved in decreasing neonatal morbidity and mortality. Recently, it has also proven that the exclusive administration of human milk can provide antibodies to protect against COVID-19. More research is recommended to explore other unidentified immunological components, which may be beneficial to the infant.

## Glossary

**Bioactive peptides** Organic substances (made up of amino acids joined by amide or peptide bonds), normally remain encrypted in the structure of the parent proteins and are released mainly by enzymatic processes. They have a significant role in human health by affecting the digestive, endocrine, cardiovascular, immune, and nervous systems. Although some exist free in their natural source.

**Chemotactic activity** Also called "chemotaxis"—the movement of an organism or entity in response to a chemical stimulus. Somatic cells, bacteria, and other single-cell or multicellular organisms direct their movements according to certain chemicals in their environment.

**Colostrum** A nutrient-rich first milk produced by a woman's breasts during pregnancy. It changes to transitional breast milk a few days after delivery of the baby.

**Growth factors** Biologically active molecules that can affect the growth of cells; stimulates cell proliferation.

**Immunostimulatory** Ability to cause stimulation of cells of the immune system.

**Immunomodulator** Medicines (molecules) that change your immune system so it works more effectively. They include treatments that increase or decrease your immune response. Immunomodulators treat various conditions, including cancer and autoimmune diseases.

**Lactoferrin** Human milk contains the bioactive protein lactoferrin, which acts as a component of the innate immune system. It has antipathogenic, immunomodulatory, and anti-inflammatory properties.

**Probiotics** Live microorganisms (bacteria and yeast) that are intended to have health benefits to our body. One can take them as dietary supplements.

## References

Adkins Y, Lönnerdal B (2003) Potential host-defense role of a human milk vitamin B-12-binding protein, haptocorrin, in the gastrointestinal tract of breastfed infants, as assessed with porcine haptocorrin in vitro. Am J Clin Nutr 77(5):1234–1240

Amezcua Lopez JA, Garcia Morales E, Solis Pacheco JR, Gutierrez Padilla JA, Zepeda Morales ASM, Angulo Castellanos E, Lopez Mincitar M, Flores Arevalo KF, Aguilar Uscanga BR (2019) Influence of the diet of Mexican women on the nutritional quality and the presence of beneficial microorganisms in human milk. Nutr Hosp 36:1139–1149

Arici M, Bilgin B, Sagdic O, Ozdemir C (2004) Some characteristics of *Lactobacillus* isolates from infant faeces. Food Microbiol 21:19–24

Bachrach VR, Schwarz E, Bachrach LR (2003) Breastfeeding and the risk of hospitalization for respiratory disease in infancy: a meta analysis. Arch PediatrAdolesc Med 157:237–243

Baldi A, Ioannis P, Chiara P, Eleonora F, Roubini C, Vittorio D (2005) Biological effects of milk proteins and their peptides with an emphasis on those related to the gastrointestinal system. J Dairy Res 72:66–72

Ballard O, Morrow AL (2013) Human milk composition: nutrients and bioactive factors. Pediatr Clin North Am 60(1):49–74

Brandtzaeg P (2003) Mucosal immunity: integration between mother and the breast-fed infant. Vaccine 21:3382–3388

Brandtzaeg P (2010) The mucosal immune system and its integration with the mammary glands. J Pediatr 156(2 Suppl):S8–S15

Buccigrossi V, de Marco G, Bruzzese E et al (2007) Lactoferrin induces concentration-dependent functional modulation of intestinal proliferation and differentiation. Pediatr Res 61(4):410–414

Carr LE, Virmani MD, Rosa F, Munblit D, Matazel KS, Elolimy AA, Yeruva L (2021) Role of human milk bioactives on infants' gut and immune health. Front Immunol 12:604080

Chang R, Ng TB, Sun WZ (2020) Lactoferrin as potential preventive and adjunct treatment for COVID-19. Int J Antimicrob Agents 56:106118. https://doi.org/10.1016/j.ijantimicag.2020.106118

Chirico G, Marzollo R, Cortinovis S, Fonte C, Gasparoni A (2008) Antiinfective properties of human milk. J Nutr 138(9):1801S–1806S

Chowanadisai W, Lönnerdal B (2002) α1-antitrypsin and antichymotrypsin in human milk: origin, concentrations and stability. Am J Clin Nutr 76:828–833

Chulada PC, Arbes SJ Jr, Dunson D, Zeldin DC (2003) Breast-feeding and the prevalence of asthma and wheeze in children: analyses from the third national health and nutrition examination survey, 1988-1994. J Allergy Clin Immunol 111:65–67

Conzelmann C, Zou M, Groß R, Harms M, Röcker A, Riedel CU et al (2019) Storagedependent generation of potent anti-ZIKV activity in human breast milk. Viruses 11:591. https://doi.org/10.3390/v11070591

Coppa GV, Coppa MD, Bruni S, Morelli L, Soldi S, Gabrielli O (2004) The first prebiotics in humans: human milk oligosaccharides. J Clin Gastroenterol 38(Suppl 2):S80–S83

Demers-Mathieu V, DaPra C, Mathijssen G, Sela DA, Jarvinen KM, Seppo A, Fels S, Medo E (2021) Human milk antibodies against S1 and S2 subunits from SARS-CoV-2, HCoV-OC43, and HCoV-229E in mothers with a confirmed COVID-19 PCR, viral symptoms, and unexposed mother. Int J Mol Sci 22:1749

Diaz-Ropero MP, Martin R, Sierra S, Lara-Villoslada F, Rodriguez JM, Xaus J, Olivares M (2006) Two Lactobacillus strains, isolated from breast milk, differently modulate the immune response. J Appl Microbiol 102:337–343

Dong Y, Chi X, Hai H, Sun L, Zhang M, Xie W-F, Chen W (2020) Antibodies in the breast milk of a maternal woman with COVID-19. Emerg Microbes Infect 9(1):1467–1469. https://doi.org/10.1080/22221751.2020.1780952

Donovan SM (2006) Role of human milk components in gastrointestinal development: current knowledge and future needs. J Pediatr 149:S49–S61

Dvorak B, Halpern MD, Holubec H et al (2003) Maternal milk reduces severity of necrotizing enterocolitis and increases intestinal IL-10 in a neonatal rat model. Pediatr Res 53(3):426–433
Fan H, Luo Y, Hong B, Wang L, Jin X, Chen Y et al (2020) The effect of whey protein on viral infection and replication of SARS-CoV-2 and pangolin coronavirus in vitro. Signal Transduct Target Ther 5:275. https://doi.org/10.1038/s41392-020-00408-z
Field CJ (2005) The immunological components of human milk and their effects on immune development in infants. J Nutr 135:1–4
Field CJ, Clandinin MT, Van Aerde JE (2001) Polyunsaturated fatty acids and T-cell function: implications for the neonate. Lipids 36:1025–1032
Forchielli ML, Walker WA (2005) The role of gut-associated lymphoid tissues and mucosal defense. Br J Nutr 93(Suppl 1):S41–S48
Fox A, Marino J, Amanat F, Krammer F, Hahn-Holbrook J, Zolla-Pazner S et al (2020a) Evidence of a significant secretory-IgA-dominant SARS-CoV-2 immune response in human milk following recovery from COVID-19. medRxiv [Preprint]. https://doi.org/10.1101/2020.05.04.20089995
Fox A, Marino J, Amanat F, Krammer F, Hahn-Holbrook J, Zolla-Pazner S, Powell RL (2020b) Robust and specific secretory IgA against SARS-CoV-2 detected in human milk. iScience 23:101735. https://doi.org/10.1016/j.isci.2020.101735
Francisco JP, Dong H, Yaqoob P (2010) In vitro immunomodulatory activityof Lactobacillus fermentum CECT5716 and Lactobacillus salivarius CECT5713: twoprobioticstrainsisolatedfrom-human breast milk. Immunobiology 215:996–1004
Gdalevich M, Mimouni D, Mimouni M (2001) Breast-feeding and the risk of bronchial asthma in childhood: a systematic review with meta-analysis of prospective studies. J Pediatr 139:261–266
Godhia ML, Patel N (2013) Colostrum—its composition, benefits as a nutraceutical—a review. Curr Res Nutr Food Sci 1(1):37–47
Goldman AS (2000) Modulation of the gastrointestinal tract of infants by human milk. Interfaces and interactions. An evolutionary perspective. J Nutr 130:426–31S
Hallgren O, Aits S, Brest P, Gustafsson L, Mossberg AK, Wullt B et al (2008) Apoptosis and tumor cell death in response to HAMLET (human alpha-lactalbumin made lethal to tumor cells). Adv Exp Med Biol 606:217–240
Hanson LA, Korotkova M, Lundin S, Håversen L, Silfverdal SA, Mattsby-Baltzer I et al (2003) The transfer of immunity from mother to child. Ann N Y Acad Sci 987:199–206
Haversen L, Ohlsson BG, Hahn-Zoric M et al (2002) Lactoferrin down-regulates the LPS-induced cytokine production in monocytic cells via NF-kappa B. Cell Immunol 220(2):83–95
Hosea Blewett HJ, Cicalo MC, Holland CD, Field CJ (2008) The immunological components of human milk. Adv Food Nutr Res 54:45–80
Ichikawa M, Sugita M, Takahashi M et al (2003) Breast milk macrophages spontaneously produce granulocyte-macrophage colony-stimulating factor and differentiate into dendritic cells in the presence of exogenous interleukin-4 alone. Immunology 108(2):189–195
Jakaitis BM, Denning PW (2014) Human breast milk and the gastrointestinal innate immune system. Clin Perinatol 41(2):423–435. https://doi.org/10.1016/j.clp.2014.02.011
Lang J, Yang N, Deng J, Liu K, Yang P, Zhang G et al (2011) Inhibition of SARS pseudovirus cell entry by lactoferrin binding to heparan sulfate proteoglycans. PLoS One 6(8):96–104. https://doi.org/10.1371/journal.pone.0023710
Lara-Villoslada F, Olivares M, Sierra S, MigueRodríguez J, Boza J, Xaus J (2007) Beneficial effects of probiotic bacteria isolated from breast milk. Br J Nutr 98(Suppl. 1):S96–S100
Lepage P, Van de Perre P (2012) The immune system of breast milk: antimicrobial and anti-inflammatory properties. Adv Exp Med Biol 743:121–137
Lievin-Le Moal V, Servin AL (2006) The front line of enteric host defense against unwelcome intrusion of harmful microorganisms: mucins, antimicrobial peptides, and microbiota. Clin Microbiol Rev 19(2):315–337
Liu W, Chen M, Duo L, Wang J, Guo S, Sun H, Menghe B, Zhang H (2020) Characterization of potentially probiotic lactic acid bacteria and bifidobacteria isolated from human colostrum. J Dairy Sci 103:4013–4025

Lönnerdal B (2003) Nutritional and physiologic significance of human milk proteins. Am J Clin Nutr 77(Suppl 1):1537–143S
Manzoni P, Stolfi I, Messner H et al (2012) Bovine lactoferrin prevents invasive fungal infections in very low birth weight infants: a randomized controlled trial. Pediatrics 129(1):116–123
Marild S, Hansson S, Jodal U, Oden A, Svedberg K (2004) Protective effect of breastfeeding against urinary tract infection. Acta Paediatr 93:164–168
McElroy SJ, Weitkamp JH (2011) Innate immunity in the small intestine of the preterm infant. Neoreviews 12(9):e517–e526
Meenakshi (2020) The human milk microbiota-origin, probiotics properties and its benefits for newborns babies. Pharma Innov J 9(10):340–342
Mirpuri J, Brazil JC, Berardinelli AJ et al (2010) Commensal *Escherichia coli* reduces epithelial apoptosis through IFN-alphaA-mediated induction of guanylate binding protein-1 in human and murine models of developing intestine. J Immunol 184(12):7186–7195
Moossavi S, Miliku K, Sepehri S, Khafipour E, Azad MB (2018) The prebiotic and probiotic properties of human milk: implications for infant immune development and pediatric asthma. Front Pediatr 6:197. https://doi.org/10.3389/fped.2018.00197
Morrow AL, Ruiz-Palacios GM, Altaye M, Jiang X, Guerrero ML, Meinzen-Derr JK, Farkas T, Chaturvedi P, Pickering LK, Newburg DS (2004) Human milk oligosaccharides are associated with protection against diarrhea in breast-fed infants. J Pediatr 145:297–303
Nanthakumar N, Fursunyan RD, Sanderson IR, Walker WA (2000) Inflammation in the developing human intestine: a possible pathophysiologic basis for necrotizing enterocolitis. Proc Natl Acad Sci U S A 97:6043–6048
Ogra PL, Rassin DK, Garofalo RP (2006) Human milk. In: Remington JS, Klein JO (eds) Infectious diseases of the fetus and newborn infant, 6th edn. Elsevier Saunders Company, Philadelphia, pp 211–243
Ogra PL, Walker WA, Lönnerdal B (eds) (2020) Milk, mucosal immunity, and the microbiome: impact on the neonate. Nestlé Nutrition Institute workshop series, Basel, Karger, vol 94, pp 1–11. https://doi.org/10.1159/000505031
Pace RM, Williams JE, Järvinen KM, Belfort MB, Pace CDW, Lackey KA et al (2020) COVID-19 and human milk: SARS-CoV-2, antibodies, and neutralizing capacity. medRxiv [Preprint]. https://doi.org/10.1101/2020.09.16.20196071
Pace RM, Williams JE, Järvinen KM, Belfort MB, Pace CDW, Lackey KA, Gogel AC, Nguyen-Contant P, Kanagaiah P, Fitzgerald T, Ferri R, Young B, Rosen-Carole C, Diaz N, Meehan CL, Caffé B, Sangster MY, Topham D, McGuire MA, Seppo A, McGuire MK (2021) Characterization of SARS-CoV-2 RNA, antibodies, and neutralizing capacity in milk produced by women with COVID-19. mBio 12:e03192–e03120. https://doi.org/10.1128/mBio.03192-20
Palmeira P, Carneiro-Sampaio M (2016) Immunology of breast milk. Rev Assoc Med Bras 62(6):584–593
Pellegrini A, Thomas U, Bramaz N, Hunziker P, von Fellenberg R (1999) Isolation and identification of three bactericidal domains in the bovine α-lactal bumin molecule. Biochim Biophys Acta 1426(3):439–448. https://doi.org/10.1016/s0304-4165(98)00165-2
Penttila IA (2010) Milk-derived transforming growth factor-beta and the infant immune response. J Pediatr 156(Suppl 2):S21–S25
Plaisancie P, Claustre J, Estienne M et al (2013) A novel bioactive peptide from yoghurts modulates expression of the gel-forming MUC2 mucin as well as population of goblet cells and Paneth cells along the small intestine. J Nutr Biochem 24(1):213–221
Quitadamo PA, Comegna L, Cristalli P (2021) Anti-infective, anti-inflammatory, and immunomodulatory properties of breast milk factors for the protection of infants in the pandemic from COVID-19. Front Public Health 8:589736
Ruiz L, Garcia-Carral C, Rodriguez JM (2019) Unfolding the human milk microbiome landscape in the omics era. Front Microbiol 10:1378

Sadauskaite-Kuehne V, Ludvigsson J, Padaiga Z, Jasinskiene E, Samuelsson U (2004) Longer breastfeeding is an independent protective factor against development of type 1 diabetes mellitus in childhood. Diabetes Metab Res Rev 20:150–157

Sakwinska O, Moine D, Delley M, Combremont S, Rezzonico E, Descombes P, Vinyes-Pares G, Zhang Y, Wang P, Thakkar SK (2016) Microbiota in breast milk of Chinese lactating mothers. PLoS One 11:e0160856

Salminen S, Isolauri E (2006) Intestinal colonisation, microbiota and probiotics. J Pediatr 149:S115–S120

Sherman MP (2013) Lactoferrin and necrotizing enterocolitis. Clin Perinatol 40(1):79–91

Spatz DL, Davanzo R, Müller JA, Powell R, Rigourd V, Yates A, Geddes DT, van Goudoever JB, Bode L (2021) Promoting and protecting human milk and breastfeeding in a COVID-19 world. Front Pediatr 8:633700. https://doi.org/10.3389/fped.2020.633700

Tuaillon E, Valea D, Becquart P, Al Tabaa Y, Meda N, Bollore K et al (2009) Human milk-derived B cells: a highly activated switched memory cell population primed to secrete antibodies. J Immunol 182(11):7155–7162

Valenti P, Antonini G (2005) Lactoferrin: an important host defence against microbial and viral attack. Cell Mol Life Sci 62(22):2576–2587

Van Keulen BJ, Romijn M, Bondt A, Dingess KA, Kontopodi E, van der Straten K et al (2020) Breastmilk; a source of SARS-CoV-2 specific IgA antibodies. medRxiv [Preprint]. https://doi.org/10.1101/2020.08.18.20176743

Van Odijk J, Kull I, Borres MP, Brandtzaeg P, Edberg U, Hanson LA et al (2003) Breastfeeding and allergic disease: a multidisciplinary review of the literature (1966-2001) on the mode of early feeding in infancy and its impact on later atopic manifestations. Allergy 58:833–843

Vassilopoulou E, Feketea G, Koumbi L, Mesiari C, Berghea EC, Konstantinou GN (2021) Breastfeeding and COVID-19: from nutrition to immunity. Front Immunol 12:661806. https://doi.org/10.3389/fimmu.2021.661806

Vidal K, van den Broek P, Lorget F, Donnet-Hughes A (2004) Osteoprotegerin in human milk: a potential role in the regulation of bone metabolism and immune development. Pediatr Res 55(6):1001–1008

Wada Y, Lönnerdal B (2020) Bioactive peptides derived from human milk proteins: an update. Curr Opin Clin Nutr Metab Care 23(3):217–222. https://doi.org/10.1097/mco.0000000000000642

Walker WA (2002) Development of the intestinal mucosal barrier. J Pediatr Gastroenterol Nutr 34:S33–S39

Walker WA (2013) Initial intestinal colonization in the human infant and immune homeostasis. Ann Nutr Metab 63(Suppl 2):8–15

Weng M, Walker A (2006) Bacterial colonization, probiotics and clinical disease. J Pediatr 149:S107–S114

MIX
Papier aus verantwortungsvollen Quellen
Paper from responsible sources
FSC® C105338

If you have any concerns about our products,
you can contact us on
**ProductSafety@springernature.com**

In case Publisher is established outside the EU,
the EU authorized representative is:
**Springer Nature Customer Service Center GmbH**
**Europaplatz 3, 69115 Heidelberg, Germany**

Printed by Libri Plureos GmbH
in Hamburg, Germany